And God Said: Let There Be Evolution

The Science and Politics of Evolution

Paul Sanghera, Ph.D.

And God Said: Let There Be Evolution
The Science and Politics of Evolution

Published by
Infonential, Inc.
A California Corporation.

Copy Editor: Susan Gilmore
Technical Editor: John Serri, Ph.D.

ISBN-13: 978-1511450294
ISBN-10: 1511450290

About the Author

Dr. Paul Sanghera, an educator, scientist, technologist, author, and an entrepreneur, has a diverse background and experience in multiple fields including physics, chemistry, computer science, and the biosciences. He holds a Ph.D. in Physics from Carleton University, Canada; a Masters of Engineering degree in Computer Science from Cornell University, U.S.; and a B.Sc. from India with triple major: physics, chemistry, and math. He has comprehensive cross-disciplinary, cross-continental diverse experience in research, teaching, and learning. He has taught science and technology courses all across the world including San Jose State University, U.S.; Carleton University and Simon Fraser University, Canada; and Indian Institute of Technology (IIT) Ropar, India. He has authored and coauthored around 150 science research papers on subatomic particles of matter published in well-reputed European, American, and international research journals. At world class laboratories, such as CERN in Europe and Nuclear Lab at Cornell, he has participated in designing and conducting experiments to test quantum theories and models of subatomic particles, thereby contributing to testing the Standard Model of the Universe.

Dr. Sanghera was a lead engineer at the first Internet company, Netscape Communications, Inc. He has been on the ground floor of several startups in Silicon Valley, California and elsewhere, and has contributed to the development of state-of-the-art technologies such as Novell's NDS, the first computer network management system; Netscape Communicator, the first commercial web browser; and one of the first commercial web storefronts at Weborder, Inc.

Dr. Sanghera is the author of around two dozen books in science, technology, and project management. His current research interests involve topics in biochemistry and molecular biophysics.

About the Tech Editor

Dr. John A. Serri has a diverse background in physics, information technology, and biotechnology. He has held a variety of positions from basic physics research to application software development. His broad industrial and academic experience enables him to take on a rigorous yet practical perspective to science and technology. Dr. Serri received his BS degree in Mathematics and Physics from State University of New York (SUNY) at Albany and a Ph.D. in Physics from Massachusetts Institute of Technology where he developed innovative techniques to probe the dynamics of intermolecular collisions. One of his mentors at MIT, Dr. William Phillips was awarded the Nobel Prize in Physics in 1997, using some of this early work as a foundation. After leaving MIT, Dr. Serri joined the research staff at Bell Laboratories where he conducted basic and applied research in surface physics and chemistry. At AT&T he also helped design and build the future systems to be used for secure communications and network management. After AT&T he joined the staff of Loral Aerospace where he became one of the lead designers and developers of the Globalstar System. At Globalstar, he led the development of systems to control the space and ground network.

Dr. Serri currently serves on the adjunct faculty of the Mathematics and Computer Science Department at the California State University East Bay, is on the faculty of the University of Phoenix, and is a technical advisor for an international biotechnology company headquartered in the San Francisco Bay Area.

Table of Contents

Why This Book?

As there are many good books on evolution out there, I need to explain what drove me to write this book. I wanted to put the big picture of evolution with all of its main aspects in one book in an easy-to-understand way. I had this book in my mind that explores the development of the theory of evolution from Darwin to modern genetics in a concise yet comprehensive fashion.

Instead of overemphasizing Darwin and fossil evidence, this book extends the treatment of the theory to a much bigger scientific context, which includes biochemistry and molecular biology including genetics. Furthermore, it also explores the connection of the theory of evolution with physics, and what quantum mechanics, the well-established theory of physics, may have in store for evolution.

As the development of practical applications of a theory provides the ultimate evidence for its validity, this book explores the applications of evolution. This book also offers a reality check that many of us need in order to break away from so-called Darwinism. Such a breakaway is good for the scientific health of our nation and the world. In addition to busting the false synonymy between Darwin and evolution, this book also busts many other myths about evolution. Scientifically, it is hard to understand what something is without understanding what it is not. Last but not least, this book includes simple questions and problems with answers and solutions in order to provide the option for the reader to be interactive while venturing through the book.

In a nutshell, this book tells the story of evolutionary theory in the bigger sociopolitical, historical, and scientific context in which it has unfolded. One of the reasons for being aware of this story and keeping the story in a bigger context is that we can learn from our past experiences and avoid the mistakes of the past. For example, such a context includes the eugenics movement, which played a significant role in the history of opposition to evolution. It is important to remember the role of eugenics so that we stay alert or on guard against the possibility of such tragedies happening again in the name of science.

The need to keep this big picture of evolution in sight is direr today than ever before. Here is why: As a society, we are entering the molecular age from both dimensions, living and non-living, by using the tools of nanotechnology and bioengineering including biotechnology. As we will be able to evolve living entities in the lab, it will be possible to control evolution and its pace in many ways and thereby manually interfere with natural biological evolution. This will have profound impact on society and our planet as a whole. This reason alone is good enough for us keep the big picture of evolution in mind.

The writing of this book is an attempt to present that big picture. How much have I succeeded? I am eagerly waiting to hear the answer to this question from you so that I can improve the future editions.

Thank you for your interest in this topic and in this book.

Preface

Physics, Biology, and Evolution: Mind the Gap

Quantum physics thus reveals a basic oneness of the universe.

Erwin Schrodinger

The further the spiritual evolution of mankind advances, the more certain it seems to me that the path to genuine religiosity does not lie through the fear of life, and the fear of death, and blind faith, but through striving after rational knowledge.

Albert Einstein

Almost all aspects of life are engineered at the molecular level, and without understanding molecules we can only have a very sketchy understanding of life itself.

Francis Crick
One of the discoverers of DNA structure

One of the most important discoveries of science to date is that everything, living and non-living, is made from the same set of about 100 atoms, which in turn are made from an even a smaller set of subatomic particles. We live our lives in two coexisting and interacting worlds: the world of living things and the world of non-living things. We, along with other living and non-living entities (or things), exist and are distributed over space and time. This obviously true and trivial-sounding statement contains two important points: First, life exists not only in space, but also in *time*; and evolution occurs in the dimension of *time*. Second, not only living entities but also non-living entities exist in space and time, which, combined with the fact that both living and non-living entities are made from the same set of atoms or subatomic particles, indicates that at a fundamental level both living and non-living entities are subject to the laws of nature, some of which may be the same or similar in both cases. The study (or science) of living entities is called biology.

Even though biologists for a long time have largely managed to do without atoms, physicists and chemists have been dealing with them for at least two centuries. Physics, at its core, is a branch of science that deals with understanding the Universe and the entities in the Universe in terms of the fundamental constituents of matter such as molecules, atoms, subatomic particles, and interactions among these constituents. Physics, the most fundamental science, deals with the fundamental principles that govern the structure and function (or anatomy and physiology, as a biologist would say) of the Universe, its constituents, and physical phenomena. These principles are at work in some form for all other disciplines of science and technology such as biology, chemistry, material science, electronics, engineering, and nanotechnology.

Humans have developed a zoo of scientific disciplines and sub-disciplines. However, nature did not create and does not need borders and interfaces between disciplines of science in order to run its affairs. These borders are the artificial creations of our own limitations. Nature is one and at its most fundamental level it runs in oneness. This is another central message of the whole scientific research performed so far in various scientific fields, especially in physics and biology.

In physics, the Universe can be explained by the unification of apparently different forces of nature into one force; and in biology the observed unity behind diversity of life is elegantly explained by evolution, as you will realize as you go through this book. Coming back to a point made earlier, life distributes (or extends) itself not only in space but also in time. Thinking in the time dimension, the life you see around you is just a snapshot of the history of life on Earth that extends back in time to about 3.5 billion years ago. It is not hard to realize or imagine that things change in time. Biological evolution, referred to as evolution in this book, drives the change in life over time, and the theory of evolution explains how this occurs.

We will offer a more specific and detailed definition of evolution and the theory of evolution in **Chapter 1**, where we explain the theory of evolution as introduced by Charles Darwin and Alfred Wallace. Also in this chapter, we explore the roots of the concept of evolution that go far back into ancient history, and we explore the context and the scientific environment in which the original theory of evolution was developed. You will see in Chapter 1 that even though most of the vocal opponents of evolution treat Darwin synonymous with evolution, Darwin was neither the first one to introduce the concept of evolution nor the only one to propose the theory of evolution.

Let me make it clear from the outset: I am neither a supporter nor an opponent of the theory of evolution. This is because I am a scientist and the theory of evolution is a scientific theory. Scientists run experiments and do not campaign on scientific problems or theories. We scientist do not support or oppose theories as believers, we test theories. As long as a theory withstands the tests of observations and experimental results, we learn from it, apply it to understand physical entities and phenomena, and make this knowledge available for other disciplines and for real-world applications. If observations or experimental results demand changes to a theory, we modify the existing theory to improve it. If a theory fails miserably to explain newly found facts or phenomena, we happily replace that theory with a better one. Even well-established theories are always under scrutiny. As a matter of fact, we take more pleasure in disproving a theory than proving it, of course by using facts and reproducible experimental results. Most of us believe that you can never prove a theory, you can only disprove it.

We don't care if it is Newton's theories of classical mechanics or modern scientists' Standard Model of the Universe, whether it is Einstein's theory of relativity or Darwin's theory of evolution. We are constantly analyzing them, putting them through tests. This is because we are scientists; factual observations and experiments are our tools; and the scientific method is the technique that we use to make discoveries or test theories. In **Chapter 2**, we explore science, the scientific method, and theories in science. We discuss the meaning and role of a theory in science by exploring how science develops, progresses, and works. This chapter addresses the notion that evolution is *just a theory*.

Most of the vocal opponents of evolution, being busy with demonizing Darwin under the assumption that Darwinism and evolution are synonymous, have perhaps missed the fact that starting from Darwin and Wallace, the theory of evolution has come a long way to find its roots in genetics. This has given rise to a whole new field called microevolution, the study of changes within a population at the genetic level, leading to changes in traits (characteristics) at the organism and population (or species) level.

It may look obvious to us because we are used to it, but it is a very significant characteristic of life on Earth that it is largely composed of discrete organisms, each occupying a finite space and time (lifetime). Discrete organisms means a whole number of organisms such as one, two, three; and not one and a half, etc. This discreteness exists at all complexity levels of life from organisms down to cells and genes in the cells. This means, for example, a multicellular organism is made of a whole number of organs, each organ is made of a whole number of cells, each of which in turn is made of a whole number of molecules. The molecules of life that largely compose a cell are of four types: carbohydrates, lipids, proteins and nucleic acids (DNA and RNA). DNA is the molecule that contains genes, the blueprints of life.

Another important property of life, related to its history, is that each organism has a unique set of characteristics or traits. This means life is diverse. You don't have to be a scientist to observe and admire the astounding diversity of life. The differences are not limited to organisms from different species, even within the same species no two organisms are identical in their characteristics; spare identical twins. Darwin and Wallace knew that evolution occurs due to this diversity or variations of traits. However, they did not know the source of these variations, which was considered the main scientific criticism or weakness of the theory of evolution during their times. Unfortunately, they were unaware of the work of their contemporary, Gregor Mendel, whose work held the answer to their burning question. Mendel's experiments had revealed that genes and different versions of genes are the origin of these variations. Mendel also discovered that genes were discrete and that some characteristics or traits, called genetic traits, are transferred from one generation to the next through genes. Mendel's work and its connection the theory of evolution as introduced by Darwin and Wallace is explored in **Chapter 3**.

It is important to know that Mendel did not know genes as we know them today. He found through experiments that some discrete particles were responsible for transferring certain traits from one generation to the next (from parents to children), and he named these particles *discrete hereditary factors*, also cited as *genetic factors* or *Mendelian factors*. The subsequent discovery of Mendel's work by other scientists, connecting it to the theory of evolution, and the discovery and study of genes extended the theory of evolution from the macroscopic world to the microscopic world. In **Chapter 4**, we explore how with discoveries in molecular biology and genetics, the theory of evolution found its roots in the microworld of genes. Also in this chapter, we explain how microevolution, evolution at genetic (that is, molecular) level, gives rise to macroevolution, the evolution observable at the levels of species and above and explained by the theory of evolution as introduced by Darwin and Wallace.

As the human species, we can boast as much as we like about our superiority to other organisms, but the fact remains that we as humans have our own limitations in understanding the Universe, as alluded to earlier. First, we as individuals can only investigate and understand so many things in our life span. Second, we can only investigate, gather information, obtain knowledge, and develop our understanding in pieces, that is, in a fractured, imperfect way. Some pieces (entities, phenomena, or information) may look different from each other at the surface level. However, they may just be different facets of the same thing at some other level. Having only partial knowledge, which is in different pieces, contributes to obscuring the big picture. Nature works the way it does without caring about our limitations; it works in a beautiful oneness at its most fundamental level. So it is important to put the pieces together to understand the big picture. It is in this spirit, that we explore the connection between evolution and physics in **Chapter 5**. In this chapter, we also explore the roots of the theory of evolution extending toward quantum physics; the well-tested and established theory of physics. In other words, we explore how the principles of quantum physics may explain some aspects of evolution and verify the validity of the theory of evolution.

Evolution has its signature or imprint on almost all phenomena of life and on the relationships between different groups of organisms. Its impact on life is so prevalent that it is impossible to understand almost any key concept in biology, the science of life, without it. Darwin and Wallace developed the theory of evolution based upon the enormous amount of data they collected. Since then, the theory has been subjected to continuous scrutiny and tests from a number of different directions, and as a result has developed as a common thread running through the whole field of biology. We discuss this topic in **Chapter 6**, where we also present multiple lines of evidence for evolution from different areas and complexity levels of life demonstrating that evolution is a common thread that runs through all life.

Evolution defined the history of life on Earth discussed in **Chapter 7**, where we explore how according to evidence, life began with a cell and evolved to what we see today, millions of species filling different environments on Earth. We present a brief account of the history of life on Earth and the role of physical laws of nature and evolution in determining and shaping this history.

All organisms of a species have to perform certain functions in order to survive and reproduce. Some of these functions are shared with other species (unity) and others are unique to the species (diversity). In both cases, the morphology (external body form) of the members of a species must support these functions. In other words, the form must fit the function. As a matter of fact, form fits function at all complexity levels of life from molecules to organisms. This issue is explored in

Chapter 8, where we also explain how theory of evolution explains or predicts this feature of life.

You may wonder if the theory of evolution is so convincing, what is the basis for the opposition not only to the theory of evolution but also to the very concept of evolution? A number of surveys and studies have cited a few major causes at the bottom of this opposition including the lack of understanding of even most basic biological concepts including genetic concepts; and misinformation about, misinterpretation, and misrepresentations of the theory of evolution. A major problem is that in large part, what the opponents of the theory of evolution oppose and what the theory of evolution actually is are two very different things. We dispel some of the most prevalent myths about the theory of evolution in **Chapter 9**, where we also present a brief history of the greatest myth about evolution: *eugenics*. We discuss eugenics in some detail for several reasons: It played a significant historical role in the history of opposition to evolution and in this process it exposed the contradiction and hypocrisy of this opposition. Furthermore, it is important to remember the role of eugenics while we are entering the molecular age, which includes genetic engineering, so that we stay alert or on guard against the possibility of such tragedies happening again in the name of science.

The history of opposition to evolution with an emphasis on the opposition in the U.S. is explored in **Chapter 10**, where we also expose the intentions behind and politics of this opposition, and its potential grave consequences at national level. In this chapter, we explore how creationism, used to oppose evolution, belongs to a belief system, whereas evolution is a natural process that occurs according to the verifiable natural laws described by the scientific theory of evolution.

A major factor that distinguishes a scientific law from a belief is that the scientific law can be verified on its own without referring to divinity, and once understood it can often be put into practical use in many cases. In other words, applications can be developed based on scientific laws. Most of the devices and facilities today that are part and parcels of our daily lives such as cars, airplanes, televisions, cellular phones, and computers are applications of the laws of physics. The development of practical applications of a theory provides the ultimate evidence for its validity. Applications of the theory of evolution in various fields are discussed in **Chapter 11**.

One of the reasons for hardcore opposition to evolution or for the denial of the reality of evolution is that some of us feel that the very concept of evolution is an attack on our religious faith. Many people of faith would agree with me that this is not necessarily true.This feeling, however, is understandable. After all, evolution is not the first scientific theory where a set of religious people felt that way. During the history of science and the progression of human civilization on this planet, we have been there before more than once. We discuss this issue in **Chapter 12**, where we also explore some aspects of and reasons for such opposition to science by focusing on the opposition to evolution. History is a witness that most of us over time find ways to reconcile our faith with scientific reality.

Paul   September 2013
(Paul Sanghera, PhD)

ONE

Evolution of Evolution

And God Said: Let There Be Evolution

I have no great quickness of apprehension or wit which is so remarkable of clever men.

-Charles Darwin in his autobiography

Ignorance more frequently begets confidence than does knowledge: it is those who know little, and not those who know much, who so positively assert that this or that problem will never be solved by science.

-Charles Darwin in *The Descent of Man*

1.1 Once Upon a Time

Once upon a time, there lived people on this planet who had great *trouble* in comprehending a law of nature operating all around them even after it was discovered: a law so elegant, yet so simple. This is one way the coming generations will remember us and a few generations that came before us. Most of the players perhaps have not yet observed that the great battle between the opponents and proponents of the theory of evolution is practically over. The results are already in and the dust is settling down. The

Figure 1.1 Humans killed by none other than humans. The morning after the battle of Waterloo, by John Heaviside Clarke, 1816.

results will be a part of common sense for coming generations (for some of us they already were and are), who will wonder why we fought over such a trivial and obvious matter. Historians of science will eventually determine the exact event or a set of events, year or years, or a period to mark the end of this battle. However, the beginning can be traced back to the publication of *On the Origin of Species* by Charles Darwin about one and a half century ago.

3

And God Said: Let There Be Evolution

So, which side has won the battle? I am leaving it to you to make this determination after reading the story in this book.

We may say that our story begins in 1831, when a young man of 22 named Charles Darwin (1809-1882) began his sail to the Galápagos Islands as the naturalist onboard a survey ship called the *HMS Beagle*. This five-year, around-the-globe journey not only put this naturalist in direct contact with the natural world in a very practical way, but also provided plenty of time away from routine daily life, which provided the opportunity to contemplate the dynamics of the natural world. This episode of experience worked for Darwin as a gateway to a scientific path that he walked on for the rest of his life. His experience on this journey included witnessing a great diversity of groups (species) of living organisms as well as discovery of fossils of extinct organisms from the distant past. For instance, the fossil of a giant sloth (a mammal) 400 miles south of Buenos Aires, Argentina, started him thinking about how those creatures might have become extinct, and how for that matter any species could go extinct, and so on: a natural course of scientific thinking or imagination.

To Darwin, during this journey, nothing was more scientifically stimulating than his five-week venture to the Galápagos Islands, where his observations catalyzed the scientific revolution that the world would witness for years, decades, over a century, and beyond. There he witnessed for the first time the geographically confined elegant exhibition of unity and diversity of life. He contemplated the species unique to an island, that is, they were found nowhere else in the world. Then there were closely related species, for example, those of mockingbirds inhabiting neighboring islands. Could it be that these closely related species had common ancestors that lived elsewhere, for example, on a nearby mainland?

And on this journey, on the side, Darwin learned about gauchos, cowboys of the Argentine pampas (plains), and also about the indigenous (pre-Columbian) people; and how gauchos killed the indigenous people. It provided him a window to see primal territorial impulses in the animal called human. Figuratively speaking, there are animals inside all of us. Yes, just like those organisms we call animals, we humans also battle our instincts. Our battles are even bigger than those whom we call animals: We battle not only over territories and resources but also over ideas and beliefs. And one of the recent battles over ideas and beliefs has been the battle of evolution. Could there be a connection between us and those we call animals?

It took centuries to understand evolution and to develop the scientific theory of evolution.

1.2 In the Beginning

In the beginning of the journey of science, when we had less information and knowledge, it appeared as if it was possible for one person to know everything. Therefore, instead of many scientific disciplines, there was only one discipline called *natural philosophy*, the study of nature and the physical universe. Today's physics is actually the modern and refined version of *natural philosophy*. The scientists then were called natural philosophers or naturalists. As mentioned earlier, Charles Darwin ventured to the Galápagos Islands as the naturalist onboard the *HMS Beagle*.

> **Note.** Natural philosophy, the study of nature and the physical universe, served as a precursor to the modern fields of natural sciences led by and based on physics. Physics aside, modern notions of science and scientists date only to the nineteenth century. Before then the word *science* simply meant knowledge and the title *scientist* did not exist. For centuries scientists were called natural philosophers, who often pursued a wide variety of scientific interests.

Scientific work on biological entities, however, can easily be traced as far back as Aristotle (384-322 BC), a noted Greek philosopher. It is clear from the writings of Aristotle that he was recording observations and then trying to make sense out of them by identifying patterns in the observations. These are essential elements of scientific research method: make observations and connect the observations with one another to identify a pattern. Aristotle was one of the first naturalists.

Many vocal opponents of evolution present their opposition in such a way as if to suggest the theory of evolution was just the invention of the mind of one evil-spirited man: Charles Darwin. It's important to realize that the concept of biological evolution is as old as the history of science and natural philosophy itself. It took centuries to develop the scientific theory of evolution, as we know it today, from the cumulative discoveries of many scientists (called natural philosophers in olden days) working in many fields such as biogeography, comparative morphology, geology; and today, almost all subfields of biology. As a matter of fact, the idea of evolution is as ancient as the very idea that natural phenomena can be understood at all.

The longing and effort to understand the Universe is one of a very few things that distinguish us from other animal species and is largely responsible for what we are today as human beings. In ancient times, most people, if not all, did not believe that nature could be understood in terms of a few consistent principles. So, some invented gods to explain phenomena and forces of nature such as earth and sky; floods and fire; earthquakes and volcanoes; love and war; and death and wisdom. The idea that it's possible to understand the Universe, that nature works in accordance with consistent principles that can be deciphered, is traced back to Thales of Miletus (624-546 BC), a pre-Socratic Greek philosopher who is known as one of the Seven Sages of Greece. The concept of evolution is traced back to Anaximander (610-546 BC), a friend and a student of Thales, who is also regarded as the father of Western philosophy and science by many. Anaximander proposed that humans could not have originated and survived as an independent line of life. The first human as an independent infant all by itself would be so helpless that it could not have survived on Earth. Therefore, Anaximander argued that humans must have gradually evolved in the protection of and from other animals. He went on to propose that all life forms evolved from fish through a process of modification and migration to land.

The purpose of this chapter is three-fold: to explore the roots of the concept of evolution as far back as they go in ancient history, explore the context and the scientific environment in which the original theory of evolution was developed, and describe the essence of the original theory of evolution as introduced by Charles Darwin and Alfred Wallace.

But first things first; before we begin exploring the issue of the evolution of life, it's fair to step back and ask: what is life anyway?

1.3 What is Life?

At a very fundamental level, all living organisms, including humans, are made from the same set of about 100 atoms from which any other material thing in our universe is made. Multiple atoms bond together through some kind of electromagnetic interaction to make molecules. For example, two atoms of oxygen bond to a hydrogen atom to make a molecule of water. Another example of a molecule is chlorophyll, a pigment molecule which makes plant leaves green. Millions of chlorophyll molecules are organized in a cell to convert sunlight into carbohydrates in a process called *photosynthesis*. Life arises at cellular level.

In our daily lives, we generally have a very good idea of what life is and what it is not. For example, even a child knows that a lush, green tree is alive and a rock is not. To be more specific, life is defined by the set of characteristics that living things have, which include the following:

Complexity of organizational levels. Life has hierarchical levels of organization (Figure 1.2) with increasing complexity to the extent not found in non-living entities. For all life, this organization begins with or before the organism level (living individual such as an animal, plant, or bacterium) and goes beyond. For a multicellular organism, there are levels of increasing complexity of life inside, beginning with a cell, the most fundamental structural and functional unit of life. Therefore, at its most fundamental level, a living organism is composed of one or more cells. In a multicellular organism, a set of

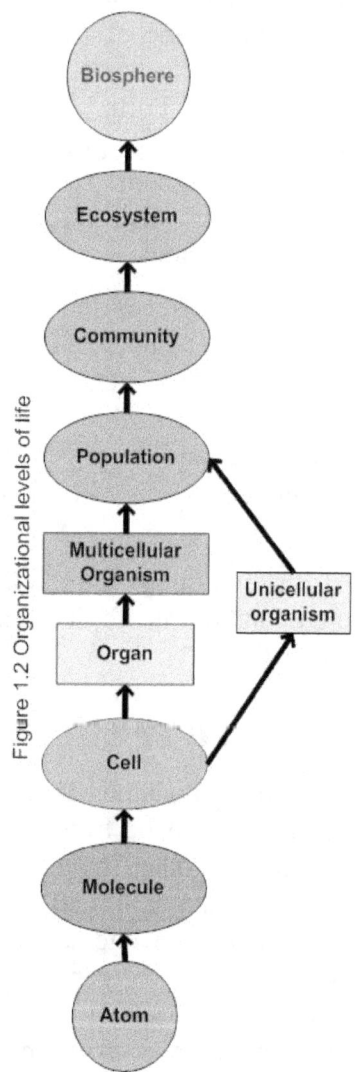

Figure 1.2 Organizational levels of life

cells organize into an array in order to perform a common function, called *tissue*. Tissues, in turn, organize into a structural unit called an *organ*, such as the heart, which perform one or more specific tasks. A number of organs interacting to accomplish one or more common tasks are called an *organ system*, such as the digestion system. A multicellular organism is composed of multiple organ systems. A *system*, in general, is a combination of components that work together to accomplish a common task.

The organizational levels of life do not stop with an organism. After all, organisms interact with other organisms and with their environment. A group of organisms that have the potential of interbreeding and producing fertile, viable offspring are called a species. A group of organisms of a given species living in an area is called a *population*. All populations of all species living in a specified area are called a *community*. One or more communities interacting with their physical environment makes what is called an *ecosystem*. The sum of all ecosystems on our planet is called the *biosphere*. Even though life begins with a cell, it's important to remember that a cell is made of molecules, which in turn are made of atoms; and that complex cellular functions can only be truly understood at molecular, atomic, and particle (subatomic) levels.

These different organizational levels of life are not merely theoretical definitions; they represent the properties and interactions of life in very practical ways as described below:

New emergent properties at each organizational level. At each level of organizational complexity, life is more than the sum of its parts. To be specific, at each level of organization, some new properties emerge which are not a part of any of the components of the system at that level. These properties emerge from how the parts of the system (such as molecules in cell or cells in an organ) at that level interact with each other

9

and how as a result of these interactions they are arranged. For example, a cell is composed of molecules, but life appears only at the cellular level and not at the level of any of the molecules of the cell. The cell performs functions which none of its molecules can perform alone.

The appearance of emergent properties, however, is not unique to life. They can also appear in non-living systems. For example, your car provides you transportation, a property which none of its parts alone can offer.

Growth and Metabolism. Organisms metabolize and grow. *Metabolism* is a set of chemical reactions which include *anabolic* (consuming energy to synthesize a complex molecule) and *catabolic* (releasing energy by breaking down a complex molecule) pathways. In order to grow, an organism takes in suitable material from the environment such as food, breaks it down if necessary, and reorganizes it into a needed structure. This process is supported by metabolism. For example, energy required to accomplish growth is obtained through metabolism. As an example, you eat a meal, digest it, and transform it into more of yourself (that is energy and molecules required to maintain life) by using metabolism.

Non-living entities do not metabolize and grow.

Response to the Environment. An organism interacts with its external environment; and senses and responds to the changes occurring in the environment internal or external to its body. Energy is generally used in this response. For example, you feel hungry and you eat. Different responses constitute different behaviors of different organisms.

The responsive behavior of a living organism is generally active as opposed to the passive (non-responsive) behavior of a non-living entity such as a book dropped down from your hands.

Reproduction. Organisms reproduce their own kind, that is, species. Reproduction is of two types: *asexual reproduction* and sexual *reproduction.* In asexual reproduction, offspring arise from only one parent and therefore inherit the genes of that one parent only; like producing a clone. In sexual reproduction, offspring acquire half of their genes from each parent. Therefore, sexual reproduction is more complex, and gives rise to more diversity and new unique properties and species over time. A molecule called deoxyribonucleic acid (DNA), the carrier of genes, is the basis of growth and reproduction. We will explore more in this direction further on in this book.

Only organisms, and not non-living entities, have DNA.

Homeostasis and Regulation. A defining feature of life is *homeostasis*, which is a steady-state physiological condition of the body of an organism that favors cell survival. For example, keeping temperature and pH value (a measure of the concentration of hydrogen (H^+) ions, which are practically protons) within a very narrow acceptable range. This steady state is accomplished by sensing and responding to changes in the environment, which includes regulating some processes. For example, *thermoregulation* in an organism helps maintain a constant internal body temperature even if the external temperature changes.

Now that we have some idea of how life is defined in science, we can explore the context in which the theory of evolution was developed.

1.4 Prevailing Belief and Scientific Environment

As discussed in Section 1.2, pre-Socratic philosopher Anaximander proposed that humans evolved. Nevertheless, the idea got lost with time. Through his contributions to many areas of science, another Greek philosopher, Aristotle (384-322 BC), had a great impact on the development of early science in the West. He viewed species (different groups) of organisms as fixed and unchanging, just like he viewed the heavens (parts of the Universe as seen from Earth such as stars and galaxies) as perfect and unchanging. Consider a ladder standing against a wall. Assume that each rung of the ladder represents a form (say species) of life and the height of the rung represents the complexity of that form of life represented by that rung. According to Aristotle, each form of life was perfect, permanent, and had a rung allotted to it on this ladder; each rung was occupied. The ladder represented the scale of increasing complexity of life from bacteria to fly, to fish, to alligator, to human. This view survived through the centuries and it was the prevailing belief at Darwin's time as well, that all species had been created in a perfect state. One of the reasons for the survival of these views was that these ideas were consistent with those found in the books of various religions, including Christianity. For example, with the literal interpretation of the Old Testament account of creation, different species were individually designed by God and therefore were perfect. So, according to the doctrine of perfect creation, all species were created by the Creator at the same time in a perfect and permanent state and have not changed since their creation. One implication of this belief is that no species can ever become extinct by natural causes because it's perfect and permanent.

During Darwin's time, this belief of *perfect creation* was being challenged by new scientific observations, which were raising some serious questions in various fields including biogeography, comparative morphology, embryology, and geology, as discussed in the following:

Questions from Biogeography. Biology, the science of life, started as a discipline of three Cs: counting, categorizing, and calling (naming). During Darwin's time and before, in the area of biology, naturalists were observing or discovering a *chain of being* running through all forms of life from the "lowest" forms all the way to human. Their first task was to catalogue each link in this chain, that is, each type (or group) of beings called species. The motivation was: discovering and understanding each link in the chain would reveal the meaning of life. When Europeans discovered that the world was a lot bigger than Europe, they embarked upon world explorations. As a result of these explorations, thousands of new plants and animals (or the knowledge of their existence) were brought home from abroad. This added more links to the known chain of life. Following this, many naturalists, including British naturalist Alfred Wallace (1823-1913), used the catalogue to identify patterns in the information of where the species lived and the apparent relationship between them. These naturalists pioneered a field called *biogeography*: the study of patterns in the geographical distribution of species.

The pioneers of biogeography discovered some intriguing patterns in the chain of life. One of these patterns was this: They identified some species living only on certain islands in the middle of the ocean, isolated from the rest of the world. To their surprise, they discovered that some of these isolated species were remarkably similar to some of the species living on the mainland, far across the vast expanses of the ocean.

These similarities were also found between species living across opposite sides of mountain ranges, which could not possibly be passed through. Yet there were some species geographically isolated from each other, which looked similar (they had some common characteristics) but were very different in many important ways. These observations raised questions such as these: How can two geographically-isolated species such as those on different continents end up looking so much alike; and why may some totally different species have some characteristics oddly similar? The simplicity of the hypothesis of perfect creation, according to which all species were created at the same time in perfect form and have not changed since then, could not answer these questions about the complexity of apparent unity behind diversity. Instead, the questions raised by these similarities and differences gave rise to a whole new field called *comparative morphology.*

Questions from Comparative Morphology. An organism's external form is called morphology. Comparative morphology is the scientific study of body plans and structures among groups of organisms. The scientists working in morphology observed that some organisms that are externally very similar are quite different internally. For example, think of porpoises (dolphin like marine mammals) and fishes. The opposite is also true: some organisms that appear different outwardly are very similar in their internal body plans. For example, the bones in a bat wing, human arm, and porpoise flipper are very comparable. So the question was: why are species that differ so much in one set of features so much alike in another set of features? The hypothesis of perfect creation could not answer this question either. It also could not answer questions like this: if the original design was so perfect that there was no need for modification over time, then why do some organisms have

14

body parts that have no function? For example, an ancient aquatic whale had ankle bones, but it did not walk. We, the human, have a coccyx, which is like the tailbone in many other mammals to support the tail.

And then there were questions from embryology.

Questions from Embryology. Embryology is the subfield of biology in which scientists study the development of an embryo from the fertilization of the egg (ovum) to the fetal stage of the embryo. The word embryo has its origin in the Greek word *embryon*, which means the unborn. The discipline of embryology also dates back to pre-Darwin era. From studies in comparative anatomy and embryology, scientists discovered that all vertebrate (animal with a backbone) species begin their early development the same way. To be more specific, they observed that embryos of different vertebrate species begin looking like embryonic fish. However, these embryos diverge in their structure at later stages of development. Here is the general pattern of this development:

1. Embryos of all vertebrate species, in the beginning, contain very similar numbers and shapes of blood vessels, nerves, and organs.

2. At a later stage, some of these blood vessels, nerves, and organs disappear.

3. Some of these structural elements go through different twists and turns in their form.

4. This later divergence in the development of initially similar looking embryos results in different adult forms of different groups of vertebrates such as amphibians, birds, fish, mammals, and reptiles.

This did not make sense at all and caused confusion among scientists. The *perfect creation* doctrine collapsed in the face of these observations.

And then there were questions from geology.

Questions from Geology: Fossils. To understand his experience during the voyage of the *HMS Beagle*, Darwin drew many of his ideas from the work of geologists studying fossils, which are the remains (or traces) of organisms that lived in the distant past. Most fossils are found in rocks called *sedimentary* rocks, the rocks that are formed from the mud and sand settled down to the bottom of lakes, seas, and swamps. In geology, the word *sediment* refers to mineral or organic matter deposited by air, ice, or water. Sediments travel through rivers to seas and swamps. Older layers of

Sediments and Sedimentary Rocks

Figure 1.3 An example of a sedimentary rock clearly showing the strata (horizontal layers). Image: courtesy of Martin M. Short, Sr, NASA.

sediment are covered by new layers and this process of compression gives rise to superimposed layers of rocks; these layers are called *strata*. In this way, the strata represent a vertical time scale; older (lower) layers formed before newer (upper) layers. An example is shown in Figure 1.3. Some strata contain fossils, which are exposed along the rocks due to changes such as lowering of the water level.

All over the world, geologists had been mapping the layers of rocks exposed by processes such as erosion and quarrying.

To their surprise and confusion, they found identical sequences of rock layers in different parts of the world. Note that as of the 1700s (that is even before Darwin), fossils in the layers were accepted as stone-hard evidence of earlier forms of life. While identifying patterns, scientists discovered another piece in the puzzle: sequence of complexity. Here is an example: Many deep layers held fossils of simple marine life. The fossils held by the upper layers were similar looking but more intricate (complex or sophisticated). The degree of intricacy gradually increased from bottom to top. At first this may sound very much like Aristotle's ladder of complexity of life discussed earlier in this section. However, realize that contrary to Aristotle's ladder, this ladder of strata has a scale of time, that is, the more complex a life form is the later it appeared. The species in the top (or higher) layers looked more like modern species. The big question was: if each species had been created in a perfect state at the same time, then what did that observed sequence of complexity mean?

Adding yet another piece to the puzzle, geologists were finding fossils of huge animals with no representatives living on Earth at the time the fossils were discovered. So were these kinds (species) of animals created in a perfect and permanent state, and if so, why did they become extinct? For example, chances are that you have seen great dinosaur skeletons in a natural history museum. Fossil studies indicate that dinosaurs were a diverse group of animals that populated the Earth for over 160 million years, from the late Triassic period (about 230 million years ago) until the end of the Cretaceous period (about 65 million years ago). The extinction of most dinosaur species occurred during the Cretaceous–Tertiary extinction event. Figure 1.4 illustrates 18 species of a group of dinosaurs called *basal ceratopsia*. The ceratopsian dinosaurs ranged in

And God Said: Let There Be Evolution

size and weight from 1 meter and 23 kilograms to over
9 meters and 5,400 kilograms.

Figure 1.4 A scaled illustration of 18 species of *basal ceratopsia* dinosaurs.
Image: public domain.

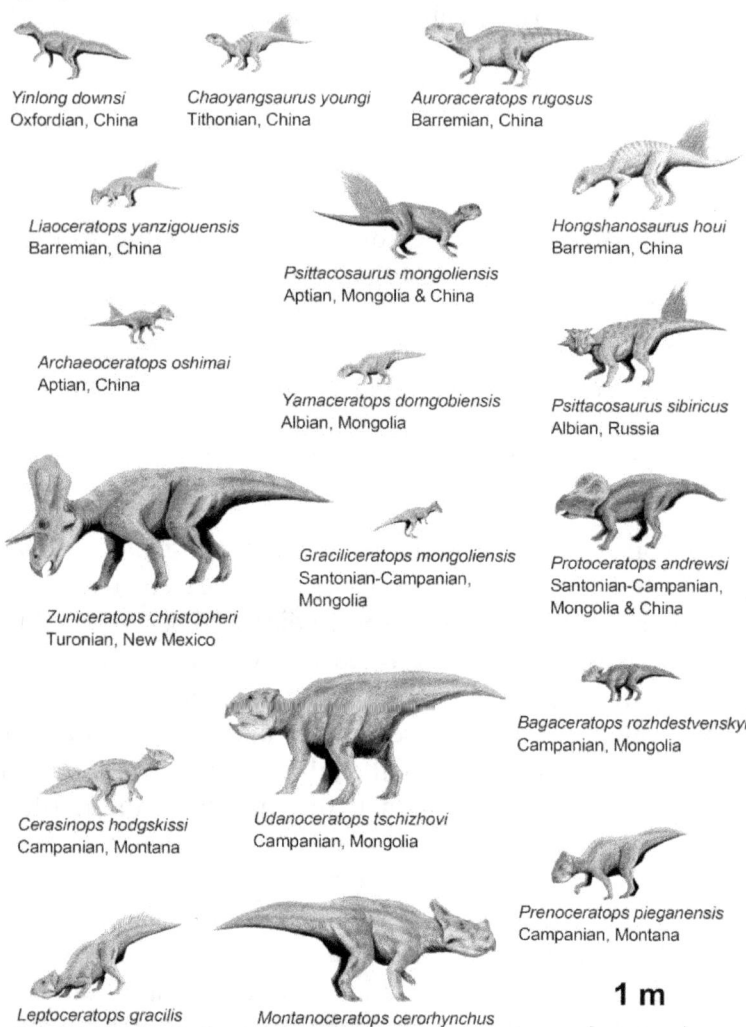

Yinlong downsi
Oxfordian, China

Chaoyangsaurus youngi
Tithonian, China

Auroraceratops rugosus
Barremian, China

Liaoceratops yanzigouensis
Barremian, China

Psittacosaurus mongoliensis
Aptian, Mongolia & China

Hongshanosaurus houi
Barremian, China

Archaeoceratops oshimai
Aptian, China

Yamaceratops dorngobiensis
Albian, Mongolia

Psittacosaurus sibiricus
Albian, Russia

Zuniceratops christopheri
Turonian, New Mexico

Graciliceratops mongoliensis
Santonian-Campanian,
Mongolia

Protoceratops andrewsi
Santonian-Campanian,
Mongolia & China

Bagaceratops rozhdestvenskyi
Campanian, Mongolia

Cerasinops hodgskissi
Campanian, Montana

Udanoceratops tschizhovi
Campanian, Mongolia

Prenoceratops pieganensis
Campanian, Montana

Leptoceratops gracilis
Maastrichtian, Wyoming &
Alberta

Montanoceratops cerorhynchus
Maastrichtian, Montana

1 m

In a nutshell, discoveries made in the nineteenth century in the areas of biogeography, comparative morphology, and geology raised questions that could not be answered by the prevailing belief of the time that each species had been created by its Creator in a perfect and permanent state. Such times of surprise and confusion in science are the most creative times indeed, when old (and established) ideas and theories are questioned and new theories are developed. In the study of life, such were the times into which Charles Darwin and Alfred Wallace were born.

1.5 Wresting with the Puzzling Body of Evidence

In the face of questions unanswered by the perfect creation doctrine, some scientists started thinking beyond this doctrine, which was widely accepted at the time. An obvious way of thinking beyond this was: If species were not created in a perfect and permanent state as the observations such as fossil records and the existence of useless body parts suggested, then could it be that species had been changing over time? This thinking contained the seeds of the theory of evolution. However, this was not the dominant way of thinking at the time. It happens so often in all great discoveries in science, that the underlying cause of some natural processes under scrutiny is not obvious. Therefore, the prevailing beliefs influence how we interpret the clues and data about these processes, and in this way we miss the mark. So, many of the eighteenth and nineteenth century naturalists bent over backward to reconcile the new evidence against perfect creation within the traditional framework of perfect creation. Citing an example, one such naturalist was a French scientist, Georges Cuvier (1769-1832), who noted that the older the fossils were the more dissimilar they were to the current life forms. He also observed that moving from one layer to the next in the strata some species disappeared, whereas some new species appeared. Instead of taking this as evidence of gradual change over time, Cuvier, in his attempt to not break away from the framework of perfect creation, advocated the hypothesis of *catastrophism*. According to this hypothesis, some events in the past occurred suddenly due to mechanisms which are not operating in the present. He also observed from fossils that extinctions had been a common occurrence in the history of life. Therefore, he had to swallow the implication that catastrophes were common in the past: each boundary between strata represented a catastrophe. He tried to make these implications less ridiculous by arguing that these catastrophes were confined to local regions that were later populated by species emigrating from other regions, and hence the appearance of new species. It is, however, noteworthy here that a whole field of the studies of fossils called *paleontology* was largely developed by Cuvier. Yet, Cuvier led the school of thought that tried to squeeze the meaning of the evidence of changes emerging from the fossil record into the traditional framework of perfect and permanent creation.

There were several naturalists, however, in the other school of thought who, in the light of new evidence, thought along the lines of gradual change over time. For example, Scottish geologist James Hutton (1726-1797) proposed that the geologic features of Earth can be explained by mechanisms causing gradual change which are still in operation today as they were in the past. These mechanisms give rise to slow but continuous change, which accumulates over time into a profound change. Charles Lyell (1797-1875), the foremost geologist of his time, incorporated these ideas into his principle of *uniformitarianism* discussed in the next section.

1.6 Forecasting Shadows of the Theory of Evolution

First, many scientists in the eighteenth century interpreted the adaptations, that is, the remarkable match between organisms and their environment, as evidence for perfect creation by the Creator. Yet, as mentioned earlier, Charles Darwin and Alfred Wallace were not the first ones to think along the line of gradual change or evolution as opposed to catastrophes. In fact, there were several naturalists during the eighteenth century, a century before Darwin, including Charles Darwin's grandfather Erasmus Darwin (1731-1802), who hypothesized that life evolves as the environment changes. Another naturalist worthy of a mention here is Georges-Louis Buffon (1707-1788), who about a century before Darwin tried to understand the similarities between apes and humans. He went as far as talking about the common ancestry of apes and man. He addressed these issues in his *Historie Naturelle*, a 44-volume encyclopedia describing almost everything known about the natural world at that time.

Note. Erasmus Darwin, grandfather of Charles Darwin, was an English physician and a naturalist who, being a freethinker, turned down George III's invitation to be a physician to the King. One of the key thinkers of the Midlands Enlightenment movement, he was also a physiologist, abolitionist, inventor, and a poet.

Although Buffon from his studies believed in the change in species over time, he failed to provide a coherent mechanism about how this change occurred. Nevertheless, he proposed that the environment acted directly on organisms through what he called *organic particles*. In another publication, *Les Epoques de la Nature (1788)*, Buffon proposed that our planet was much older than the 6,000 years proclaimed by the church. Regarding change, he introduced concepts that were very consistent with the principle of *uniformitarianism*, which Charles Lyell proposed 40 years later. The principle of uniformitarianism, which incorporated Hutton's ideas (mentioned in the previous section), stated that change occurred gradually and that the mechanisms of change were constant over time; meaning that the geological processes of today are the geological processes of the past, operating at the same rate as in the past. In other words, change occurs slowly and gradually as opposed to the sudden change proposed by the principle of catastrophism. These ideas existed before Darwin, and he was profoundly influenced by them. What separated Buffon from most of his contemporaries was his empirical and philosophical pursuit of causes and explanations beyond the prevailing and well-accepted (or believed) explanations of his time. He did this in the face of two thousand years of dogma sponsored by the religious establishment of the time. His daring way of looking at the natural world paved the way for other revolutionary thinkers, who are largely responsible for much of what we now know about the natural world.

> **Note.** The two thousand years of dogma faced by many naturalists at the times of Buffon was composed of two church doctrines that provided sweeping biblical explanations for most questions about biological diversity. The first doctrine was that all creatures were created independently of one another by God and then organized into a hierarchy (chain of being) with Man occupying the most elevated rank beneath God. The second doctrine was that of the *Young Earth*: The age of the planet Earth being 6,000 years.

Note that the point here is not that catastrophes do not occur at all or that catastrophes have no significance; the point is that catastrophes are not responsible for all (even most) change represented by the fossil evidence. You will see in Chapter 4 how the standard theory of evolution accommodates change through catastrophes.

Charles Darwin was strongly influenced by the ideas of Buffon, Hutton and Lyell. In the framework of these ideas, he realized that if geologic change resulted from slow and gradual mechanisms as opposed to catastrophes, then the Earth must be much older than its widely accepted age of a few thousand years proposed by the creation doctrine. However, Darwin was not the first to apply the idea of slow but continuous change to biological evolution. There was one predecessor of Charles Darwin who actually presented a testable mechanism for how life gradually changed or evolved over time in response to changes in the environment. He was French biologist Jean-Baptiste de Lamarck (1744-1829), who devised his hypothesis based on his comparison of living species with fossils. His hypothesis had two parts: evolution of new characteristics, and inheritance of these characteristics by offspring. The first part, the principle of *use and disuse*, stated that the parts of the body that are used extensively become larger and stronger, whereas the unused parts deteriorate over time. In other words, use it or lose it. The second part of his hypothesis stated that an organism passes the acquired new characteristics to its offspring, a principle called *inheritance of acquired characteristics*. Lemarck cited the example of how the long neck of giraffes evolved out of the need or desire to reach the branches higher up in the trees. He argued that giraffes stretched their necks during their lifetime out of the need to reach food, and a long neck acquired this way was passed down to the offspring. Little by little, increases per generation in neck length would accumulate to give rise to a very long neck over several generations.

According to Lamarck's hypothesis, life was created in its simplest form, and it keeps improving gradually due to its drive for perfection. This drive representing internal need along with environmental pressure brings about changes in the body form and functionality that can be passed to the offspring. What largely distinguishes Lamarck's hypothesis from the standard (or modern) evolutionary theory (developed later) is the proposal that the macro-characteristics (phenotypes) acquired during an organism's lifetime could be inherited by its offspring: *inheritance of acquired characteristics*. This is also the major problem with Lamarck's mechanism, as even simple observations will counter this claim. For example, if you develop large muscles through weight training, your children will not automatically grow up as individuals with large muscles. As another example, a tree during its lifetime can be trained to grow as a dwarf (bonsai) by pruning and shaping. Nevertheless, seeds from this tree would produce trees of normal size as opposed to the size of a bonsai. So, the idea of inheritance of acquired characteristics was ridiculed at the time and has been getting bad rap ever since. This idea also had another implication which has been proven to be wrong, that is that an individual evolves, whereas evidence suggests that it's the population and not the individual that evolves. We will discuss this issue further on in the book in Chapter 4.

So, Lamarck's hypothesis, along with catastrophism, faded away due to lack of support by scientific observations and experimental results. However, Lamarck's greatest contribution to biology is his visionary recognition that patterns in fossils suggest gradual evolution of life over time.

Unfortunately, he is remembered not for this vision and not for being the first in offering a testable hypothesis for evolution, but for presenting the *incorrect* mechanism for evolution, the emphasis on *incorrect*. In many ways, Lamarck laid the foundation for the theory of evolution originally presented by Darwin and Wallace. Furthermore, Lamarck made many other contributions to biology. For example, not many naturalists at that time considered invertebrates (animals without a backbone) worthy of study. Not only did he begin the study of invertebrates, Lamarck coined the word *invertebrates*. With a diverse background in botany and zoology, he believed that all living things should be studied as a discipline and he was one of the very first few naturalists who used the term *biology*. He took on the enormous challenge of creating the new field of biology through learning and researching: yes, he coined the term *biology*. But his recognition of the evolution of life remains his most important contribution to biology.

> **Note.** To test Lamarck's hypothesis of inheritance of acquired traits, in the 1880s, August Weismann cut the tails off mice and found that their offspring developed tails anyway. He repeated his experiment over multiple generations in a sequence and found that the offspring always continued to develop tails. This way, he falsified Lamarck's hypothesis of inheritance of characteristics acquired during the lifetime of a parent organism.

To summarize: though pragmatically the mechanism proposed by Lamarck had been shown not to be true, it was a predecessor to the idea of natural selection. Lamarck's ideas contributed to progress in biology by challenging the belief in the immutability of species, the belief that all species were created separately and permanently. This way Lamarck set the stage for the theory of evolution.

And God Said: Let There Be Evolution

Jean-Baptiste Lamarck was born on August 1, 1744 in the village of Bazentin, located in the Picardie region of Northern France. He was born as the eleventh child in an impoverished aristocratic family with a centuries-old tradition of military service. Lamarck entered a Jesuit college (seminary) in Amiens in the late 1750s. In 1761 he joined the French army where he showed great physical courage on the battlefield in the Pomeranian War with Prussia. During his military post in Monaco, he stumbled onto *Traité des plantes usuelles*, a botany book by James Francis Chomel.

Jean Baptiste Lamarck (1744-1829)

After the war was over in 1763, Lamarck left the army due to an accidental injury. With a reduced pension of only 400 francs a year, but determined to pursue a profession of his choice, Lamarck began to study medicine and botany while supporting himself by working in a bank office. After studying medicine for four years and French flora (as part of his botany studies) for ten years, he published his work in 1778 on the plants of France. This work appeared in three volumes titled *Flore Française*. The book earned him praise from many scholars including Buffon, one of the top French scientists of his time, who mentored Lamarck and helped him gain membership into the French Academy of Sciences in 1779. Also due to the wide acceptance of *Flore Française*, Lamarck was appointed an assistant botanist at the royal botanical garden, the Jardin des Plantes, which was also a center for medical education and biological research.

Later, Lamarck was appointed professor of zoology at the Muséum national d'Histoire naturelle when it was founded in 1793. As a result of his diverse background and research interests, Lamarck worked on various topics and published a series of books on invertebrate zoology and paleontology. He even wrote papers on topics in physics, including meteorology. Although he won praise from some noted scholars about some of his books, overall Lamarck's work never became popular during his lifetime. For instance, he never won the respect or prestige that his colleague Cuvier and his patron Buffon enjoyed. While Cuvier respected Lamarck's work on invertebrates, he used his influence to discredit Lamarck's ideas on evolution. However, Charles Darwin, Haeckel, Lyell, and other early proponents of evolution acknowledged Lamarck as a great zoologist and as a forerunner of research in evolution. Yet today, the name of Lamarck is largely associated merely with a discredited hypothesis of heredity: *the inheritance of acquired traits*.

On the financial side, Lamarck's life was a continuous struggle against poverty. Around 1818 he began to lose his eye sight and spent his last years completely blind. Upon his death on December 28, 1829, in Paris, he received a poor man's funeral (although his colleague Geoffroy Saint-Hilaire gave one of the orations). Because Lamarck was buried in a rented grave, after five years his body was removed. Now, no one knows where the remains are of this founder of the field of biology.

Lamarck published his hypothesis in 1809, the year Charles Darwin was born.

1.7 Here Comes Charles Darwin

We described in Section 1.4 how experimental data and observations in multiple fields were challenging the prevailing belief of perfect creation. Nevertheless, despite all the evidence against it, in the beginning of the nineteenth century, *perfect creation* was still the prevailing belief: species had remained unchanged since their creation. In 1831, in the midst of this confusion created by questions unanswered by the perfect creation doctrine, a young boy of 22 named Charles Darwin was trying to figure out what to do with his life. Charles, born on February 12, 1809 in Shrewsbury, England, was fascinated with all things life. Since his childhood, he had a keen and consuming interest in nature with hobbies including reading books on nature, fishing, collecting plants and insects, and hiking and exploring fields around his home. Charles Darwin, baptized on November 17, 1809 at St. Chad's church in Shrewsbury, regularly attended services at the Unitarian church on High Street with his mother. Even though his mother passed away when he was eight years old, he continued to go to church on a regular basis with his sisters.

In 1825, at his physician father's insistence, Charles became a student at the University of Edinburgh to study medicine. However, the crude and painful surgical procedures used on patients, a standard practice during those times, sickened him. Eventually, bored with his studies, he quit medical school. Directed by his father again, he went to Cambridge University to become a clergyman. Many science scholars during those times in England belonged to the clergy. It may come as a surprise to many that at Cambridge, Darwin actually earned a degree in theology:

Charles Darwin (1809-1882) in 1854

the study of spirituality, religious faith, and God. However, as part of this four-year degree program, he also studied geology and other natural sciences with keen interest. A leading geologist, Adam Sedgwick, took Darwin with him on a geology field trip to Wales in Southern England. Impressed with Darwin's interests and abilities, Professor Sedgwick predicted that Darwin would make his mark in science. Also while at Cambridge University, Charles Darwin became the protégé (the receiver of mentorship) of Reverend John Henslow, a botany professor there.

After finishing his degree in theology, Darwin had no eminent drive to become a minister to preach gospel. His direction in life was still unclear when Professor Henslow (Darwin's mentor at Cambridge), who had perceived Darwin's real interests, actively recommended him to Captain Robert Fitzroy, who at the time was preparing the HMS Beagle, a survey ship, for a long voyage around the world. Fitzroy selected Darwin as a naturalist on the Beagle due to his education and also because they belonged to the same social class and were about the same age.

1.8 Here Comes Beagle: The Mother Ship of the Theory of Evolution

I call the Beagle (Figure 1.5) the mother ship of the theory of evolution because it was the observations made and the data collected during the voyage of the Beagle that, in light of the already lingering questions discussed earlier, led Charles Darwin to formulate the theory of evolution. However, note that the primary official purpose of the voyage was to chart the poorly known stretches of the coastline of South America. So, at a time when Western Europeans were exploring and charting the whole world, HMS *Beagle*, a *Cherokee* class (10-gun class) of brig-sloop of the Royal Navy, set sail for South America on December 27, 1831. During the Atlantic crossing, Charles Darwin read the first volume of Charles Lyell's *Principles of Geology*, the book that he received from Professor Henslow as a parting gift. The expedition, originally planned to last for a duration of two years, lasted for almost five years—the *Beagle* returned on October 2, 1836.

Figure 1.5 HMS *Beagle* in the seaways of Tierra del Fuego, painting by Conrad Martens during the voyage of the Beagle (1831-1836). From *The Illustrated Origin of Species* by Charles Darwin, abridged and illustrated by Richard Leakey. The abbreviation HMS stands for His/Her Majesty's Ship.

During the five-year-long voyage of the Beagle, nature hit Darwin with all kinds of biological data from all sides. While the ship's crew was working surveying the coast, Darwin was busy on the shore observing and collecting animals and plants. This way, Darwin spent most of his time exploring: overall three years and three months on land, and eighteen months at sea. He took notes of diverse species living in diverse environments that ranged from the humid forests of Brazil to the expansive grasslands of Argentina, to misty mountains of the Andes, and to sandy shores of remote islands. He found a good number of fossils, many of which looked unusual at the time: different from living species, yet distinctly local in their resemblance to the living organisms of that continent. He collected thousands of different varieties of animals and plants.

Darwin was fascinated by the unusual organisms that he found at the cluster of volcanic islands called the Galápagos, located near the equator about 540 miles west of South America. The Galápagos proved to be a mine of biological data. There he collected several kinds of mocking birds. Some of them, even though they looked similar at first glance, were different enough to be categorized as different species. Darwin noticed that most of the animals there were uniquely local, and were not known to exist anywhere else in the world. Some of them were even localized to individual islands, and others were found only at two or more neighboring islands. From these observations, Darwin hypothesized that this cluster of islands, the Galápagos, had been populated by organisms whose ancestors were strayed away, say by a storm, from the mainland of South America in the distant past. After colonizing the Galápagos, they multiplied, diversified, and gave rise to new species to fill the environmental niches available there without competition. This type of speciation is called *adaptive radiation*. Speciation, in general, is a process of splitting one species into two or more new species.

In a nutshell, during the voyage of the Beagle, Darwin observed clear evidences of adaptations, that is, inherited characteristics of organisms that enhance their survival and reproduction in specific environments. As mentioned earlier, finding connections and patterns is the core element of scientific thinking. Darwin started to see the connection between adaptation to the environment and the origin of new species. He started pondering: could it be that a new species arises from an ancestral species by the gradual accumulation of adaptations to a changing or different environment?

Along with the Beagle, Darwin returned to England with an enormous amount of data that included 1529 species in bottles of alcohol, 3907 dried specimens, live tortoises, hundreds of pages of zoology and geology notes, and a 770-page diary filled with observations and thoughts based on those observations.

Scientists make sense of observations and data by using an existing theory or by developing (or proposing) a new one if the existing theories fail to explain the data at hand.

1.9 Here Comes the Theory of Evolution

After his return (onboard the Beagle) to England in 1836 with an enormous amount of data and samples, Darwin started putting pieces together to recognize patterns. For example, stored in the cargo hold of the Beagle, came home a group of preserved bird specimens, finches from the Galápagos Islands. Darwin had learned the skill of preserving bird specimens during his studies at the University of Edinburgh. He originally thought the finches in his group of specimens belonged to different species because they had different morphological features such as different beak sizes and shapes. John Gould (1804-1881)–an artist and ornithologist–developed illustrations of finches in this group (four of these illustrations are shown in Figure 1.6). Gould determined the following:

1. All of these birds were closely related species or sub-species of finches.

2. Yet, these species were so similar that all of them could belong to a related group of buntings.

3. Although these species were unique to the Galápagos Islands and found nowhere else, they had their look-alikes in South America, the nearest mainland, 900 km away.

Based on Gould's work, Darwin began theorizing how the beak size (Figure 1.6) of finches may have changed over the generations owing to their different environments, such as differences in the size of seeds and insects available to them for consumption on the various islands. He knew from Gould's work that most of the finch species lived nowhere else than the Galápagos Islands, yet they shared characteristics (traits) with some species on the mainland. Could it be that all these finch species were diversification from a common ancestral species that somehow in the distant past made it to the islands from the mainland? It's not impossible to imagine that a strong storm may have blown some mainland birds into the sea: some of them obviously died, and a few of them made it to the islands alive. This scientific line of thought could then be applied to other species of organisms that Darwin brought home from the voyage.

An immediate result of the analysis performed by Darwin and Gould was the publication in 1839 of *The Voyage of the Beagle* in which Darwin wrote, "Seeing this gradation and diversity of structure in one small, intimately related group of birds, one might really fancy that from an original paucity of birds in this archipelago, one species had been taken and modified for different ends." This publication can be considered as the first milestone toward developing the full-blown theory of evolution, which was eventually presented in *On the Origin of Species* (popularly called *Origin of Species*), published in 1859, 20 years later. What took him 20 years? Largely because, like any good scientist, he wanted to ensure that the arguments in his theory were firmly based on facts and were beyond reproach. However, by the 1840s, Darwin had worked out the major features of his evolutionary hypothesis; and in 1844, he summarized his hypothesis in an essay that he did not publish immediately, apparently because he anticipated that there would be uproar against his hypothesis or theory once published. This is easy to realize if you note that with a few exceptions, all people were creationists before Darwin and Wallace. Almost all people including scientists believed in the natural theology of life. As mentioned earlier, Darwin himself earned a degree in theology and was sent to Cambridge for this purpose with the intentions of becoming a minister. In spite of delaying publishing his work, Darwin continued working on his evolutionary ideas and compiling evidence in support of them.

1. Geospiza magnirostris.
2. Geospiza fortis.
3. Geospiza parvula.
4. Certhidea olivacea.

Figure 1.6 An illustration of finches from Galápagos Archipelagos by John Gould. Appeared in The *Voyage of the Beagle* by Charles Darwin.

Remember Charles Lyell whose book Darwin read during the voyage of the Beagle? Darwin explained his hypothesis to Lyell and a few other scholars. Even though Lyell himself was not yet convinced of evolution, he suggested to Darwin to publish his ideas before someone else arrived at the same conclusion and published it first. Low and behold, Lyell's prediction came true in 1858, when Darwin received a manuscript of a paper from Alfred Russel Wallace, a British naturalist working in the East Indies. Wallace had developed a hypothesis of *natural selection* similar to Darwin's: you will learn about natural selection further on in this chapter. Wallace requested Darwin review his paper and forward it to Lyell if it was worth publishing. Darwin forwarded the paper to Lyell, accompanied by a note that Darwin wrote to Lyell: "Your words have come true with a vengeance…I never saw a more striking coincidence…so all my originality, whatever it may amount to, will be smashed." On July 1, 1858, when Lyell and a colleague presented Wallace's paper to the Linnean Society of London, they also included extracts from Darwin's unpublished essay of 1844.

After this event, Darwin quickly put his book together titled: *On the Origin of Species by Means of Natural Selection*, commonly referred to as *The Origin of Species*. The book was published the next year, 1859. However, as mentioned earlier in this chapter, Darwin was not the first scholar to present the concept of evolution. He certainly was the first one to convince most biologists, through his book, that evolution is the cause of the diversity of life. Due to his sound arguments supported by data, biologists started taking the concept of evolution seriously. Darwin had presented his ideas based on an avalanche of evidence and in the form of a hypothesis that could be tested and falsified. In other words, his ideas fit into the scientific method, discussed in the next chapter.

> **Note**. You may find it interesting that in the first edition of *The Origin of Species*, Darwin never used the word *evolution*, except that the final word of the book is "evolved". Instead of *evolution*, he rather used the phrase *descent with modification*.

Nevertheless, as Darwin expected, the publication of his book triggered a storm of controversy. This is because just like before Thales of Miletus, the world did not believe that nature could be understood in terms of a few principles; before Darwin, the world did not believe that life could be understood in terms of a few natural principles. This book single-handedly lifted the mysteries and phenomena of life from mythology or theology and placed them into science for inspection.

Alfred Russel Wallace was born in Kensington Cottage near Usk, Monmouthshire, England (now part of Wales) on January 8, 1823. He was one of the nineteenth century's greatest naturalists, who made remarkable contributions to various fields such as biology, anthropology, and biogeography. Even though he is the co-discoverer of evolution by natural selection, his role is much undervalued in the modern literature about evolution, including most of the biology textbooks. Due to his pioneering work on the geographical distribution of animal species, he's also considered by many to be the father of biogeography.

Alfred Russel Wallace (1823-1913)

After a few years of school education, when Wallace's father passed away, he had to drop out of his school and join his elder brother's surveying business. Through this four-year experience, he acquired a talent for observation and detailed recordings which helped him immensely later on in his research work. In 1843, due to a slump in surveying work, Wallace found a position at the Collegiate School in Leicester as a master to teach arithmetic, drafting, surveying, and English. Leicester had a good library, and he made good use of it to study several important works on natural history. During the year of 1844, he made the acquaintance of a young, amateur naturalist named Henry Walter Bates, who was already an accomplished entomologist (zoologist focusing on insects).

In 1845, Wallace moved to Neath, Wales, where he first read Robert Chambers' controversial book *Vestiges of the Natural History of Creation*, which got him keenly interested in the topic of evolution, then known as transmutation. In 1847-1848, inspired by W. H. Edward's book *A Voyage Up the River Amazon*, Wallace convinced Henry Walter Bates to travel to Brazil to collect specimens of insects, birds, and other animals; both for their private collections and to sell to collectors and museums in Europe. However, Wallace's primary task during this expedition was to collect data to test ideas about evolution as an attempt to discover its mechanism. So, on April 26, 1848, the two young men (Bates of 23 and Wallace of 25) set off by ship from Liverpool (England) to Belém, Pará (Brazil). However, this was only the first of many projects to collect data. For example, during his collecting expedition to the Malay Archipelago (now Malaysia and Indonesia), he spent nearly eight years in the region, and undertook sixty separate journeys which totaled to around 14,000 miles of travel. During this expedition, he visited each important island in the archipelago at least once, and collected almost 110,000 insects, 8050 bird skins, 7500 shells, and 410 mammal and reptile specimen. His collections included several thousand species new to science at the time. The idea of natural selection

> dawned on Wallace in February of 1958, when he was suffering from fever in the village of Dodinga on the remote Indonesian island of Halmahera.
>
> Alfred Wallace died in his sleep at the age of 90 on November 7, 1913 at Broadstone, England, in his country house that he had built a decade earlier, named *Old Orchard*. News of his death was widely reported in the media. For example, *The New York Times* called him "the last of the giants belonging to that wonderful group of intellectuals that included, among others, Darwin, Huxley, Spencer, Lyell, and Owen, whose daring investigations revolutionized and evolutionized the thought of the century."

The lingering questions in biology and the new data, discussed in Section 1.4, were explained by Darwin and Wallace with principle of natural selection and the tree of life, discussed next.

1.10 Recognizing the Tree of Life: Change and Adaptations

Analyzing the collected data and observations during his voyage, in the light of the existing knowledge at that time, Darwin saw a pattern pointing to the unity of life despite its great diversity. He attributed this unity behind diversity to the descent of new species from previous species and the descent of all organisms from a common ancestor (or ancestors) that lived in the distant past. To present this hypothesis, he used the phrase *descent with modification* in his book *The Origin of Species*. His argument went like this: As the descendants of ancestral species lived through the changing environment, they, over multiple generations, accumulated diverse modifications to adapt to specific ways of life in order to fit their specific environment. Over a very long period of time that includes many generations, descent with modification eventually led to new species and therefore the diversity of life that we see today. Yet all these different species (diversity) started from a common origin (unity), that is, a common ancestor; therefore, they still share some common characteristics: the underlying unity behind diversity. *Adaptation* is the key in this process.

From the Galápagos Islands to mountains, plains, and deserts, anywhere in the world, adaptations are everywhere. The beak variations among the Galápagos finches (Figure 1.6) are actually adaptations to specific diets available. For example, the cactus ground finch (*Geospiza scandens*) uses its long, sharp beak to tear and ingest cactus flowers and pulp, whereas its narrow, pointed beak helps the green warbler finch (*Certhidea olivacea*) to grasp and ingest insects. The large beak of the large ground finch helps it to crack seeds fallen from plants to the ground. The result of these different adaptations is more than a dozen closely-related species of finches, some of them populating the same island. Somewhere else in the world, the Namib Desert along the coast of Southwestern Africa hosts a high richness in species of beetles, many of which evolved as a result of adaptations in terms of different methods of condensing fog as a source of water. Even though the desert does not have any rainfall, fog is abundant. For example, the head-standing beetle (*Onymacris unguicularis)* creeps to the crest of a dune when fog is present. It then faces into the wind that blows fog across the dunes and stretches its back legs to tilt its body forward with head down. With this peculiar behavior, the beetle drinks the water necessary for its survival as fog precipitates onto its body and runs down into its mouth (Armstrong 1990). Any theory of life must explain the striking ways in which different organisms adapt to their specific environments.

And God Said: Let There Be Evolution

As shown in his original 1837 sketch presented in Figure 1.7, Darwin conceived the history of life as a tree: multiple branches emanating from a common trunk all the way out to the tips of the youngest twigs. The branching represents the evolution of diversity in general, and the twigs represent the diversity of organisms or species living today. Branches diverging from a specific point represent new species evolved from a common ancestor species represented by the common trunk from which the branches emanate. As mentioned earlier in this chapter, this process of splitting one species into two or more new species is called *speciation*. In other words, speciation is the evolution of a common ancestor into at least two different groups that cannot interbreed, that is, cannot exchange genes. How does it really happen that an ancestor splits into two new species? As an example, consider two different populations of the ancestor species living in two different places. These two populations would evolve differently while adapting to different environments as the process of natural selection operates on them. Over time these differences would grow large enough to result in two different species.

Natural selection. A process in which organisms with certain inherited traits that fit the environment better have better chances to survive and reproduce. This process, by definition, always improves the fit between the population and its environment.

Each fork point in a tree (inverted) in Figures 1.7 and 1.8 represents an ancestor of all the lines of evolution represented by the branches that subsequently branch underneath it and can be linked (traced) back to that point (ancestor). The common ancestors, not living today, are also called missing links to the descendant species. The evidence for some of these missing links have been found in fossils. We will explore this issue in further detail in Chapter 6.

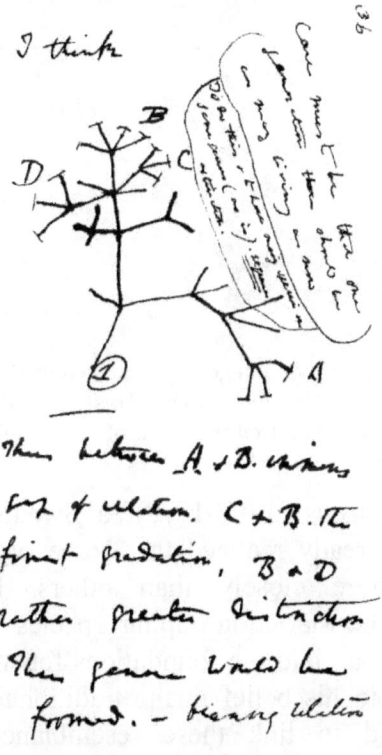

Figure 1.7. Charles Darwin's 1837 sketch; his first diagram of an evolutionary tree from his First Notebook on Transmutation of Species (1837) on view at the Museum of Natural History in Manhattan. Interpretation of handwriting: "I think case must be that one generation should have as many living as now. To do this and to have as many species in same genus (as is) requires extinction. Thus between A + B the immense gap of relation. C + B the finest gradation. B+D rather greater distinction. Thus genera would be formed. Bearing relation," (next page begins), "to ancient types with several extinct forms"

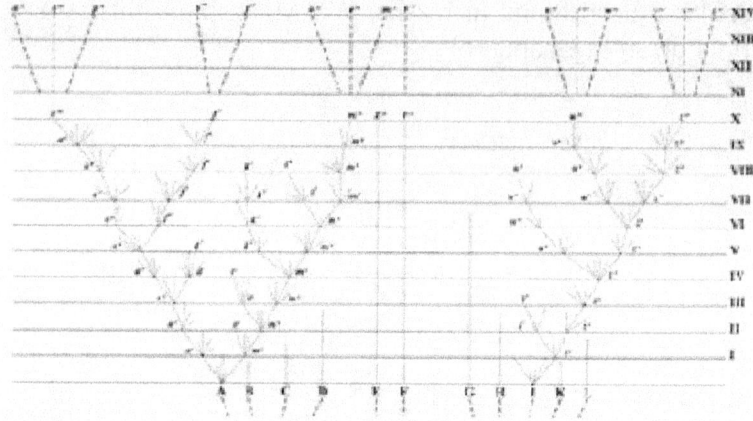

Figure 1.8 The *Tree of Life* image that appeared in Darwin's *On the Origin of Species by Natural Selection*, 1859. This is the only figure in *On the Origin of Species*. Darwin used this Tree of Life (TOL) as a model for the theory of evolution.

Carolus Linnaeus (1707-1778), a Swedish physician and botanist, had already realized that some organisms resemble each other more closely than others. Based on these resemblances, he started grouping species in a hierarchical order and this way laid the foundations for modern taxonomy. However, due to his belief in the traditional thinking of that time, he failed to link these resemblances to evolution. Nevertheless, by recognizing that the great diversity of living organisms could be organized into groups and subordinate groups, Linnaeus helped pave the way toward discovering the theory of evolution. His model meshed well with the approach taken by Darwin, to whom the Linnaean hierarchy of organisms reflected the branching history of the tree of life. By exploring further the Linnaean idea of hierarchy of life, Darwin realized that organisms at the various levels of the hierarchy are related through a single principle: *descent with modification from common ancestors*.

To summarize, diverse species arose descendant from a common ancestral species with modifications facilitated by the process of natural selection. *Descent with modification* gave rise to diversity, descent from a *common ancestor* remained the source of the underlying unity behind diversity, and natural selection facilitated the accumulation of diverse sets of modifications that fit the specific environments in which different groups of organisms lived generation after generation. These modifications, which are different for different groups that lived in different environments, are called adaptations.

What is natural selection?

1.11 Evolution by Natural Selection

Darwin perceived the diversity of life as an outcome of evolution, a process through which all life from the same origin went over the ages due to the changes in the environment. The changes affected the variations of traits among organisms, which in turn caused evolution resulting in diversity, and it happened gradually over ages. A central piece of the theory of evolution is *natural selection*, a hypothesis that Darwin (and Wallace) proposed to explain the observable patterns of evolution. As Darwin was possibly aware of the uproar his theory would cause because of its departure from traditional thinking, he introduced the hypothesis of natural selection in his book very carefully and diligently. He started with discussing familiar examples of selective breeding of domesticated animals and plants. The idea of selection should not be foreign to us now, living in the age of genetically modified food, which is equivalent to performing selection at genetic level. Moreover, we have modified species of organisms such as dogs over many generations by selecting and breeding individuals that possess desired traits. We know that plants and animals artificially selected and bred bear little resemblance to their wild ancestors. This is *artificial selection.* When nature does it, it's called natural selection, and Darwin was able to perceive this by analyzing the data.

Think About It!

What is the smallest unit of evolution?

    A. Species

    B. Population

    C. Organism

    D. Gene

Answer: Population

Natural selection operates at the organism level resulting in evolution at population level over generations. A species may be living in more than one population with different environments and hence different selection pressure. Evolution takes generations to produce results. Therefore the smallest unit of evolution is population and not an organism or species.

To present his hypothesis, Darwin cited four observations:

1. All species of organisms have inherent capability of producing more offspring than their environment can support.

2. Because resources such as food are limited, there comes a competition for survival, and as a result many overproduced offspring do not survive.

3. Members of a population often vary in their traits.

4. Individuals of a population inherit certain traits from their parents.

> **Note**. A population is a group of organisms of the same species living in one geographical location. Individuals belonging to the same species, by definition, means they have the potential to interbreed and produce a fertile offspring.

The key points in these observations are: *overproduction*, struggle (or competition) for *survival* due to limited resources, *variation* in traits, and *inheritance* of traits. It's easy to understand that reproduction capability and inheritance of traits link one generation to another, and that the nature and quality of traits plays a role in survival and reproduction.

So, considering this, it won't be difficult to make the following two inferences from the four observations cited above:

1. All traits are not equal: In a given environment, certain traits are better than others at helping an individual that possess them to compete more effectively for resources, and therefore improve their chances for survival and reproduction in that environment. Therefore, the individuals with favorable traits will leave behind more offspring than those who do not possess favorable traits.

2. This unequal ability to survive and reproduce differentiated by traits leads to accumulation of favorable traits in the population over generations. At some point, this trait shift will give rise to a new species.

These are the inferences that Darwin made from his observations and which constitute the hypothesis of natural selection according to which individuals in a population with favorable traits survive and reproduce at a higher rate than other individuals, and as a result the favorable traits (which are passed down to the next generation) prevail over generations. It's important to realize the generation connection here. An individual organism's traits can influence not only its own chances of survival and reproduction but also how well its offspring will perform in the given environment. For example, certain inherited favorable traits can give the offspring advantage over others in gathering resources, escaping from predators, and tolerating the harsh physical conditions of a given environment. Such advantages will increase the number of offspring with favorable traits. As a result, over generations, the favorable traits will dominate the population. So, natural selection this way operates as a differential in choosing one trait against the other (along with the trait-bearing organisms), and over ages this leads to the disappearance of old species and the appearance of new species.

Think About It!

By using artificial selection, you can bring about a dramatic change to create a new kind of organism in a relatively short period of time, whereas a new species through natural selection evolves over multiple generations, which takes a long time. Darwin stressed this point in his book.

Q1. Is it true that evolution is always slow?

Q2. Is natural selection the only mechanism of evolution?

A1. The speed of evolution through natural selection depends on the selective pressure by the environment.

A2. As we will discuss in Chapter 4, evolution can also occur through mechanisms other than natural selection.

Here are the connections between three central concepts of natural selection:

1. Variations. Individuals within a population vary with respect to their traits. Individuals with more favorable traits

survive and reproduce more than others. Over time this gives rise to the domination of the individuals with favorable traits in the population.

2. **Change.** When the environment changes or the population moves to a new environment, some traits adapt better to the change than others. This causes a change in dominant traits within a population over generations. This way, change in environment flows to change in the population.

3. **New species.** Over time, the fit between the new environment and the survived traits becomes gradually better. In this continuous improvement, some old species disappear and new species appear.

To avoid misunderstanding, it's important to realize the following points:

1. **The basic unit of evolution is population.** Because natural evolution is slow in most cases and it takes generations to give rise to new species, it is the population and not the individual organisms that evolve. In other words, evolution occurs at the population level over generations and not at an individual level within one lifetime. However, it is the individual that is selected for a given environment, but it is the population that evolves.

2. **Evolution is based on variations among heritable traits only.** Natural selection can increase or decrease the frequency of *only heritable* traits in a discriminative way within a population. An organism can modify its many traits during its lifetime, but only the heritable traits will contribute to the evolution of a population.

3. Favorability of traits depends on the environment. Traits that are favorable in one environment may be unfavorable or neutral in another environment.

The final point to make here is that what we call the theory of evolution today, the standard theory of evolution, is based on the theory of evolution by natural selection (that we have just discussed) originally introduced by Darwin and Wallace, and not on Lamarck's hypothesis discussed earlier in this chapter. According to that hypothesis, characteristics, let's call them macro-characteristics, such as a long neck, are built by efforts made in order to adjust to the environment to fulfill a need; and these characteristics acquired during a lifetime are then passed on to the offspring. The suggestion here is that an organism could change its structure during its lifetime because it felt the need to do so, and then that change can be passed down to the offspring. As discussed earlier in this chapter, there is no evidence that this is true. So, a natural question that arises here is: what characteristics (traits) are passed on and how. It took decades to find answers for these questions, which we will discuss in the next chapters.

> Caution! Many vocal opponents of evolution, while ridiculing theory of evolution, often confuse it with Lamarck's hypothesis and fail to distinguish between the two. So when you are reading about the theory of evolution in this book, make sure you can distinguish it clearly from Lamarck's hypothesis.

One important component missing in Lamarck's hypothesis is the principle of natural selection. It's important to note that most biologists had no trouble accepting that evolution is true, whereas many of them did have trouble with accepting natural selection as a cause for the evolution. They accepted the evolution part of Darwin's theory, but it took them awhile to accept natural selection as the mechanism of evolution. The point here is that Lamarck's hypothesis did not emerge in total isolation; there were many biologists at that time who were exploring different reasons for evolution, such as inner drive, that is, the inner drive made species develop (or evolve) in a certain prescribed direction.

According to the standard theory of evolution, organisms do not evolve within a lifetime based on their desires and needs. This theory recognizes that even within the same species, variations among traits exist. In the changing environment, natural selection operates on these variations to select the organisms with traits better adapted to the new environment. Evolution occurs not at the organism level (in the lifetime of an organism) but at the population level over generations. An organism that does not have traits that adapt well to the environment has fewer chances of surviving and reproducing.

Think About It!

Explain how Lamarck's hypothesis and Darwin's theory of evolution will explain the evolution of the long neck of a giraffe from its short-neck ancestor.

According to Lamarck:

The short-neck giraffes, during their lifetime, stretched their necks to reach food, and handed down the resulting long-neck trait to their offspring. The later generations kept increasing the neck length to meet their needs and kept handing down this increase to their offspring. Key points: a characteristic acquired during a lifetime to meet needs is handed down to the offspring.

According to Darwin and Wallace:

Within the populations dominated by short-neck giraffes, there were a few long-neck giraffes. When the environment changed so that only the long-neck giraffes could reach the food and most of the short-neck giraffes starved and died off, most of the few long-neck giraffes survived, reproduced, and proliferated. Eventually, after multiple generations, all (or almost all) giraffes had a long neck.

1.12 Theory of Evolution: Putting It All Together

Any theory of life must explain the three key observations of life: adaptations, the rich diversity of life, and yet the unity (shared traits) behind this diversity. Adaptations are the inherited characteristics of an organism that improve its chances for survival and reproduction, and gives rise to the striking ways in which different organisms fit their specific environments. Adaptation is discussed in a bigger context in Chapter 8.

As discussed in this chapter, the concept and original theory of biological evolution took centuries to develop from the cumulative discoveries of many scientists (naturalists) working in many fields such as biogeography, comparative morphology, and geology. In this chapter, we explored the context and the scientific environment in which the theory of evolution was born. Although Charles Darwin (and Alfred Wallace) got it more right than others, the original discovery of the concept and theory of evolution was a result of efforts of generations of scientists. What was original to the theory of evolution proposed by Darwin and Wallace was the mechanism that drives evolution: natural selection.

In a nutshell, here are the three core elements of the theory of evolution introduced by Darwin and Wallace:

1. **Tree of life.** Multiple species emerge (or diverge) from a common ancestor species and evolve along separate pathways. For example, based on collected data and facts, Darwin argued that three species of mockingbirds he observed on the Galápagos Islands could be traced back and related to a single colonization of species that he had observed in Latin America. The underlying idea here is, as Darwin put it, descent with modification from a common ancestor, and that explains unity (common ancestry) behind diversity (descent with modification).

   This hypothesis of common ancestry of different species gives rise to a branching tree of life. We will explore more examples of trees of life in Chapters 6 and 7. A key point here is that offspring inherit traits along with variations from the parents.

2. **Change.** All living things continually undergo a slow change from generation to generation: offspring are not identical to their parents in terms of their traits, that is, characteristics. This uneven change causes variations among organisms of the same generation as well: all siblings are not identical except identical twins. Some of these trait differences or variants, which are beneficial to the organisms in a given environment, accumulate over multiple generations through natural selection. A key point here is that variations of traits exist and flow along the branches of tree of life.

3. **Natural selection.** Now that we have variant traits, a given environment selects those organisms that are best adapted by their trait variants to the existing environmental

conditions. You can imagine that in a given variety (in traits) some kinds of organisms (with specific trait variants) survive and reproduce better in a specific environment than do others. Such organisms, due to their reproductive success, become more common over time and are said to be naturally selected. These selected organisms have been passing along, from generation to generation, the beneficial traits, which are called adaptations. A key point here is that natural selection acts upon the variations.

> **Note.** The tree of life represents the core idea of evolution: descent with modification from common ancestry. Natural selection explains how evolution occurs. Variations are the raw material on which natural selection operates.

So, the three essential elements of evolution are variations, inheritance, and natural selection: variations are inherited and natural selection operates on variations.

Some contemporaries of Darwin shared the view that each lineage of organisms (plants and animals) arose by spontaneous generation from some kind of inanimate matter and then progressed (evolved) toward greater complexity and perfection. In this hypothesis, no species can arise from another species. Darwin's theory rejected this straight-line progression in favor of a branched tree of life in which multiple species can diverge from a single common ancestor species.

> **Caution!** It is very common practice among biologists to omit Wallace's name when referring to the original theory of evolution by referring to it as Darwin's theory of evolution. However, bear in mind that the core mechanism of the original theory of evolution, natural selection, was independently discovered by both Wallace and Darwin.

An implication of natural selection is this: if the existing environmental conditions change, the group of organisms in a population that happen to possess the most traits, or characteristics, adaptive (suitable) to the new environment will come predominate over time, even if they are in small number to start with. This is where the core strength of the theory is: to predict. Furthermore, Darwin's breakthrough insight lies in the compelling argument that evolution became inevitable for the organisms whose environment changed non-randomly and whose effectiveness to reproduce depended on inherited traits. Simply put, previous generations of these organisms flourished (successfully survived and reproduced) under the old environment with the old set of traits. With a non-random change in environment, there would be a new set of traits that would fit the environment better, and the organisms with those traits would more successfully reproduce. Therefore, even if these organisms were in small numbers in the olden days, now in the changed environment they will multiply over generations and define the face of the population; hence, the evolution.

Think About It!

A population consists of 1026 birds of the same species. Among this population there are two birds which have relatively longer and sharper beaks and are considered defective or abnormal. There were a few dozen of these abnormal birds a few generations ago. However, they had trouble gathering food with their abnormal beaks and some of them died before regenerating, and hence, their numbers have been decreasing. Now there are only two of them left (a male and a female).

Due to some natural events, the environment suddenly changed and the only kind of food left is the one that can be better picked with long, sharp beaks of the abnormal birds. As a result, only half of the normal birds survive to regenerate, whereas the abnormal

---

birds live a happy, sexually-productive life. If each abnormal bird produces two baby birds and each normal bird produces only one baby bird, what will be the number of normal and abnormal birds after ten generations?

Answer:

The initial number of normal birds is 1024. Only half of them, 512, will survive to have one offspring each, this is, a total of $512 \times 1 = 512$; and out of these 512 birds only half, $512/2 = 1024/4 = 1024/2^2 = 256$, would survive to have an offspring of 256, and so on. After 10 generations the number of normal birds will be reduced to $1024/2^{10} = \frac{1024}{1024} = 1$.

The second generation of offspring for the abnormal birds will be $4 = 2^2$, the third generation will be $8 = 2^3$, and the tenth generation will be $2^{10} = 1024$.

The abnormal birds will become the norm for this population over ten generations.

It's almost always true that it helps to make new discoveries and inventions if you understand the existing ones. Great scientists often see the future with the help of their contemporaries and in Newton's words, by *standing on the shoulders of giant* figures that came before them. Charles Darwin was no exception. For example, as mentioned earlier, during his voyage, he read two volumes of *Principles of Geology* by Charles Lyell. In these volumes, Charles Lyell challenged catastrophism: the prevailing idea of that time, which endorsed the belief that sudden, violent events driven by some supernatural forces had directed the shaping of the landscape. Lyell argued for *uniformitarianism,* according to which processes such as erosion, sedimentation, and volcanic activities on average occur at more or less the same rate over ages. From his readings of Lyell, Darwin took the idea of gradual change in geological landscape and applied it to living organisms. This kind of analogical thinking plays a great role in scientific research. Darwin was also influenced by the work of Thomas Malthus (1766-1834), a British scholar in political economy and demography, which helped him to realize that populations tend to grow quickly because they have the capability of producing more offspring than the environment can support. This thinking lead to a core principle of his theory: natural selection.

One important conclusion from this chapter is that Charles Darwin was neither the first one to introduce the concept of evolution nor the lone founder of the theory of evolution. The theory was introduced on July 1, 1858 to the Linnean Society in London; Charles Darwin and Alfred Wallace were both represented. Figure 1.9 shows both sides of the medal provided by the Linnean Society to Wallace in 1908 on the 50th anniversary of presenting the papers by Wallace and Darwin on natural selection.

Figure 1.9 Both sides of the Darwin-Wallace medal awarded to Alfred Wallace at the 1908 Linnean Society meeting celebrating the 50th anniversary of the reading of Darwin and Wallace's papers on natural selection. Image: courtesy of the Linnean Society.

1.13 Summary

The concept of biological evolution is as old as the history of science and natural philosophy, but it took centuries to develop the standard theory of evolution, as we know it today, from the cumulative discoveries of generations of scientists. The idea of evolution can be traced back to pre-Socratic Greek philosopher Anaximander (610-546 BC), a friend and student of Thales (624-546 BC), who proposed the very idea that natural phenomena could be understood in terms of a few consistent principles. Anaximander argued that humans gradually evolved in the protection of and from other animals. Nevertheless, before Darwin, almost all (if not all) scientists believed, under the influence of religious beliefs, in the doctrine of perfect creation: that all species were created by the Creator independent of each other in a perfect and permanent form, and have not changed since their creation. However, during Darwin's time, biological questions from the accumulated observations and data collected in various scientific fields challenged this untested doctrine of perfect creation. In this chapter, we discussed the questions raised by biogeography, comparative morphology, embryology, and geology/fossils that could not be answered by the doctrine of perfect creation. In Chapter 6, we will revisit these questions and show how the theory of evolution answers these questions.

Jean-Baptiste de Lamarck (1744-1829) was the first scientist to propose a mechanism, in 1809, for how life evolves. According to this proposed mechanism, characteristics acquired by an organism during its lifetime could be passed down to its offspring; which did not withstand the test of even simple observation. Subsequently, in 1858, Charles Darwin (1809-1882) and Alfred Wallace (1823-1913), in an attempt to explain the wealth of accumulated data and observations, presented the theory of evolution based on the principle of natural selection.

Although the theory of evolution originally presented by Darwin and Wallace has withstood scientific scrutiny, it has come a long way to develop into its modern (standard) form, as research over about one-and-a-half century after its discovery has contributed to its improvements and details. This book presents the theory of evolution in its entirety as it stands today, referred to as the standard theory of evolution in this book. In this chapter, however, we have only discussed the original theory as presented by Darwin and Wallace. Figure 1.10 illustrates how the theory of evolution was originally realized from observations.

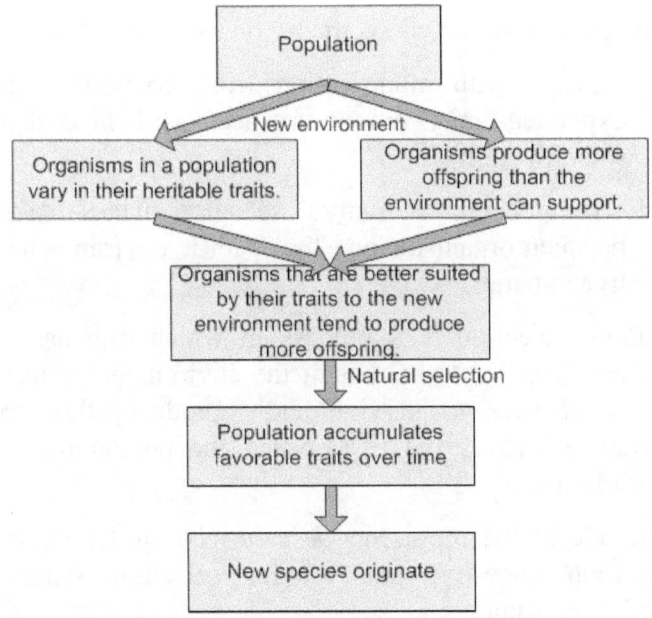

Figure 1.10 An illustration of the argument used in developing the original theory of evolution by Darwin and Wallace.

And God Said: Let There Be Evolution

In a Nutshell

✓ Charles Darwin was not the first one to think about evolution; the idea was proposed, in one or another form, by naturalists from the times of the ancient Greek philosophers.

✓ Lamarck was correct in suggesting that species evolve, but the mechanism of evolution proposed by him was not supported by scientific evidence. Nevertheless, he was the first one to propose a testable mechanism for evolution.

✓ Charles Darwin was not the lone founder of the original theory of evolution either: Alfred Wallace and Charles Darwin both independently discovered the core mechanism of evolution called *natural selection.*

✓ Charles Darwin based his theory on two main ideas:

1. Descent with modification from common ancestors explained unity (common ancestor) behind diversity (modification).

2. Over generations, natural selection makes the match between organisms and the specific environments they live in better.

✓ Natural selection is a process in which organisms with certain inherited traits that fit the environment better have better chances to survive and reproduce; this process always improves the fit between the population and its environment.

✓ The tree of life represents *descent with modification* from common ancestry, and natural selection states how evolution occurs.

✓ Only heritable traits are passed down to offspring.

✓ When the environment changes, natural selection operates on variations in traits to select the traits most favorable to the new environment. Over generations the favorable traits accumulate, and this may give rise to new species.

✓ The three essential elements of evolution are variations in traits, inheritance, and natural selection: variations are inherited and natural selection operates on variations.

✓ Because evolution is based on inherited characteristics and it takes generations to give rise to new species, it is the population and not the individual that evolves. Individuals in the population are selected for and the population evolves.

✓ Theory of evolution is a scientific theory supported by considerable amount of evidence, even at the time it was first introduced in 1858; it was developed to explain the existing data.

✓ The idea of evolution is as old as science itself. It goes back to the times of the Greek philosophers such as Aristotle and Anaximander. However, the theory of evolution originally introduced by Darwin and Wallace has been the first theory capable of withstanding rigorous tests of scientific scrutiny starting from the nineteenth century until today.

1.14 Behold

A careful reader should still have three questions from this chapter unanswered:

1. What really are the heritable traits? In other words, what kinds of traits are passed down to the offspring and how?

2. For natural selection to operate there has to be trait variants. What is the origin of these trait variants? For

example, in talking about the trait of eye color, why does one person have blue eyes and another one brown eyes? For instance, different versions (variants) of the same trait originate from the environment or from within?

3. Where does the modification or change in an existing trait come from? Due to this change or modification, old traits may disappear and new traits may appear during the course of evolution. For example, does this change come from the environment or from within?

These questions are addressed in Chapters 3 and 6. For now, for your peace of mind, the short one-line answer to question 1 is: heritable traits are the traits that are based on and therefore can be passed through genes; the short answer to question 2 is that variations come from alternative versions of genes called alleles; and the short answer to question 3 is: modifications in traits occur due to changes in the genes, called mutations. Of course these answers were not available at the times of Darwin and Wallace. However, do not forget that environment also plays its role in evolution as discussed in this chapter. The role of the environment in the case of genes will also be discussed in upcoming chapters. Also note that in this chapter, our presentation of the theory of evolution has been limited to its original version as introduced by Darwin and Wallace. As it will become clear in the rest of the book, the theory has been improved and extended to accommodate the new discoveries and studies after Darwin and Wallace.

Before we explore these issues, let's deal with another issue that some readers might be struggling with by now. One of the biggest attacks on the theory of evolution by its opponents is: it's just a theory. This one-liner attack can be countered with one-liner defense: It's not *just a theory*, it's a scientific theory. All right, evolution is a scientific theory, but what is a scientific theory anyway? How do you know that a given theory is a scientific theory or *just a theory*?

We address this issue in the next chapter.

And God Said: Let There Be Evolution

TWO

Evolution
Is
Just a Theory:

Or
Is It?

And God Said: Let There Be Evolution

It is a good morning exercise for a research scientist to discard a pet hypothesis every day before breakfast. It keeps him young.

Konrad Lorenz

The great tragedy of science — the slaying of a beautiful hypothesis by an ugly fact.

Thomas Henry Huxley

2.1 Once Upon a Time

An incredible thing happened during the spring of 2009 when my wife and I and our son of twelve were participating in a class organized at our church that lasted for a few Sundays. The topic was *Christian World View*, and the class was based on a video by Rick Warren (author of the *Purpose Driven Life*) and Charles Colson (in the past commonly named as one of the *Watergate Seven*). The sessions were led and moderated by a fellow I used to call Uncle Stalin in my own narrow circle due to his dogmatic views, in my opinion, on a few topics.

Figure 2.1 Straw man; a rather ineffective scarecrow.

One of the sessions was devoted to *beating* on the theory of evolution. Rick and Chuck (Charles) basically mischaracterized the theory and then mocked it, a style very typical of vocal opponents of evolution. I have great respect for Rick Warren, for his writing and communication skills, but what he and Chuck did in that video is very typical of the opponents of evolution. After the video, *Uncle Stalin* reinforced the message including the mantra: it's *just a theory*, why should we believe it? I am usually an ideal

65

audience and do not express my disagreements in church meetings (and pretty much anywhere else as well) unless they are solicited. But being a scientist, I could not let this one go quietly. So I stood up and respectfully made a very brief speech on what do we, scientists, mean by theory in science. My 12-year-old son was able to counter the other arguments presented by Rick and Chuck because he had read about the theory of evolution in his biology book and had discussed it with me.

The point here is that if you create a straw man (as they say in philosophy) and then beat it, you are beating your own creation not the opposing viewpoint: This is critical thinking 101. A biologist church member in the class added to my voice. To my surprise the next Sunday, Uncle Stalin walked over to me and thanked me for "my contribution" to the previous Sunday class. Disturbed by my comments, he had spoken to some other people in the meantime. "Thank you, now I can live with both: evolution and my faith," Ron said, concluding his compliment.

I never referred to him as Uncle Stalin ever again. Science or not, in order to logically condemn a concept, idea, or a theory; you first need to know it properly in its own terms.

2.2 In the Beginning: Doing Science

The word *science* has its origins in a Latin verb that means *to know*. Science is a discipline that consists of the body of knowledge, and the techniques and efforts to develop and advance that body of knowledge. This is the body of knowledge of nature, organized in a rational and verifiable way. To obtain this knowledge, science involves the investigation of physical entities and natural phenomena through observation, theoretical explanation of the observations and data, and experimentation to test the predictions of the explanations.

By nature, we mean the whole physical Universe with all of the phenomena occurring in it and all the physical entities, living or non-living. So, nature (also referred to as the natural world) consists of large and diverse varieties of physical entities (also referred to as just entities in this book) and phenomena, which can be studied at different levels of detail and from different perspectives. Starting from physics (which was also called natural philosophy in ancient times), different branches of science developed over time to study these entities and phenomena. Talking physics, there are about 50 billion (one billion=10^9) galaxies and 10^{21} stars in our visible universe. Talking biology, there is a microscopic living universe right on our own planet: at any given moment, there are more than 10^{31} living microorganisms distributed among about 2 to 3 billion species on Earth. However, while each subject matter is vastly different, the fundamental way of making investigations and acquiring knowledge is the same in all branches of science; this is called the *scientific method*, which we will explore further on in this chapter.

Note. The power of science is that it is always scrutinized and is always open to revision; quite a contrast to belief systems which are often based on some fixed dogmas.

Be it biology, physics, or any other field, the beauty of science is that you can theorize and fantasize as much as you wish, but at the end of the day, your theory must withstand the test of experiments or it will be shown the way out.

In this chapter, we discuss the meaning and role of a theory in science by exploring how science develops, progresses, and works. For this purpose, we explore three avenues: scientific method; relationship between hypothesis, theory, and law; and materialization of the truth of scientific theories and laws into applications.

2.3 Theorizing in Science

To dismiss the theory of evolution, one of the arguments that is often given is: well, it's *just a theory* (implying or suggesting that it is an untested speculative idea) and it has so many loopholes. First, let me nip this myth in the bud by pointing to the fact that as you saw in Chapter 1, Darwin and Wallace came up with the theory of evolution based on the body of evidence they collected and based on the evidence that existed before them. Furthermore, Darwin held back from publishing his theory for about twenty years after the voyage of the Beagle, where an abundance of data was collected. What was he doing during those twenty years? Not much else than amassing more data and evidence which turned out to be for evolution and not against it.

However, there is more to the notion of, "oh it's just a theory." This argument is based on a prevailing misguided notion of the concept of a *theory*. By definition, a scientific theory is not *just a theory*; it is a well-tested explanation of observed phenomena. Also by definition, speculation (*just a theory*) is an opinion or a belief that may be based on a personal conviction and is not necessarily supported by verifiable evidence. For example, ancient Greek philosophers talked about atoms, the basic building blocks of all matter. That was a speculation. The world had to wait until the twentieth century for the theories of atoms that could lend themselves to and withstand scientific scrutiny (tests). Only such a discovery and understanding of atoms would be sound enough to begin the consumer electronic revolution.

Think About It!

Note the difference between a belief and a theory: a theory once tested often gives rise to real world objective applications whereas a belief does not. In this modern age, it's hard to see anything in our daily lives which is not an application of science: transistors to televisions, microphones to cellular phones, computer chips to the Internet, and drugs and other cures for diseases are only a few applications of theories of science to count. This is a key point that we need to appreciate.

Just like the Greek philosophers and naturalists who speculated about atoms, naturalists had begun speculating about evolution as well centuries before Darwin and Wallace. As discussed in Chapter 1, just like atoms, the idea of evolution goes back to the times of the ancient Greek philosophers. However, all the proposals (except the one by Lamarck) before Darwin and Wallace that life evolves failed to present any testable mechanism of evolution. In contrast, theory of evolution put forward by Darwin and Wallace was the theory that could be put through rigorous tests of scientific scrutiny. Due to its predictive power, the theory, once presented, started the Cambrian explosion of evolutionary thought. An essential feature of a scientific theory is that it must be falsifiable, for instance, it must make predictions that could be tested as true or false.

In science, a hypothesis or a set of hypotheses often precedes a theory. A *hypothesis* is a tentative description or explanation of specific observables about a physical entity or phenomenon, or a set of entities or phenomena. It answers the well-framed questions regarding the entity or the phenomena. The key to science is not belief but research, a scientific query that begins with defining a problem and collecting information and includes data experiments, data collection/observation, and analysis.

> **Think About It!**
>
> Some Greek philosophers presented the idea that everything is made of atoms. Would you refer to this proposition as atomic theory?
>
> Answer. No. Just a statement that things are made of atoms does not qualify as a theory in science. An atomic theory would explain how atoms behave and interact with one another to make those things that are made of atoms. And it would make testable (verifiable) predictions. The atomic theory that was presented centuries later, did meet these criteria.

As illustrated in Figure 2.2, scientific research can be divided into two interrelated stages: empirical research and fundamental research. Empirical research is data-based research describing results which can be reproduced by other scientists through observations or experiment. Fundamental research is the research that discovers the fundamental principles behind the observed phenomena or increases our understanding about the already discovered principles. Empirical research and its results

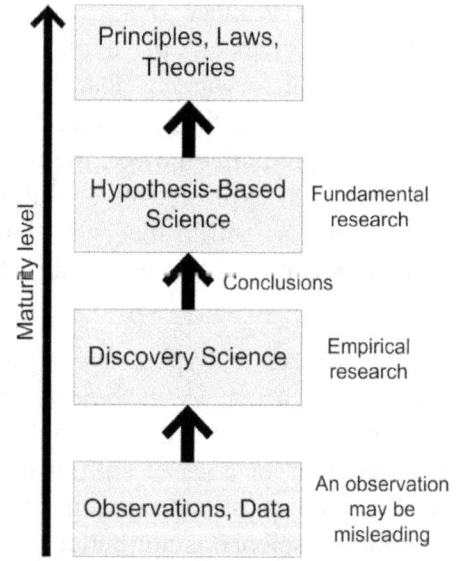

Figure 2.2 Conceptual logical flow from observations to principles and theories.

are also called discovery science and fundamental research and its results are also referred to as hypothesis-based science. Logically speaking, the stage of discovery science precedes the stage of hypothesis-based science.

Discovery Science. Discovery science is the description-based science that describes an entity or a phenomenon based on observations and data. Discovery science is focused on answering what and how questions. For example, what an entity is composed of, or how a phenomenon works, are questions in the realm of discovery science. Aristotle observed that the Sun rises in the east and sets in the west. He also observed that the Sun revolves around the Earth and completes its cycle in 24 hours. This is discovery science. After the invention of the microscope, scientists observed that living organisms consist of cells. This is also discovery science, and so is deciphering the genetic makeup (genome) of different species.

Another term that comes in the realm of discovery science is phenomenological research, which is an inductive and descriptive research approach based on the structure of the experience or the phenomenon which a person or an object goes through. This approach is developed from phenomenological philosophy which looks at the world in terms of the phenomenon that a person or object experiences as opposed to the interactions between objects in the Cartesian coordinates. It is more qualitative than quantitative.

However, sometimes observations in discovery science may be misleading, that is, lead you to wrong conclusions. For example, Aristotle's observation of the Sun rising in the east every morning and setting in the west every evening led him to incorrectly conclude that the Sun revolves around the Earth and completes one cycle in 24 hours. This is why science does not stop at the discovery stage. But what is the next step?

The best products of discovery science are conclusions reached by induction. For example, that the Sun always rises in the east and sets in the west is a conclusion reached by induction after a few days of observation. Similarly, after studying a variety of organisms, scientists concluded: all organisms are made of cells. This is also a conclusion of discovery science reached by inductive reasoning. Note that these inductions are very close to being hypotheses in the sense that they have some built-in predictive power attributed to their generalization.

One problem (or limit) of discovery science is that you may produce overwhelming amounts of data without any central principle that unifies it all. For example, these days we have tons of data available from genetic research such as DNA sequencing and genomics. However, this data may be used to discover fundamental principles of nature by using hypothesis-based science.

You enter the realm of fundamental science or hypothesis-based science when you look behind the findings of discovery science to find the answers to *why*. Why does the Sun always rise in the east and set in the west? Why are all organisms made of cells? Sometimes the difference between why and what is how you ask the question? For example, you can ask: why does the Sun always rise in the east? Or you can ask: what are the causes for the Sun to always rise in the east? The point here is that when you start looking for the natural causes for the observations or the data, then you enter the realm of fundamental research or hypothesis-based science. In this way discovery science often leads to hypothesis-based science, or you can say that empirical research leads to fundamental research.

Hypothesis-Based Science. Hypothesis-based science is usually science based on fundamental research that seeks the causes for observed phenomenon, which may subsequently lead to discovering fundamental principles. Here is one way how you proceed: In search of causes behind a phenomenon, you ask a well-framed question and formulate a hypothesis as an answer. Hypotheses will make predictions that you test by running an experiment.

Hypothesis-based science uses deductive reasoning as opposed to the inductive reasoning used in discovery science. While inductive reasoning makes a conclusion by generalizing a set of observations, deductive reasoning works in the reverse direction. Here is an example of deductive reasoning: because all living organisms are made of cells, both humans and the banana tree are made of cells. Deductive reasoning can be used to make predictions of a proposed hypothesis or a principle.

> **Caution!** We have presented here the conceptual logical flow from observations to principles, laws, and theories. In the practical research world, things don't occur in strictly the same way. For example, scientists may hypothesize from the data and keep doing discovery science rather than moving to fundamental science; and sometimes discovery science and fundamental science may overlap in certain aspects.

A hypothesis or a set of hypotheses develop into a theory only after withstanding years of scientific scrutiny and tests. So, in science, when we call something a theory that means it is well tested, that it has withstood all (or most) of the scientific tests performed so far. Furthermore, a single observation or result that is not consistent with a theory opens up that theory

And God Said: Let There Be Evolution

for revision: determine the limitations of the theory, improve it, or replace the theory with an alternative theory that explains all the previous observations and is not challenged by the new observation. Even well-established theories cannot escape continuous scientific scrutiny.

Think About It!

As you learned in Chapter 1, Lamarck was the first scientist to propose a mechanism for evolution, which was not supported by evidence. Can you call his mechanism a scientific hypothesis?

Answer. Yes, the mechanism proposed by Lamarck was a hypothesis by the very fact that it was verifiable and testable.

Consider the case of Newtonian physics, also called classical physics, based on well-tested theories resulted from the centuries-long work of several scientists such as Aristotle, Copernicus, Kepler, Galileo, Newton, and Maxwell. These theories explain well the phenomena occurring in the macroworld we live in, ranging from the motion of planets to the motion of your car. Applications such as your car, satellites, and airplanes are developed based on these theories. However, experimental observations at the atomic level toward the end of nineteenth century challenged Newtonian physics. As a result, the limitations of Newtonian physics uncovered at atomic level were identified and a new theory called quantum theory (also referred to as quantum mechanics or quantum physics) was developed. It does not mean, however, that classical physics is wrong and we can throw it away; all it means is that classical physics is a good approximation in the macroworld of more accurate theory, quantum physics; but this approximation breaks down in the microworld, the world of atoms and subatomic particles. It also does not mean that quantum mechanics is a 100% true theory and physicists can go home now. Quantum physics is still under test like any other theory in science. I, for example, earned my Ph.D. in physics by testing quantum theory of subatomic particles using the data that we collected from experiments run at the particle physics lab called CERN near Geneva, Switzerland (Figure 2.3).

Figure 2.3 Even well-established theories are under continuous scientific scrutiny. The author of this book tested some aspects of quantum theory by participating in an experiment set up in the largest particle accelerator/collider in the world: CERN's Large Electron-Positron (LEP) collider. This figure shows the aerial view of the collider as it straddles the French-Swiss border; the collider is 17 miles in circumference and 100 m under the ground. This accelerator was later converted to the large hadron collider (LHC), featured in the 2009 movie *Angels and Demons* (Columbia Pictures).

In the macroscopic world of cars, buildings, satellites, and planets, there is no need to use quantum theory, although in principle you could. Therefore, Newtonian physics is still a nearly perfect approximation, and is used in building applications in the macroworld, whereas quantum physics is used to build applications in the microworld. So, we as a society enjoy the applications of both classical physics and quantum physics. These applications of science form the fields of engineering and technology, which put the laws and theories of science, especially of physics (classical and quantum), to work. The point here is: No theory is an absolute truth because it's not possible to test them under all circumstances and conditions, because at a given point in time we do not know all the circumstances that may appear in the future. However, a well-tested theory such as quantum theory, gravitational theory, and yes, theory of evolution is as close to the *truth* as a scientist can reach at the time with the given tools. This is why most scientists intentionally avoid using the word *truth*. If, in the future, any evidence turns up against the theory of evolution, biologists will revise it or discard it for an alternative theory that explains everything that the original theory has explained so far and the new observation that the old theory did not. Anyone can attempt to disprove evolution and if you can scientifically disprove it, or even better replace it with a better theory, that would be a scientific achievement. Scientists are not closed to the idea of disproving evolution, or any other theory, as long as it can be done within the framework of the scientific method. Such a practice of modifying or even discarding well-established theories is the strength of science and not a weakness.

So, it's important to realize that theories for scientists are not an end in themselves, they are means to an end; and the end is acquiring knowledge about entities and phenomena in and around us. Scientific theories, laws, principles, and hypotheses are part of the scientific method used to acquire knowledge.

2.4 Scientific Method

Scientific method is a technique (or process) used in science for acquiring new knowledge and amending the existing knowledge through verification and testing. As illustrated in Figure 2.4, scientific method is composed of three basic elements: observation, hypothesis, and experiment, which are in a triangular relationship with each other. These three elements are explained in the following:

Observations. This element involves making observations about a certain aspect of nature such as an entity or a phenomenon. You may need to do this to scientifically define (or redefine) a new (or an already existing) problem. The word observation here has a broader meaning, which goes beyond just observing with your naked eyes. Observation is the process of gathering information (data) by using your five senses and other tools. For example, observations include data collected by a microscope or by any other tool during an experiment. Observations may be qualitative (descriptions) or quantitative (numerical measurements). Data is simply a recorded set of observations. Defining the problem based on your observations includes framing questions which you will answer by proposing a hypothesis or a set of hypotheses, which in turn will explain your observations or answer your questions.

Before proposing a hypothesis, you check out the work of other scientists on the subject at hand. By the way, this is one perspective of looking at how science has become a set of specialties or expertise; people have specialized in certain areas because of the sheer amount of existing knowledge that has to be mastered (checked out and understood) before proceeding to the next step in research and development.

Hypothesis. A hypothesis is a proposition or a set of propositions that present an explanation for an observed phenomenon or a group of phenomena. It provides a tentative answer to well-framed questions related to observations of the phenomena under study. The word hypothesis has its roots in the Greek verb which means *to suppose*. A hypothesis is an explanation on trial, meaning the explanation lends itself to tests and therefore is falsifiable. A hypothesis usually makes some predictions. The hypothesis (along with its predictions) needs to be tested through an experiment or a set of experiments.

Because all predictions of your hypothesis are potentially false, you devise a method or an experiment to test them.

Think About It!

Can a hypothesis be a speculation?

Answer. Yes, it could be a speculation, but it must be a falsifiable, that is, testable speculation. A speculation which is not falsifiable or testable is not a hypothesis.

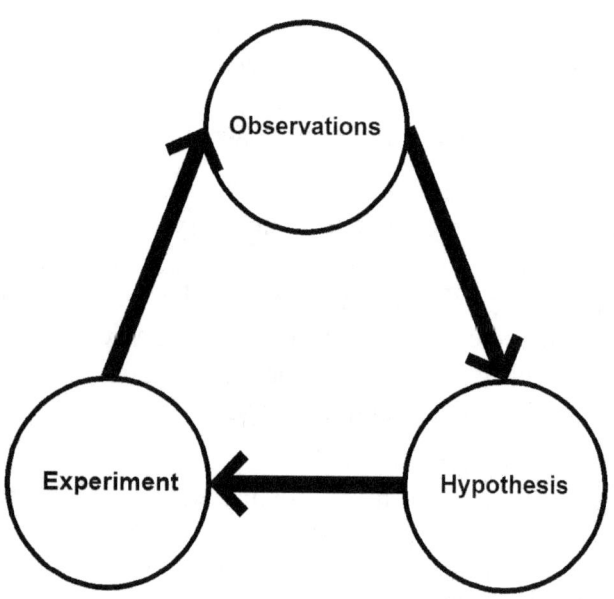

Figure 2.4. Triangular relationship between the three elements of scientific method: observations, hypothesis, and experiment.

Experiment. An experiment is an act to test (that is to confirm, verify, or reject) a claim or a set of claims such as a hypothesis and its predictions. Experiments can also add to your set of observations including data to complete the triangular relationship between observations, hypothesis, and experiment shown in Figure 2.4. During an experiment, you may also make an observation, such as collect data, that may lead you to a new and different problem. Many great scientific discoveries were made due to such accidents.

The scientific method is illustrated in Figure 2.4 in terms of the three concepts discussed in the previous paragraphs. The details of how the scientific method, in general, is implemented by scientists and researchers include the following steps:

1. Make observations about an entity or a phenomenon.

2. Define a problem based on observations.

3. Use your experience and knowledge to make sense of the problem. Gather as much information about the problem as you can, which includes checking out the work of others.

4. Set forth a testable hypothesis or a set of hypotheses that offer an explanation for the observations, answers to questions, and solution to the defined problem.

5. Work out the details of the hypothesis and its testable predictions.

6. Plan an experiment to test the hypothesis and its predictions.

7. Perform the experiment and collect data.

8. Go to Step 1 to analyze the experimental data, make observations, and draw conclusions. You may have to

go through this cycle many times, which may include modifying your hypothesis or to propose a new hypothesis.

9. Publish your work to such detail that other researchers should be able to conduct the same experiment and verify your results.

Think About It!

A hypothesis in science must be testable and falsifiable. According to these criteria, is Lamarck's hypothesis discussed in Chapter 1 scientific?

Answer. Yes, it is scientific because it can be tested and is open to be falsified. Lamarck's *principle of use and disuse* has predictions about organisms adapting to a new habitat. These predictions can be tested, for example, by using the fossil record. The principle of inheritance of acquired characteristics can be tested by comparing living organisms with their offspring. These tests have proven Lamarck's hypothesis wrong.

Note that scientific method is the general framework for performing scientific research. Although the three core elements of the scientific method (observation, hypothesis, and experiment) are always true at a big picture level, practically speaking, scientists do not always follow the method exactly the way described here. For example, in this age of expertise, a theoretician may propose a hypothesis and leave the task of testing it through an experiment to experimentalists. This is just a fact of life. Doing science requires great skill and knowledge and time-consuming research to the point that there are now those that focus on experiments and those that focus on theories. There are only a few brave souls who can do both.

Also, sometimes scientists stumble onto information or observation that they did not plan for and take it from there. Some biologists do surveys and work without a hypothesis, but tomorrow some biophysicist, for example, may come along, take that information and propose a hypothesis based on that information or observation collected by the biologists during the surveys. Nevertheless, generally speaking, at the big picture level, science does follow the scientific method. In science, a hypothesis must be testable and falsifiable. It means that there must be some way to check the validity of the hypothesis, and there must be some conceivable event that if observed would prove that the hypothesis was false. A hypothesis or a set of hypotheses extensively tested by many scientists becomes a theory. Even a theory remains open to scientific scrutiny. So, you can replace hypothesis in Figure 2.4 with *hypothesis and theory*.

To summarize, a hypothesis is a principle-like statement made as an explanation of a phenomenon, generally based on current or previous observations, extensions of existing scientific theories, or both. The scientific method requires that a hypothesis must be verifiable, that is, you must be able to test it.

Note. It's crucial to the scientific method that your hypothesis and its predictions must be testable and you must give enough information in your publication of results so that other scientists are able to duplicate the experiment and verify your results.

You must have heard of physical laws or scientific laws. What are they? A physical law, also called a law of nature, or a scientific law, is a generalized conclusion or a set of generalized conclusions based on observations of a physical entity, set of physical entities, or behavior. These generalized conclusions are reached through repeated observations and scientific experiments and are therefore generally accepted

81

within the scientific community. While a theory has its range of validity and exceptions, a law usually is universal and does not have any exceptions.   An example is the law of conservation of energy, which states that matter and energy combined can neither be created nor destroyed; they just change forms. For example, you can convert mechanical energy into electrical energy, but you cannot create electrical energy (or any other form of energy) from scratch. Similarly, you can convert matter into an equivalent amount of energy and vice versa. Also you can create one kind of matter from another kind of matter, but not from scratch (nothing). This is a universal law, true anywhere in the Universe, both for living and non-living entities, and both at macro- and microscale. Other laws include the first law of thermodynamics, second law of thermodynamics, Coulomb's law of electrostatic force; and the list goes on.

> Caution! If you don't know the specific scientific laws mentioned here, don't worry because you don't need to. The main point here is that a law is more universal than a theory.

A hypothesis may turn into a law through repeated confirmation by scientific experiments. For example, the hypothesis about gravity is such a well-tested hypothesis that not only is it part of the theory of gravity, it also exists in the form of a physical law, written precisely as a mathematical formula, called Newton's law of universal gravitation, which states that every entity with mass in the Universe attracts every other mass with a force which is directly proportional to the product of the two masses and inversely proportional to the square of the distance between them. However, here is a word of caution: Even though a law can be universal in some criteria, it may not be universal in some other criteria. For example, the law of gravitation is universal in the sense that it applies to both living and non-living matter, and it's true anywhere in the

Universe. However, at the micro-level this law breaks down and the force at work is quantum gravity, which scientists are still trying to understand. Nevertheless, a law is generally more universal and without exceptions as compared to a theory. So, you can replace hypothesis in Figure 2.4 with *hypothesis, theory, and law*. As summarized in Figure 2.5, a law has a higher maturity level than a theory, and a theory has a higher maturity level than a hypothesis.

A theory usually has a much broader scope than a hypothesis or a law. A set of well-tested hypotheses, laws, or both can be compiled into a theory. For example, quantum theory (also called quantum mechanics or quantum physics) explains how nature works at the atomic level, that is, at the level of atoms and molecules. If you read the history of the development of quantum theory, you will learn that experiments at the atomic scale were explained by presenting different hypotheses in the beginning. Some of these hypotheses and laws withstood the scientific scrutiny of the experiments. Examples are de Broglie's hypothesis of wave-particle duality, Planck's law of

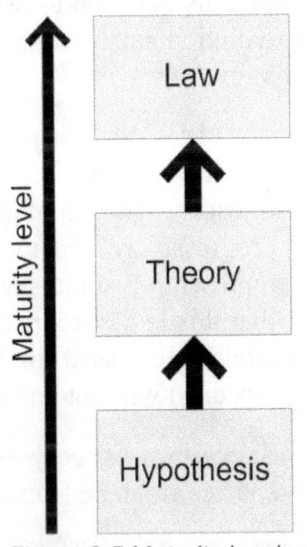

Figure 2.5 Maturity levels of hypothesis, theory, and law

quantized radiation, Heisenberg's uncertainty principle, and Einstein's hypothesis of quantized light energy. Again, explaining these laws is not the purpose here; the point is that all of these laws and hypotheses merged into what we call today quantum theory. Similarly, the law of natural selection and many principles at the genetic level are parts of the theory of evolution.

The scientific method described in this section is applicable to all branches of science, also referred to collectively as sciences. One of the essential characteristics of a scientific theory is to make predictions, testable predictions.

2.5 Predictions of the Theory of Evolution

In science, the strength of a theory lies in the spectrum of predictions it makes. A scientific theory makes predictions which can be tested through experiments or observations. So, an obvious question to ask is: what predictions does the theory of evolution introduced by Darwin and Wallace make? Here are a few.

♦ **A ladder of complexities.** Theory of evolution predicts that life started off as simple and then evolved into different lineages. This should give rise to an overall ladder of progressively increasing complexities with lineages representing some species changing over time. Remember that this ladder represents change, in contrast to Aristotle's ladder discussed in Chapter 1, which has fixed rungs (species) without any change.

♦ **Speciation.** Theory of evolution predicts that some new species emanate from the old ones.

♦ **Missing links.** There are some species not living today but which lived in the distant past that acted as common ancestors for the species living today.

♦ **Imperfection.** If all species were consciously designed by a creator, they would be perfect as the perfect creation doctrine discussed in Chapter 1 suggests. However, if species evolved from other species and not created by a creator, they must have some imperfections developed during evolution, for example, some remnants from their ancestor species which have no use today.

♦ Natural selection. We should be able to see natural selection, discussed in Chapter 1, operating on wildlife. A salient prediction of natural selection is that the fit between the population and its environment should improve over generations.

A careful reader will notice that considerable evidence in support of these predictions has already been presented in Chapter 1 because, after all, the theory of evolution was introduced to explain the existing data and observations. Even more rigorous proof and account of how these predictions have come true will be presented in the forthcoming chapters. Also, Darwin and Wallace only introduced the classical part, or the macro part (or organism and species level), of the theory of evolution. In the next three chapters, we will see how this classical theory extends to include genetics (the micro part). With this will come more predictions. We will explore how those predictions compare to the facts.

2.6 Scientific Theories: Putting It All Together

To put it all together, here is a typical scenario how scientific theories are born and how they are dealt with by scientists. When we (scientists) see some experimental observations that cannot be explained by the existing body of scientific knowledge, we hypothesize. A sound hypothesis usually makes predictions. We test the predictions of the hypothesis and modify it to accommodate any new information that came from our experiments. If we gather more (and enough) related experimental information outside the realm of the existing hypothesis, we make another hypothesis. Eventually, over time, with enough experimental information, the big picture starts emerging. Subsequently, after extensive testing, we try to fit all the information and inter-related positively-tested hypotheses together into what is called theory and base it on a smaller number of assumptions. The beauty of such a theory lies in the fact that all the previously scattered hypotheses and principals can now be derived from the theory based on a very few assumptions.

> Note. From its beginning, science has been growing into more and more a many-person endeavor, gradually becoming more complex. As a result, big picture discoveries usually emerge slowly and gradually, with an occasional big breakthrough by someone with a larger and deeper view who brings the facts together. Isaac Newton and Albert Einstein are good examples of this.

No, we don't stop even at this point. We keep on challenging the theory with new facts to see if they fit into the theory in order to keep building our confidence in the theory. One of the strengths of science is that theories are always under scrutiny and are up for revision or even a one-way trip to the junkyard. Such is the power of science; it holds no loyalty to a specific theory or authority and always welcomes improvements and advances. This underlines the difference between science and a belief system such as a religion or today's astrology. We scientists thrive on discovering contradictions and we do not shy away from saying goodbye to an existing theory if the facts can be better explained by a new theory. As a matter of fact, scientists on one hand develop theories and on the other hand spend their lives to prove them wrong. This is the story of Earth-centric versus heliocentric theories in cosmology, and this is the story of classical mechanics and quantum mechanics, and the story of the theory of evolution of life.

Unlike scientific theories, many beliefs and opinions that we hold cannot be tested. Unless something lends itself to a test, there is no way to disprove it, so it's not falsifiable, and hence does not belong to science. Although personal convictions usually have tremendous value in our lives, they should not be taken as or confused with a scientific theory. Also, in contrast to scientific theories, beliefs, opinions, and dogmas do not give rise to tangible physical applications. For example, what has astrology done for us lately or ever in terms of physical applications?

Table 2.1 Examples of scientific theories

Theory or principle	Brief Description
Cell theory	All organisms are made of one or more cells, the basic building blocks of life.
Gravitational theory	Any two massive objects attract each other with a force that is directly proportional to their masses and inversely proportional to the square of the distance between them.
Theory of evolution	Living species of organisms evolved from common ancestral species through the process of descent with modification.
Germ theory	Germs cause infectious diseases.
Quantum theory	Explains the behavior of physical entities at microscopic level in terms of wave-particle duality: each physical entity can behave as a particle or as a wave depending on the physical conditions around it.

2.7 Summary

I see the power of ignorance at its strongest and ugliest display when some opponent of evolution, such as a politician, condemns theory of evolution as *just a theory*. This person obviously does not realize the irony that the camera, the television, satellites, the satellite dish, and the cable networks through which his or her words are reaching the audience are all based on theories in science, and evolution is also a theory in science. Figure 2.6 sums up the hierarchical relationship between hypothesis, theory, law, and applications, in a nutshell.

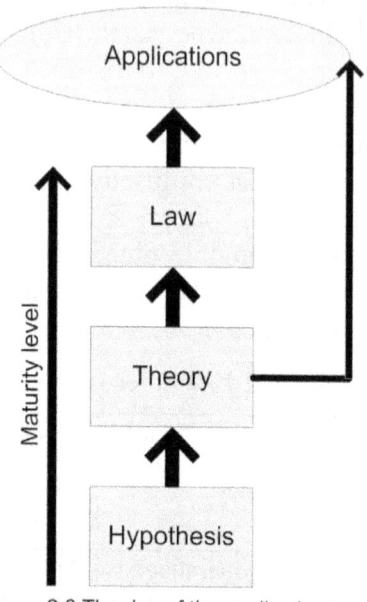

Figure 2.6 The rise of the applications that define our lives in the modern age.

The fields of engineering and technology use scientific theories and laws not only to develop applications for use in routine life but also to develop research tools to advance science.

In a Nutshell

✓ **Just a Theory?** In science, a theory, also called a scientific theory, is not *just a theory* or a speculation, but is a set of principles or explanations well supported by evidence and reproducible tests.

+ By definition, a scientific theory must be falsifiable, that is, lend itself to reproducible tests.

+ A scientific theory is continuously under scientific scrutiny and is subject to modification or rejection.

✓ Scientific method. It is a technique (or process) used for acquiring knowledge through verification. Scientific method is composed of three basic elements: observation, hypothesis, and experiment; which are in a triangular relationship with each other.

✓ Hypothesis. A hypothesis is a falsifiable and testable proposition or a set of propositions, which present a tentative explanation for a set of entities or observed phenomena. It's an educated guess based on observations.

✓ Theory. A theory is a well-refined set of principles or explanations at the next stage of maturity to hypotheses on the path of scientific scrutiny, and may integrate multiple well-tested hypotheses and laws.

✓ Scientific Law. A scientific law is a statement that describes a natural phenomenon or a set of phenomena that invariably or universally holds true under specific conditions or circumstances.

✓ Applications. Endless arrays of applications that define our living today are coming from the fields of engineering and technology, which themselves along with applications are directly based on scientific theories and laws.

✓ Evolution. Just like many other theories in science, such as quantum theory or theory of gravitation, evolution is not *just a theory*, but a scientific theory.

2.8 Behold

In this chapter, we explored how science develops, progresses, and works through hypotheses, theories, and applications. In science, a theory is not just a speculation; it's a well-tested set of principles or explanations about entities or phenomena. Therefore, sooner or later, theories often find their applications directly or indirectly. You can ask: if evolution is a well-tested theory, are there any applications of it? The answer is yes, and in this book we have a devoted a whole chapter, Chapter 11, to explore the applications of evolution.

The introduction of the theory of evolution, for example in *Origin of Species,* was a starting point for an endless series of questions that even today continue to inspire scientists. This is a sign of a successful scientific theory. The first and foremost question at the time that Darwin himself faced was: what is the source of variations in the traits of organisms that drive evolution? This question had to wait for longer than a century for a complete answer.

In the next chapter, however, we shall see that the answer was beginning to take roots at about the same time the question was being asked. The next chapter also demonstrates the power of the scientific method through the story of how an obscure Austrian monk became the father of modern genetics, largely through his study of pea plants.

THREE

Theory of Evolution: The Monastery Connection

And God Said: Let There Be Evolution

My scientific studies have afforded me great gratification; and I am convinced that it will not be long before the whole world acknowledges the results of my work.

Gregor Mendel

The value and utility of any experiment are determined by the fitness of the material to the purpose for which it is used, and thus in the case before us it cannot be immaterial what plants are subjected to experiment and in what manner such experiment is conducted.

Gregor Mendel

3.1 Once Upon a Time

In its early days, the biggest support for the theory of evolution that put the theory on firm ground came from a monastery where a monk named Gregor Mendel (1822-1884) performed experiments to study inheritance in the Abbey garden between 1856 and 1863. Gregor Mendel, a contemporary of Charles Darwin, is now recognized as the founding father of modern genetics. He was born on July 20, 1822 in Heinzendorf, Austria, which is now part of Czech Republic. Although he grew up in poverty on his parent's small farm, Mendel did not let financial hardship and illness to stop him from excelling in high school and later at the Olmutz Philosophical

Gregor Mendel (1822-1884)

Institute. Mendel, who was baptized two days after his birth, entered the Augustinian monastery of St. Thomas at the age of 21. At this monastery, which was a religious as well as a scientific center in

Brno, Austria; Mendel was pronounced a priest in 1847. In 1851, he was sent by the monastery to the University of Vienna under the sponsorship of Abbot C.F. Napp to study physics and chemistry for two years. There, during his development as a scientist, Mendel learned the value of experimentation and mathematics to explain natural phenomena, under the influence of scientists such as Christian Doppler, a physicist famous for his physics discovery known as the *Doppler Effect.* His botany professor Franz Unger encouraged Mendel's interest in finding the causes of variations in plants. In 1853, he returned to the monastery as a teacher, principally in the field of physics, and was assigned to teach at a local school, where several of his colleagues were enthusiasts of scientific research. Furthermore, his fellow monks shared fascination with the breeding of plants. By 1867, Mendel replaced Napp as Abbot of the monastery.

Figure 3.1 Mendel's church. Courtesy: American Philosophical Society.

What did Mendel have to do with evolution? Well, as you learned in Chapter 1, trait variations, trait inheritance, and natural selection are three essential ingredients of the theory of evolution. How does inheritance happen? That was the grey area that Darwin struggled with, and Mendel at that time was working on studying inheritance.

3.2 In the Beginning

Sir John Herschel (1792-1871), England's most famous living scientist at the time, was one of Darwin's scientific heroes, whose work on natural philosophy had inspired Darwin in part to get involved in the field of science. Herschel himself was interested in the natural causes of the origin of species, which he referred to as *mystery of mysteries.* While he called Lyell's uniformitarianism, discussed in Chapter 1, an open bold speculation on the origin of species, he believed that Darwin did not really have a theory that would explain the origin of species. His biggest objection was that Darwin could not explain the source of variations in the traits of organisms. Furthermore, he could not see how variations in traits could possibly give rise to new useful traits and hence new species. He dismissed this line of argument as the "law of higglety-pigglety". However, in Herschel's criticisms, there was one question that was also haunting Darwin and that was: what are the causes of variations?

Actually there are two interrelated questions here; let me decouple them for you. For natural selection (discussed in Chapter 1) to operate, there must be variations in traits, which have been (and still are) there without a doubt. For example, when you see a blond-haired woman and a brunette-haired woman, you are seeing a variation in the trait of hair color. The question is: what is the source of these variations? The second related question refers to the appearance of new traits in the course of evolution, and the question is: where do these new traits come from, or what is their origin? This chapter deals with both of these questions.

In this chapter, we explore three avenues: Mendel's experiments, how these experiments answered the two questions discussed above, and how linking the Darwin-Wallace theory of evolution with Mendel's work extended the theory from the macroworld of populations and organisms to the microworld of genes.

Mendel's experiments were aimed at studying inheritance.

3.3 Inheritance: The Grey Area

Recall from Chapter 1 that the three essential elements of evolution described by the theory put forth by Darwin and Wallace are trait variations, inheritance, and natural selection. Among a population, trait variants are inherited and natural selection operates on these trait variants. When we say Darwin's theory of evolution correctly describes the unity and diversity of life, we neither mean that Darwin's account of evolution was complete nor do we mean that it was completely correct to the letter. For example, as we have already mentioned, Darwin could not explain where the trait variations came from; nor could he explain how the new traits spread to the subsequent generations: an inheritance question, and inheritance was a grey area at that time. Surely Darwin knew of these shortcomings in his theory. Part of the reason for these shortcomings was that Darwin and Wallace were ahead of their time to make this discovery. So was Mendel, whose discovery of genetic principles was ignored for about half a century before it was rediscovered and linked to Darwin and Wallace's theory of evolution. Not much scientific work had been done by that time to understand inheritance, an essential ingredient of evolution. So, the theory of evolution originally could not explain how inheritance works. As you will see later in this and the forthcoming chapters, it was important to understand inheritance in order to find the origin and causes of variations?

The knowledge base of inheritance available to Darwin at the time contained some opinions, beliefs, and untested hypotheses such as the blending inheritance hypothesis.

3.4 Blending Inheritance: The Dead End

In Chapter 1, you saw how some naturalists (or scientists) got stuck with the belief (or hypothesis) of catastrophism. The hypothesis of blending inheritance is another example of how sometimes scientists, just like people outside the realm of science, get stuck with an untested and unproved belief or hypothesis. Needless to say, when it happens, it impedes the progress of science, and thereby the progress of society or human kind.

Hardly anybody would doubt that certain characteristics (traits) such as hair color, eye color, and height (being short or tall) are inherited by the offspring from their parents (and ancestors through parents). The issue had been the unresolved mystery of how exactly inheritance worked. Prior to the discovery of genetics, biologists and other scholars generally believed in the hypothesis of *blending inheritance*. According to this hypothesis, the hereditary material from both parents blended like a fluid at fertilization in a uniform fashion. Think of red wine mixing with grape juice, or vodka mixing with orange juice. This means that the offspring should acquire the characteristics intermediate between their parents. However, this hypothesis has a very vivid prediction: After many generations, a freely (randomly) mating population will give rise to a population of individuals that have very uniform characteristics, that is, all individuals look alike, and variations tend to cease. But the evidence against this prediction and therefore against the blending hypothesis in general are all around. Even children of the same parents do not exactly look alike unless they are identical twins. Furthermore, breeding experiments have also contradicted this prediction.

Blending hypothesis also fails to explain many other facts of inheritance such as reappearance of some traits after skipping generations. For example, a child may not get the attached earlobes of her mother, but the grandchild may. Also, if the traits are truly blended then any new variant (when it's rare) will be progressively and quickly diluted by breeding with the great majority of organisms who did not yet have that trait. Therefore, even if a new trait appeared, it would never become dominant in the population and would soon disappear. This prediction goes against the factual history of the development of life on our planet.

Although never proved, the blending hypothesis was so prevalent at Darwin's time that even physicists like H.C. Fleming Jenkins (known to the world as an inventor of telpherage) believed in it. As discussed earlier, blending would wipe out variation, and without variation natural selection would have nothing to select from, and hence there would be no adaptive evolution through natural selection. Darwin struggled with this dilemma in attempt to understand the cause of variations and how variations are inherited.

> Note. Telpherage is a transportation system in which carriers such as cars are suspended from or run on wire cables or the like, especially one operated by electricity.

Although Darwin never fully supported the hypothesis of *blending inheritance*, it made its way into his theory in a limited form. In the 1868 edition of *The Origin of Species*, Darwin proposed that small particles, *gemmules* as Darwin called them, were responsible for transmitting characteristics to the offspring. These gemmules, thrown off by parts of the body such as organs, were carried through the bloodstream and congregated in the sexual (reproductive) cells of the organism.

Because these gemmules come from all parts of the body to constitute the germ (sexual) cells, the germ cells also contain the characteristics developed during the lifetime of the organism. This way the characteristics (traits) at the individual level and variations at population level are transmitted to the offspring through the gemmules of the germ cells. The gemmules, by collecting the changes and variations created during the lifetime of an organism, would offset the loss of variations by blending.

As it turns out, in presenting this mechanism for the cause of inheritance of variations, Darwin was right on one count and wrong on another. He was correctly suggesting that the transmission of inheritable traits occurs through some sexual entities at atomic or molecular level, smaller than cells. As you will see in this chapter and in the forthcoming chapters, these sexual entities are genes that Mendel called *genetic factors*. Darwin went wrong, however, by proposing that these gemmules were collected from all parts of the body and hence contained the characteristics developed during the lifetime of the organism. This brought Darwin's argument close to Lamarck's problematic hypothesis of inheritance of acquired characteristics.

There were two sources of Darwin's problem: First, he was accepting the blending hypothesis and then defining gemmules in a way to offset the loss of variation by blending. The second source of his problem was that he, like all other scientists of his time, was thinking in the direction from macro-characteristics to the microparticles that would transmit the characteristics, and not the other way around. In other words, the approach was top-down and not bottom-up. This error is very natural when we, macro-beings, approach the phenomena of the microworld.

3.5 Journey from Macroscopic World to Microscopic World

The problem with Lamarck's hypothesis was that it was focused on phenotypes, that is, the macro instead of the micro. While Darwin made the right advance by thinking micro in terms of gemmules, his direction was wrong: from macro (phenotypes) to micro, and not the other way around. Lamarck and Darwin were not alone; all scientists (including physicists) of their times were either thinking macro or thinking from macro to micro. Today, we know that we need to think from micro to macro because of the simple fact that the macro is built from the micro.

> **Caution!**
> **Characteristic, trait, phenotype, genotype: mind the gap.**
> In this book, we use characteristics and traits synonymously as features of organisms that may belong to their form, function, or behavior. Strictly speaking, traits are variants of a characteristic (also called character). For example, eye color is a characteristic, and blue eyes and green eyes are traits. Inheritable traits are called *phenotypes*. We refer to phenotypes as macro-characteristics in the sense that they belong to an individual organism as opposed to its genes. A genotype is a set of alleles (versions of a gene) that give rise to a phenotype. A genotype is also called genetic makeup.

Being human, our greatest blessing and our greatest curse has been that we are macro-beings. Due to this fact, even though the Universe is built from the bottom up, we started investigating it from the top down. And there is a reason for that: we did not have a choice. Obviously, the first things we saw and tried to understand around us were the macro items: our eyes laid upon galaxies and stars, and not upon molecules or atoms. This is why the history of physics started from planets, or other celestial and terrestrial bodies, and eventually made its way to molecules, atoms, and subatomic particles. Similarly, the discipline of biology started from the study of living organisms, such as plants and animals, and made its way to cells and the molecules within cells. So, even though the Universe is built from micro to macro, we have studied it from macro to micro. In other words, we had it backwards, and in this sense we have been very good reverse engineers.

> **Caution!**
> When physicists use the term micro as compared to macro they usually mean everything small: micro (10^{-6}) and smaller including nano (10^{-9}). In this book, we use the term micro in the same sense.

Nevertheless, one of the most important lessons from the entirety of physics research is that nature does its most important and fundamental design work on a microscopic scale. The Universe and the systems, living and non-living, in it are designed at the microscopic level, and are built from the bottom up. This fact, for instance, is reflected by how a human is developed: Two half-cells (sperm and egg) join to make a zygote, which divides and develops into an embryo. Embryonic cells differentiate into different specialized cells to make different parts (tissues and organs) of the body, and subsequently we have a complete human being within months. Therefore, the reality of the macroworld that we experience on a daily basis has its roots in the microworld of atoms and molecules. There are many things that atoms and molecules do which cannot be understood in terms of what objects in the macroworld do; whereas there is nothing that macro-objects do that cannot be understood in terms of what micro-objects (atoms and molecules) do.

And God Said: Let There Be Evolution

This is what we know today but did not know then. Interestingly enough, when Darwin and other biologists were struggling with their macro-micro problem, physicists were running into their own. However, problems in science are blessings. The macro-micro problem in physics led to the development of quantum physics, whereas this problem in biology led to the development of genetics and molecular biology: very interesting parallels indeed.

Being the most fundamental of all the sciences, physics has profoundly affected all scientific fields. In this book, you will witness the interaction between physics and biology in three ways:

1. Physicists or the scientists whose original or major expertise was in physics contributing to research in biology.

2. Analogy between physics theories and biological theories.

3. Physical laws directly at work at microscopic and macroscopic levels to make biological evolution happen.

As mentioned earlier, we had two problems in dealing with the micro and macro aspects of the Universe and the systems in it: we started from understanding the macro and made our way toward micro, and then we tried to understand the micro in terms of our experience with the macro. There was another limitation that we faced: we can only investigate and understand so many things in our short lifetime. Given the enormity of information that we have to grasp in order to make an advance in research, we can only investigate, gather information, obtain knowledge, and develop our understanding piecemeal. We always need somebody to put the pieces together, and sometimes pieces do get ignored or get lost, especially in olden times when we were not living in the information age, the age of the Internet that we live in now. This is what happened to Mendel's research.

3.6 Quantized or Discrete Inheritance: The Undiscovered Proof Right Under Their Noses

At about the same time Darwin and Wallace presented the theory of evolution, Gregor Mendel was gathering experimental data that would not only support the theory but also help answer the big question unanswered by the theory: what is the origin of the change or variations in traits that causes evolution? The first answer to this question came from studying not animals but plants, a reflection of how evolution is the unifying principle in biology. However, so prevailing were the existing beliefs at that time that Mendel's work went unnoticed, and it took scientists about half a century to rediscover it and connect the dots between his work and theory of evolution introduced by Darwin and Wallace.

So, when Darwin was struggling with this problem in the 1868 edition of his book, little did he know that an obscure Austrian monk had already discovered and published the right solution to his problem supported by experimental results two years earlier. When Darwin took his last breath in 1882, he was unaware of the fact that the discovery made by Mendel's experiments that would rescue his theory was secure in the same library, the Linnaean Society library, where his theory was released to the world. Making matters worse, Mendel died in obscurity in 1884 with his work (which would later revolutionize biology) ignored, unknown, or forgotten all the same. This was partly a reflection of the popularity of the *blending hypothesis*, discussed in Section 3.4.

And God Said: Let There Be Evolution

While working on the inheritance of trait variations in plants, three botanists—Carl Correns, Erik von Tschermak, and Hugo de Vries – rediscovered Mendel's work in 1900. Once brought to light, Gregor Mendel's now famous breeding experiments with peas eliminated the confusion and dilemma created by the idea of blending inheritance. The actual experiments by Mendel were performed in late 1850s and early 1860s in his monastery's two-hectare experimental research garden (Figure 3.2), which was originally planted by Abbot Napp in 1830. From 1856

Figure 3.2 Mendel's research garden. Courtresy: American Philosophical Society.

to 1863, Mendel cultivated and tested about 29,000 pea plants. Mendel's garden displayed obvious morphological differences between different types of peas: smooth versus wrinkled seeds, short versus tall stems, purple versus white flowers, and so on. The main results from these experiments directly challenged the blending hypothesis, as shown in the following:

- When two contrasting types such as those with purple flowers and those with white flowers were crossed, the offspring looked like one of the two parents, not an intermediate: either purple flowers or white flowers.

- With further crosses of an offspring with itself, both of the types could reappear in undiluted form in the subsequent generations: purple and white.

As an example, Figure 3.3 illustrates the results of a crossing between white- and purple-flowered peas. Mendel crossed true breeding purple-flowered pea plants with true breeding white-flowered pea plants. The true-breed generation is called the parent (P) generation. In the first offspring generation (F_1), all pea plants turned out to be purple-flowered. When Mendel self-pollinated F_1 plants, that is, crossed an F_1 plant with another F_1 plant, the offspring generation (F_2) turned out to be about 75% purple-flowered and 25% white-flowered. In addition to the flower color, Mendel observed this pattern for many other traits as well such as plant height and the morphology of the plant seeds.

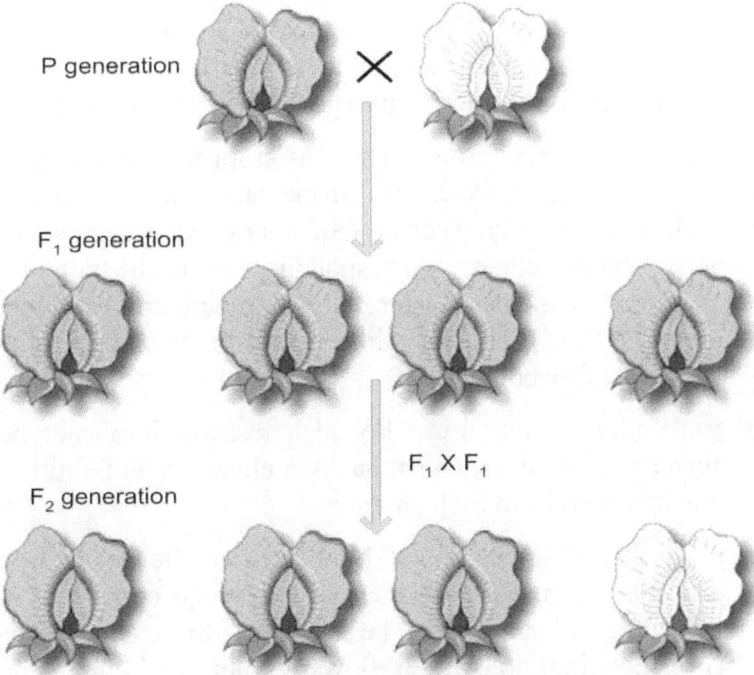

Figure 3.3 Crossing of pea plants: an example of Mendel's experiments

And God Said: Let There Be Evolution

As you can see, these experiments rejected the idea of blending variants in favor of discrete variants which are passed from parents to offspring, but which are not always visible. They may disappear in one generation and reappear in a following generation. Mendel called these discrete variants "genetic factors," also cited as discrete hereditary factors or Mendelian factors, without knowing anything then about what we now know of genes. Based on these experimental results, Mendel developed a model to explain the inheritance pattern that he observed in his experiments, presented as follows in terms of modern terminology of genes and alleles (versions of genes):

1. Traits are linked to genes. Inheritable traits arise from genes (which Mendel referred to as *genetic factors*) and are transmitted from generation to generation through genes.

2. Alleles cause variations. The reason for trait variations in inherited characteristics lies in the fact that an organism, such as a pea plant, consists of alternative versions of a gene, called alleles, corresponding to each trait. For example, for the pea flower color trait there are two alleles (gene versions), one for white flower color and the other for purple flower color.

3. Both parents contribute. For each macro-trait (phenotype) such as color, an organism has two alleles (micro-entities), one inherited from each parent.

4. Dominance and recession. Now, the two alleles, one from each parent, may be identical, for example correspond to the same color, or may be different, for example, one corresponding to white flower color and the other corresponding to purple flower color. If the two alleles are different, one of them is dominant over the other, called the

recessive allele, which has no effect in determining the trait. For instance, if a pea plant inherited a white flower allele from one parent and a purple flower allele from the other parent, its flowers will be purple because the purple flower allele is dominant. But the white flower allele is still there, just not expressed genetically in color. But if you now cross these purple flower plants with each other, some of the offspring will get two recessive (white color) alleles and the white flower color will reappear, as shown in Figure 3.3.

Recall that each parent has two alleles corresponding to, say color, and the offspring will inherit two alleles, one from each parent. Which of the two will be inherited from each parent during the sexual reproduction process and how? To understand the answer to this question, you will need to understand the process called meiosis, in which sperms and eggs (collectively called gametes) are produced. First, the inheritance patterns of Mendel's discrete genetic factors were soon discovered to be mirrored by the behavior of the gene-containing entities in the cell nucleus, called chromosomes. A chromosome is a structure that carries a single large DNA molecule in the cell. Different species of animals and plants carry different numbers of chromosomes. For example, a human has twenty-three pairs of chromosomes in each of its cells; each member of a pair comes from one of the two parents; say a red chromosome is inherited from the male parent and blue chromosome from the female parent (Figure 3.4). Such cells that contain chromosomes in pairs are called *diploid cells*. A special kind of cell called germ cells, also called sex cells or reproductive cells, are responsible for reproduction. Figure 3.4 illustrates an example in which a germ cell with one pair of chromosomes produces four gametes through a special kind of cell division called meiosis: First, the chromosome pair is replicated. The two duplicate copies of each replicated chromosome in the pair are called sister chromatids. Subsequently, the attached pairs of sister chromatids separate and each sister chromatid, now a chromosome, is separated into its own cell, called a gamete. What is shown in Figure 3.4, happens to each pair of chromosomes in a cell. The resulting gametes are called *haploid cells* because they contain half of the number of chromosomes that diploid cell of an organism contain. A male gamete is called a sperm and a female gamete is called an egg. A sperm fuses with an egg to make a diploid cell called a zygote through a process called fertilization. The full organism develops through cell division from the zygote.

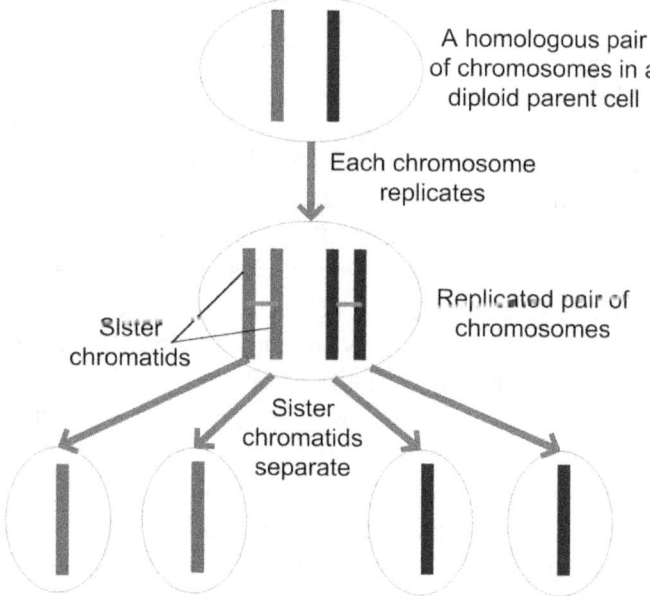

Figure 3.4 Production of gametes through meiosis.

Now consider a heritable trait that rises from two alleles (version of a gene), one on each member of a chromosome pair called homologous chromosome pairs. Now look at Figure 3.4 again and realize that the two alleles corresponding to each heritable trait separate and each of the two alleles go into a separate gamete (sperm or egg). This separation of two alleles corresponding to a trait into two sperms (in case of a male) or two eggs (in case of a female) is called *Mendel's law of segregation.* Therefore each sperm or egg gets only one of the two alleles that are present in each cell of each parent. These haploid gametes (sperm and egg) fuse during fertilization to make a diploid cell, the zygote, that gets back the two alleles for each characteristic (one from each parent) and ultimately gives rise to an organism. In the fertilization process, which of the two alleles from the male will meet with which of the two alleles from the female to make a pair of alleles in the zygote? This is where statistics enters the picture in terms of probability or chance. Opponents of evolution usually mock the theory evolution at this point by saying human life cannot be a matter of chance. However, it's not a mystery here, but a simple mathematical probability of which of the two alleles from one parent will pair with which of the two alleles from the other parent in the zygote during fertilization. This probability (or chance) is part of science and is a fact of life.

Think About It!

How many chromosomes are in a human sperm (or egg)?

Answer

Number of chromosomes in a human diploid cell = 23 pairs = 46
→

Number of chromosomes in a human haploid cell (sperm or egg) = 46/2 = 23

And God Said: Let There Be Evolution

So it was clear from Mendel's experiments that the discrete factors (genes) carry trait information and two copies of each gene are inherited by the offspring, one from each parent. Although both copies were present, one would dominate the other and produce the corresponding visible trait. But if in future generations, an offspring gets both recessive copies of the allele, the corresponding lost trait in the previous generation will reappear. Also note that the only traits that are heritable through genes are transmitted from one generation to another and therefore, only variations in those traits play a role in evolution.

In a nutshell, the root cause for variations in traits that Darwin was looking for can be found in the following three facts:

1. Genes give rise to certain traits, and it's those traits that are transmitted from parents to their offspring. Different species have different sets of genes, which gives rise to variations among species.

2. Each organism contains two copies of a gene corresponding to a trait. These alternative versions of genes (alleles) are the source of variations in traits within the same species.

3. An organism inherits only one of the two copies of a gene from each parent to make its own pair of genes to derive a trait from. This is how variations are transmitted from generation to generation.

In the light of Mendel's model, now let's take another look at how Darwin was struggling to find the cause for variations in traits. As said earlier, in the 1868 edition of *The Origin of Species*, Darwin proposed that small particles, say *gemmules*, were responsible for transmitting the characteristics acquired during the lifetime of an organism. These gemmules, thrown off by parts of the body, were congregated in the sexual (reproductive) cells of the organism. So Darwin called in the discredited Lamarck hypothesis of *inheritance of acquired characteristics* to cancel the *blending* effect. The right thing to do was: instead of cancelling the effects of two wrong hypotheses (blending and Lamarck), throw away both. Given that Lamarck's hypothesis of acquired characteristics was already discredited, there were two chasms to cross:

1. Break out of the blending hypothesis and think fresh.

2. Realize that not all characteristics of an organism would be transmitted to the offspring; only those that arise from genes (gemmules) would be transmitted.

Then, what was stopping Darwin and others from taking this apparently obvious step? The answer is: another chasm related to the second chasm mentioned above, a much deeper and wider chasm, the chasm that divides the macroworld that we live in from the microworld of gemmules or genes that they were tinkering with. The chasm existed in their thinking in terms of top-down approach, as mentioned earlier, that is, thinking from macro to micro as if micro is from macro. For example, thinking that macro-characteristics are developed during our lifetime and then are transmitted through the microparticles that Darwin called gemmules. The chasm lies between this thinking and the reverse thinking, bottom-up thinking, the thinking from micro to macro, that macro is from micro. For example, macro-characteristics arise from the microparticles (that now we call genes); then all you need to transmit to the offspring are genes. Although in the beginning of twentieth century, due to the rediscovery of Mendel's work, the cause for the variation of traits was realized, it took another 50 years to cross this chasm between the macroworld and the microworld, with the discovery of DNA structure in the 1950s.

And God Said: Let There Be Evolution

Mendel's experiments discovered that characteristics are not transmitted directly; it's not the characteristic per say, it's the genes that are transmitted, and characteristics arise from the transmitted genes. So in thinking micro to macro, and not the other way around, we can spare Darwin and other biologists in experiencing difficulty in crossing this chasm. Interesting enough, at around the same time, scientists working in the field of physics, the most fundamental science, were also stuck in fundamentally the same (or similar) chasm. As mentioned earlier, the reason is that we are macro-beings and look at the rest of the Universe from the macro-perspective. First we put Earth at the center of the physical universe. It did not work. Similarly, we put ourselves at the center of the Universe in the sense that we looked at things of all sizes from the macro-perspective. In terms of science, this is equivalent to saying that we were trying to apply laws of physics developed by studying phenomena in the macroworld, to understand the behavior of atoms, the micro-entities. The results from experiments being performed at atomic level were challenging the laws of the physics of the macroworld, called classical physics.

So when biologists were struggling with the blending hypothesis, physicists were struggling with their own blending laws called classical physics or Newtonian physics. Interestingly enough, the root of both the problems lie in continuum thinking, which works in the macroworld but breaks down in the microworld where discrete or quantum thinking is necessary to get things right. For example, a body capable of emitting heat (or energy) up to 50 Joules can emit 12.0 Joules, 12.1 Joules, 12.15Joules, 12.156 Joules and so on; any value smaller than or equal to 50 Joules. Another example of the continuum view is the blending hypothesis in biology according to which the hereditary material from both parents blends like a fluid at fertilization in a uniform fashion. That means the genetic materials from two parents can blend in any proportion on the continuum scale. In top-down thinking it's a reasonable hypothesis, because after all we can find organisms with variations in their traits in a continuous fashion. For example, I can find individuals with height 8.5 ft, 8.56 ft, 9 ft, 9.1 ft, 9.15 ft and so on. There is no such restriction that a person has to be 6 ft, 7ft, 8 ft, or 9 ft and nothing in between. The continuum approach works well in the macroworld but breaks down in the microworld, where certain physical variables such as energy and momentum for certain systems are quantized, that is, can acquire only certain discrete values, and not just any value on the continuum. This is an underlying principle of quantum physics called *quantization*, from which quantum mechanics (or quantum physics) draws its very name.

Caution! Note that we are not by any means suggesting that Mendel applied (or discovered) quantum physics to interpret the results of his experiments, nor are we suggesting that the study of genetics is equivalent to quantum mechanics or quantum biology. The scope of the discussion here is more that of an analogy or parallelism.

From this discussion, you can see how Darwin and Wallace were much ahead of their time because their theory of evolution pushed biology into a chasm at the time that even physics had not encountered yet, let alone crossed. Mendel was even farther ahead of his time because he crossed this chasm in an empirical fashion through his experiments before anyone else did.

In Chapter 1, we discussed how the theory of evolution explains the unity behind diversity of life. Let's explore how Mendel's genetic factors explain the same fact of life.

3.7 Mendel's Genetic Factors: Unity Behind Diversity

Mendel's genetic factors, genes as we call them today, represent the unity behind diversity of life in the sense that genes are the blueprint of all life on our planet, including plants and animals. For instance, proteins, which play a crucial role in determining the structure and function of various parts of an organism, are produced according to chemical instructions in the genes. Mendelian genetic laws or models discovered from pea plants are equally valid for animals including human. As an example, Table 3.1 presents some human traits that can be understood in terms of gene alleles within the framework of the Mendelian genetic model discussed in the previous section. Therefore these traits are called Mendelian traits.

Table 3.1 Examples of Mendelian traits in human.

Trait	Pair of Alleles	Dominant/recessive
Cheek dimples	dimples/no dimples	Dominant/ recessive
Chin cleft or chin dimple	cleft/no cleft	Dominant/ recessive
Earlobes	free hanging/ attached	Dominant/ recessive
Thumb	Hitchhiker's/ straight thumb	Dominant/ recessive
Tongue rolling	able/unable to roll tongue	Dominant/ recessive
Widows peak	have it/does not have it	Dominant/ recessive

> **Note.** If you are not familiar with the term, *widow's peak* refers to a distinct point in the hairline in the center of the forehead. The term has its roots in the superstition that hair growing to a point on the forehead is an omen of early widowhood.

Mendel focused studies on traits that are based on two alleles: one clearly dominant and other clearly recessive. It turns out that this is the simplest case and that the expression of genes for some traits is more complex. However, the underlying general principles that Mendel discovered remain true. For example, there are three, instead of two, possible alleles that determine your blood type: two dominant alleles (A and B), and one recessive allele (O). A given individual will only have two out of three alleles, that is, one out of six possible combinations: AA, AB, AO, BO, BB, and OO

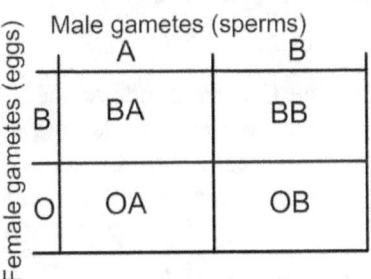

Figure 3.5 Punnett square showing all possible allele combinations for the offspring that result from an AB X BO cross.

(keep in mind that AB is identical to BA, AO is identical to OA, and BO is identical to OB). So, according to Mendelian genetics there will be only four blood types: O corresponding to OO, A corresponding to AA and AO, B corresponding to BB and BO, and AB. This is because O is recessive, and A and B are dominant. As said earlier, an individual can have only two of these three alleles. As an example, Figure 3.5 shows the cross between a male parent with the blood genetic makeup (genotype) of AB (two blood alleles) and a female parent with the blood genetic makeup of BO. Such a diagram is called a Punnett square, which shows all possible genetic makeups (pairs of alleles) for the offspring: BA, BB, OA, and OB. You can say from the Figure that there is 25% probability that the offspring will have blood AB, 25% probability for the blood type A (due to OA), and 50% probability for the blood type B (due to BB or OB).

Think About It!

John's wife is expecting. John and his wife both have a cleft chin. What is the maximum and minimum probability that their child will have a cleft chin?

Answer:

Assume C corresponds to the allele for the presence of the cleft chin, the dominant allele, and c corresponds to the allele for the absence of cleft chin, the recessive allele. Because cleft chin is a dominant trait, a parent or a child may have a cleft chin as Cc or CC.

Maximum probability:

The best chance for the child to have cleft chin is when each of the parents have two dominant alleles corresponding to cleft chin: CC. In this case the probability for the child to have cleft chin is 1, 100%.

Minimum probability:

The minimum probability for the child to have a cleft chin corresponds to the situation when each of the parents has the gene pair: Cc. The Punnett square for this inheritance is shown in Figure 3.6. The possible combinations the child could have are: Cc, cC, CC, and cc.

The three combinations (CC, cC, and Cc) will result in cleft chin and cc will result in non-cleft chin. Therefore the minimum probability for cleft chin is 0.75 or 75%.

Also, there are traits for which neither the dominant allele is fully dominant, nor the recessive allele is fully recessive. This is called incomplete dominance. All these variations from the simplest example are well understood in terms of genetic principles and contribute to the variations of traits and hence the diversity of life.

The unity behind the diversity of life is explained in terms of the fact that all life, with all of its diversity, is based on the same universal code composed of genetic factors (genes). Genes play their roles as discrete particles as opposed to blending with each other as described earlier.

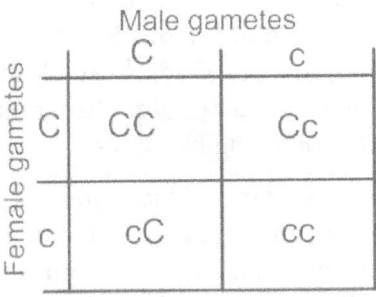

Figure 3.6 Punnett square for all possible pairs of alleles corresponding to the chin cleft for an offspring from the Cc X Cc cross.

3.8 The Quantum Rescue of Biology and Physics

Biology made its quantum jump from the macro- to microworld in terms of Mendel's model of inheritance, whereas physics made its quantum jump from macroworld to microworld in terms of quantum mechanics, a theory based on the work of many physicists including Max Planck, Einstein, Louis de Broglie, Werner Heisenberg, and Erwin Schrödinger. Two elements common to both Mendel's model and quantum physics are interesting enough to note:

1. **Quanta.** According to Mendel's model of inheritance, heredity material is transmitted in discrete units called

genetic factors (or genes) and not in continuous fluid as suggested by the blending hypothesis. In physics, any bound particle such as an electron in an atom bound to the nucleus emits or absorbs energy in discrete units called quanta. Even more generally, and closer to the gene scenario, any light source such as your electric bulb emits light in discrete units called photons. It may only emit a whole number of photons and not a fraction. Similarly, at an analogical level, you can think of genes as quanta of heredity, as opposed to continuous fluid of heredity as in the blending hypothesis.

2. Uncertainty. Uncertainty is one of a few basic unsettling principles that even the founders of quantum mechanics (also called quantum physics) had trouble believing in. Einstein, one of the founders of quantum mechanics, for instance, expressed his disbelief through his famous quote: *God does not play dice.* According to the uncertainty principle in quantum mechanics, we cannot measure the position and momentum of a particle simultaneously accurately but only in terms of probabilities. We have already discussed in this chapter how probability (and hence uncertainty) entered biology through Mendel's model of inheritance. We will explore this topic further in Chapter 5.

However, note that the comparison here is only at an analogical and empirical level. Quantum physics is a fundamental theory that still remains to be applied to biology at molecular and atomic level. The similarities that we have noted here may be an indicator of a deeper connection between quantum mechanics and biology, including evolution. We will explore this further in Chapter 5 in the context of evolution, of course. For now, let's complete the story of how the theory of evolution introduced by Darwin and Wallace found its roots in genetics.

3.9 Theory of Evolution Finds Its Roots in Genetics

The inheritance patterns of Mendel's discrete genetic factors were soon discovered to be mirrored by the behavior of the gene-containing entities in the cell nucleus, called chromosomes. We will discuss this topic further in Chapter 4. For now, realize that Mendel's experiments discovered the cause of variations in traits by refuting the hypothesis of blending inheritance. This happened toward the end of nineteenth century and the very beginning of the twentieth century, which coincides with the time when classical (Newtonian) physics began collapsing at the microscale; and quantum mechanics ideas, hypotheses, and principles started appearing to explain experimental results.

Mendelian genetics and Darwin's theory of evolution were fused together in the beginning of the twentieth century. As discussed earlier in Section 3.6, Mendel's model of inheritance provided the answer as to the cause of trait variations. But there was still a question of how the traits changed with time and new traits came into being. Once we know that traits and their variations originate from genes and alternative versions of genes (alleles), it's easy to realize that modification and change in traits originate from modifications of genes, called gene mutations. Hugo De Vries (1848-1935), one of the three botanists who rediscovered Mendel's work, as mentioned earlier in Section 3.6, published the first volume of his mutation theory in 1901, in which he proposed that evolution occurs due to mutations in the Mendelian genes, which give rise to alternate versions of genes, called alleles. Thus Mendelian genetics along with mutation made the theory of evolution complete by answering the question where trait variations among a population come from and how the existing traits change. Note that the answer to the question raised in the macroworld of phenotypes came from the microworld of genes. In the process, theory of evolution extended from the macro-scale to the micro-scale.

However, at that point in biology, in the beginning of twentieth century, a fundamental question remained unanswered: what really are these discrete hereditary factors or genes? What are they made of, how do they store and transfer the inheritable characteristics, and how do they modify in the course of evolution? In 1901, nobody had any clue to the answers of these questions. It took another 50 years or so to make a breakthrough in this direction with the discovery of DNA structure. By the 50-year anniversary of the publication of *On the Origin of Species*, genetic information was being explored and scientists were realizing that this information existed in form of some kind of threads inside the cell nucleus. By the 100th anniversary of the publication of *On the Origin of Species*, the solution to the mystery of mysteries–as put by John Herschel–had been discovered in 1953 by James Watson and Francis Crick in form of a large acidic-polymer called deoxyribonucleic acid (DNA). The discovery of DNA and its structure had stunning implications for our understanding of the origin of life, diversity of life, variation of traits, and heredity. It opened up a whole new field called molecular biology which includes molecular genetics. Does molecular genetics support or refute the theory of evolution? The scrutiny of the theory of evolution with molecular genetics is discussed in Chapter 4.

3.10 Summary

Gregor Mendel demonstrated through experiments that heritable characteristics (traits) are transmitted through discreet genetic factors, now called genes. Based on experimental results, he proposed that traits are inherited through genes and variations in a heritable trait originate from the alternative versions of the gene, called alleles, corresponding to that trait. Hugo De Vries, who was one of the scientists who fused Mendelian genetics with the theory of evolution, proposed that the change in traits and the rise of new traits occur due to the modifications in genes called mutations. This basically completed the standard theory of evolution originally proposed by Darwin and Wallace by answering the unanswered questions.

So, by now we know that there are at least four founders of the complete theory of evolution: Charles Darwin, Alfred Russel Wallace, Gregor Mendel, and Hugo de Vries.

In a Nutshell

✓ According to Mendel's model of inheritance, supported by experiments, heredity material is transmitted in discrete units called genes and not in continuous fluid as suggested by the blending hypothesis. In other words, it is genes that are passed down from generation to generation.

✓ Because genes are the origin of only some specific traits, only those traits are passed down from generation to generation. Therefore variations only in gene-based traits, the heritable traits, play a role in evolution.

✓ The variations in a trait, necessary for natural selection to operate on, are caused by alternative versions of the gene, called alleles, corresponding to the trait.

✓ New traits arise due to modifications in the genes, called mutations, discussed in Chapter 4.

✓ This way, theory of evolution developed from the data collected in the macroworld of organisms and species found its roots in the microworld of genes. This adds at least two more founders to the theory of evolution: Gregor Mendel and Hugo de Vries.

3.11 Behold

In Chapter 1, we explored what is the theory of evolution and how it was originally discovered. We also identified a question that the theory of evolution, as introduced by Darwin and Wallace, could not answer: what is the cause of variations in traits and change in traits over time, which are responsible for evolution? As you learned in Chapter 1, *natural selection* is the *how* of evolution: how does evolution occur? But what is the *why* of evolution? In other words, why does variation and change in traits occur? What brings this variation and change? Darwin could not answer it. This was the main reason that a prominent scientist of Darwin's era, John Herschel, who had the greatest influence on Darwin, doubted the theory of evolution as presented in *On the Origin of Species*. He (and Darwin) did not know that the answer to this question had emerged in terms of genes (genetic factors) from the experiments performed by Mendel. It would take about another century to really understand the answers in terms of DNA. In this chapter, we have explored how Mendel's model for inheritance answered that question in terms of discrete genetic factors now known as genes. How this answer was understood or realized in terms of the DNA or molecular genetics is discussed in the next chapter.

As discussed in this chapter, as macrobeings, we the human faced three limitations in our attempts to understand the Universe and the systems in it: First, given our short lifetimes and the enormity of the information we need to grasp to make advances in research; we can only investigate, gather information, obtain knowledge, and develop our understanding piecemeal; and then those pieces need to be put together to develop the big picture. Second, we studied the Universe and systems in it from macro to micro, whereas they are built from micro to macro. And third, when we developed the tools to penetrate to the microlevel, we tried to understand the microworld in terms of our experience with the macroworld, whereas the reality is the other way around: the properties and phenomena in the macroworld have their roots in the microworld.

All these three limitations combined kept Darwin and other scientists of his time from discovering the causes of variations, which were central to the theory of evolution. They discovered macroevolution and unknowingly were trying to make their way from macroevolution to microevolution. What needed to be realized was: microevolution is the origin of macroevolution, and not the other way around. We explore this idea in the next chapter.

And God Said: Let There Be Evolution

FOUR

Discovering the Roots of Evolution in the Microworld

And God Said: Let There Be Evolution

Everything important is, at bottom, utterly simple.

John Archibald Wheeler

It's necessary to be slightly underemployed if you are to do something significant.

James D. Watson

Almost all aspects of life are engineered at the molecular level, and without understanding molecules we can only have a very sketchy understanding of life itself.

Francis Crick

4.1 Once Upon a Time

a)   b)

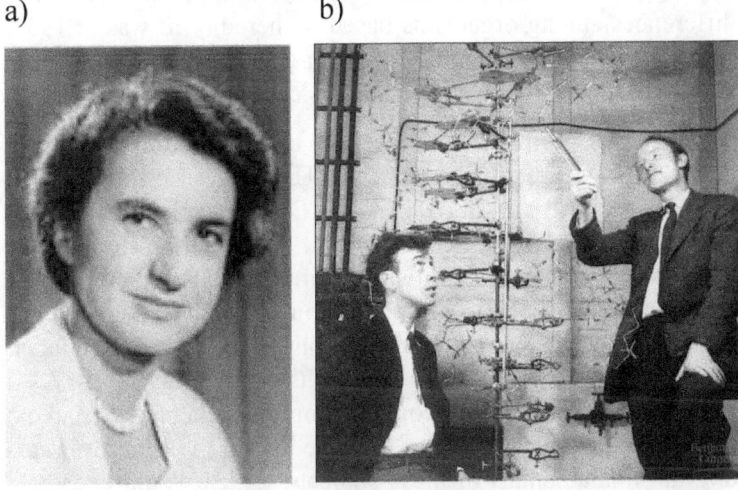

Figure 4.1 a) Rosalind Franklin, and b) James Watson (sitting) and Francis Crick (standing).

In 1962, Francis Crick and James Watson received their Nobel Prize in Physiology or Medicine for their discoveries on DNA, which they shared with Maurice Wilkins, a colleague of Rosalind Franklin. Franklin did not share in the Nobel Prize as she had unfortunately passed away in 1958 at the young age of 37, and the rules of the Nobel prizes stated that a recipient must be alive to receive the prize. James Watson, a zoologist and biologist, had grown an interest in

And God Said: Let There Be Evolution

biochemistry, especially in genes, during his Ph.D. work at Indiana University in the U.S. To pursue his research interests in biochemistry, he went to Europe as a postdoctoral fellow. There in Europe, he met Francis Crick at Cambridge University in 1951. This meeting was the beginning of the smallest and most effective research team that the field of biology had ever witnessed.

Rewind: James Watson, born in Chicago, Illinois, U.S., on April 6, 1928, received a B.Sc. degree in Zoology from University of Chicago in 1947, and a Ph.D. in Zoology from Indiana University in 1950. After completing his Ph.D. research on the effect of X-rays on bacteriophage (a virus) multiplication, he continued pursuing his interests in genetics; the scientific study of heredity, and similarities and differences among organisms based on heredity. It was in 1951, during the course of working on topics in genetics as a postdoctoral fellow, when he met Francis Crick, a physicist turned biologist.

Francis Crick, born on June 8, 1916, in Northampton, England, obtained a B.Sc. degree in Physics from University College London in 1937. He started working on his Ph.D. in physics, which was interrupted in 1939 by the Second World War. During the war, he worked as a scientist for the British Admiralty, mainly in connection with magnetic and acoustic mines. Inspired by the book *What is Life?* by one of the founders of quantum physics, Erwin Schrödinger, he started studying biology after leaving the Admiralty in 1947. At that time, Crick knew no biology, no organic chemistry, and no crystallography. He spent most of his time during the next few years learning the fundamentals of these subjects, which would later help him to make a breakthrough discovery of DNA structure in collaboration with James Watson. So, he became a research student for the second time in 1950, when he was accepted to Caius College, Cambridge.

When James Watson and Francis Crick met in 1951, the race for the discovery of DNA structure was already in full speed. Crick took a side trip (which turned out to be a Nobel trip) from his Ph.D. work to join forces with Watson to figure out DNA structure. He obtained a Ph.D. in 1954 on a thesis entitled *X-ray diffraction: polypeptides and proteins*. Before he completed his Ph.D., however, he had already published his paper with James Watson in 1953 on DNA structure, a paper that earned him the Nobel Prize in 1962 that he shared with Watson and Wilkins.

Even before Francis Crick, physics, being the most fundamental of all sciences, had directly or indirectly contributed to biology in many ways. For example, as you learned in Chapter 3, Gregor Mendel, originally trained in physics, applied probability theory to genetics. However, physics entered (or re-entered) the field of evolution and therefore biology in a big way when according to a legend, Francis Crick and James Watson walked into the Eagle Pub in Cambridge, England, on February 28, 1953, and Crick announced to the pub crowd, "We have found the secret of Life." What they had really figured out was the now famous double helix structure of a DNA molecule, shown in Figure 4.3. More than half a century after that event at the Eagle Pub, the DNA molecule remains the biggest celebrity in biology.

4.2 In the Beginning

Recall from Chapter 1 that at the core of the theory of evolution introduced by Darwin and Wallace is the idea that nature selects variations which are favorable (or adaptable) to the existing environment. However, the origin of these variations was unknown (or unrecognized) at Darwin's time, even though it was already discovered by his contemporary, Mendel. In Chapter 3, we explored the cause of the variations in traits within a species and among different species. We also briefly discussed the origin of new traits that arise in the course of evolution. So there were two questions: what is the cause of variations in traits, and what is the origin of the new traits? By now you know the basic answers to these questions from Chapter 3. The one word answer to both of these questions is: genes, the micro-entities that exist in all living organisms. The source of variations among traits is the alternative or multiple versions (called alleles) of a gene corresponding to a trait. This was the answer provided by Mendel's experiments. The origin of a modification to a trait or the rise of new traits are modifications to the genes themselves, called mutations, according to the mutation theory introduced by Hugo de Vries in 1901.

Let's ask some more questions. How do we know that the mutation theory is right? And what are genes anyway and how do they work? How do mutations occur and how do we know that mutations are responsible for modifying traits and giving rise to new traits? In this chapter, we explore the answers to these questions in order to reveal the roots of the theory of evolution in the microworld. In the process of doing so, you will be introduced to some concepts in molecular biology and genetics in an attempt to make this material self-contained.

The answers to the core unanswered questions about evolution were resting in a molecule of life called deoxyribonucleic acid (DNA).

4.3 DNA: The Hoofprint of Evolution

Now, about 150 years since the introduction of the theory of evolution by Darwin and Wallace, the key questions about *how exactly traits are passed down to the subsequent generations and how traits evolve* have already been resolved by remarkable developments and progress in molecular biology and genetics. Note that there are traits, such as eye color, and there are trait variants (variations in traits), such as blue eyes and green eyes. Now we know that the origin of variations in traits of organisms is basically the same as the origin of traits themselves: the genes. Genes are the origin of traits; alternative or multiple versions of a gene (alleles) corresponding to a trait are the origin of variations in the trait; and the changes in genes called mutations are the origin of the modifications in a given trait or the rise of a new trait. Overall, genetic variants produced by alleles and gene mutations give rise to trait variants. Genes are part of a molecule called deoxyribonucleic acid (DNA).

A careful reader must have realized the following very important point by now: If it is the genes (or DNA) that are the origin of heritable traits and variations in those traits, and if it's the genes that are passed down from one generations to the next generation in terms of DNA molecules, then by comparing the genes (or the DNA molecules) of different species, we should be able to determine if evolution occurred or not. So, you can see that the field of genetics is the biggest and the ultimate test bed ever for the theory of evolution. It's a make or a break for the theory and the concept of evolution. A person with an open mind about the validity or invalidity of the theory of evolution ought to be very excited at this juncture.

You would think, and correctly so, that by using discoveries and experimental results in genetics, we should be able to either find the presence or absence of the hoofprint of evolution from species to species. To explore this remarkable story or history of life, however, you need to understand that just like life, the genetic material is organized into multiple structures of different complexity levels, as illustrated in Figure 4.2. Genes exist as regions of a DNA molecule, which itself is packaged inside another structure called chromosome, which in turn exists in the nucleus of a cell. The collection of all the chromosomes inside a cell of an organism is called the genome of that organism. These structures are described in the following:

Genes. Mendel's discrete genetic factors discussed in Chapter 3, which we now know as genes, are the origin of traits. *Discrete genetic factor* means that a gene is the smallest unit of heredity. Although Mendel knew nothing about genes as we know them today, based on his experimental results, he hypothesized the existence of genetic particles that he called genetic factors.

Alleles. As discussed in Chapter 3, alleles are alternative or multiple versions of a gene that produce variant traits, for example, a cleft chin and a chin without a cleft.

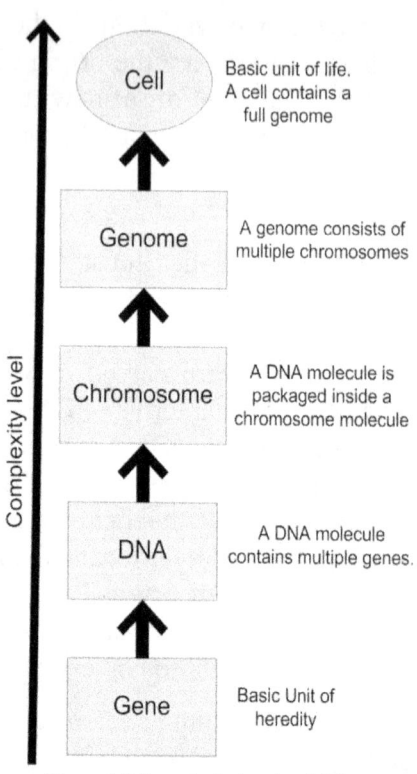

Figure 4.2 Complexity levels of different genetic structures

DNA (deoxyribonucleic acid). The deoxyribonucleic acid molecule has two strands twisted around each other into a double helix. Genes are parts (regions) of the DNA molecule. The DNA double helix is formed by *base pairs* attached to a sugar-phosphate backbone as shown in Figure 4.3. A base can be any of the four chemicals: adenine (A), guanine (G), cytosine (C), and thymine (T). These DNA bases pair up with each other, A with T, and C with G, to form units called base pairs. Each base is attached to a sugar molecule and a phosphate molecule. The triplet of base, sugar, and phosphate together is called a nucleotide. Chains of nucleotides are arranged in two long strands that form a spiral in the shape of a double helix. All of this pairing and shaping happens spontaneously due to the bonds created by different kinds of electromagnetic interactions between different atoms, smaller molecules, and regions of the DNA molecule. The information in DNA is stored as a code made up of the way base pairs (or nucleotides) are sequenced.

As illustrated in Figure 4.3, the structure of the double helix is somewhat like a ladder, with the base pairs forming the ladder's rungs and the sugar and phosphate

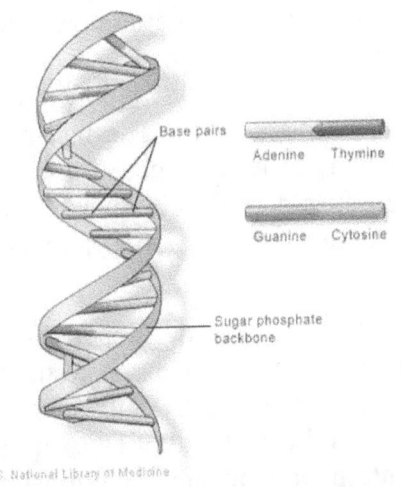

Figure 4.3 An illustration of the double helix structure of the DNA molecule. Image: courtesy of the U.S. National Library of Medicine, National Institute of Health.

molecules forming the vertical sidepieces of the ladder, called the *sugar-phosphate backbone.* To give you an idea, you, like any other human, have about 3 billion base pairs organized into 46 DNA molecules in each of your cells (except mature red blood cells where one chromosome holds one DNA molecule). Furthermore, get this: 99.9 percent of your DNA is identical to all other humans. This means only 0.1 percent of your DNA provides you uniqueness. In other words, 99.9 percent of our DNA, which is the same in all people, unites us together as one species, the human species, and 0.1 percent of our DNA is responsible for variations among us, that is, diversity.

And God Said: Let There Be Evolution

As mentioned earlier, the bases (or base pairs) in a DNA molecule are sequenced in a certain order. Two DNA molecules are identical if their sequences are identical. The sequence, or order, of bases determine the information available for building and maintaining an organism with specific traits; similar to the way in which letters of the alphabet appear in a certain order to form words and sentences which we use to convey specific messages or ideas; or the orders of bits (1s and 0s) in a computer code, which specifies a machine instruction. For instance, the bits (0 and 1) in a computer code arranged in different order will code for different information. Similarly, the same set of base pairs or nucleotides sequenced in different orders will code for different information.

Another important property of DNA is that it can self-replicate, that is, make copies of itself, which is necessary if, starting with a zygote, cells have to multiply to develop a full organism. Each strand of DNA in the double helix can serve as a pattern or template for duplicating the sequence of bases. This is critical when cells multiply because each new cell needs to have an exact copy of the DNA present in the original cell. Each cell has its own set of chromosomes (DNA molecules): 46 in a human cell.

Note. Cells multiply by using a process called *cell division*: one cell grows to a point where it splits into two, and the two cells subsequently split into four, and so on. As you can see, practically speaking, it's really cell multiplication.

Chromosomes. A chromosome is a structure into which a DNA molecule is organized. It also contains some proteins which serve to package the DNA and control its functions. Each chromosome contains one very long DNA molecule and the proteins associated with it. As described in Chapter 3, chromosomes come in pairs, one member of the pair originally from one parent and other member from the other parent: this is how you get two versions of a gene (alleles) one on each of the two chromosomes. So, in each of your chromosome pairs you have one chromosome from your father and the other from your mother. Which allele from a parent you receive depends on which of their two chromosomes you received. This transfer happens during meiosis and fertilization as discussed in the context of pea plants at the level of alleles in Chapter 3. A human cell has 23 pairs of chromosomes holding about 25,000 genes: 22 pairs of non-sex chromosomes (called autosomes) and one pair of sex chromosomes giving a total of 46 chromosomes per cell. The 22 pairs of autosomes, shown in Figure 4.4, look the same in both males and females. The 23rd pair, the sex chromosomes, differ between females and males as females have two copies of the X chromosome (one from each parent), while males have one X and one Y (X from the mother and Y from the father).

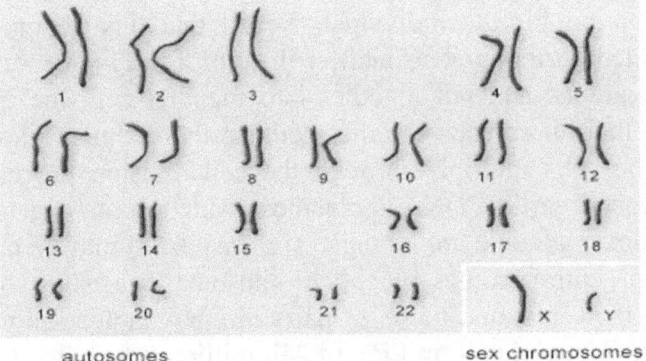

Figure 4.4 23 chromosome pairs in a human cell. The 22 pairs of autosomes (non-sex chromosome pairs) are numbered in descending order of size. The other two chromosomes, X and Y, are the sex chromosomes related to reproduction.

And God Said: Let There Be Evolution

> **Think About It!**
>
> Do the chromosomes shown in Figure 4.4 belong to a male or a female?
>
> **Answer:**
>
> It belongs to a male because a female sex chromosome pair consists of two X chromosomes, and only a male has a Y chromosome.

Genome. A genome is an organism's complete set of DNA. For example, the 23 pairs of chromosomes in each of our cells is our genome. In other words, each of our cells (except mature red blood cells) consists of a complete genome. This in general is true for all organisms. Genomes vary widely in size across species: the smallest known genome for an autonomously living organism (a bacterium) contains about 600,000 DNA base pairs, whereas human and mouse genomes have about 3 billion base pairs. As said earlier, each cell consists of a complete genome, that is, a complete set of genes for the individual, which means the entirety of the hereditary information for the individual. Recall that this genome was compiled during meiosis and fertilization from the genomes of both parents, and put together into the first cell, the zygote. From there it spreads to other cells as the organism develops and the cells multiply through the cell division process. As mentioned earlier, DNA molecules (which contain genes) in the human genome are arranged (or organized) into 23 pairs of distinct chromosomes, each chromosome consisting of one DNA molecule; and these 23 pairs of DNA molecules vary in length from about 50 million to 250 million base pairs. Certain parts of the DNA sequence that consist of hereditary information are called genes. Each gene consists of a chain of multiple base pairs of the DNA molecule. Overall, the human genome consists of about 20,000 to 25,000 genes. Table 4.1 compares the genome sizes of organisms from various species.

Think About It!

In a human cell, there are 23 pairs of chromosomes. How many total chromosomes there are in a human genome?

Answer

$2 \times 23 = 46$

Because a cell contains the full genome.

Table 4.1 Comparative genome sizes of human and organisms from other species. Information source: The research journals *Nature* and *Science*.

Organism	Estimated Size of the genome (number of base pairs)	Estimated Number of Genes in the genome	Total Number of Chromosomes in the genome
Homo sapiens (human)	3.2 billion	25,000	46
Mus musculus (mouse)	2.6 billion	25,000	40
Drosophila melanogaster (fruit fly)	137 million	13,000	8
Arabidopsis thaliana (plant)	100 million	25,000	10
Caenorhabditis elegans (roundworm)	97 million	19,000	12

Saccharomyces cerevisiae (yeast)	12.1 million	6,000	32
Escherichia coli (bacteria)	4.6 million	3200	1
H. influenzae (bacteria)	1.8 million	1700	1

Cell. A cell is the smallest structural and functional building block of life; all living organisms are made of cells. The human body contains about 100 trillion (a trillion is a million times a million, or one thousand billion). A cell that can replicate itself contains a complete genome.

Gamete. As introduced in Chapter 3, a gamete is a sex half-cell, also called a haploid reproductive cell, such as sperm or egg in an animal, and pollen and an egg in a plant. During fertilization, a gamete from a male (sperm) unites with a gamete from a female (egg) to form a full (diploid) cell called a zygote.

Think About It!

In some cultures, a woman is given much grief if she only gives births to baby girls and no baby boy. From genetic viewpoint, if we have to blame the father or the mother for producing only baby girls and no boy, who should we blame?

Answer

The father.

Both of the sex chromosomes in a mother's cell are X chromosomes, whereas the father has both X and Y chromosomes. A zygote develops into a boy if the zygote contains a gamete with Y chromosome, which can only come from the father.

Now that you have some idea about the various microscopic structures or entities that make up and run life, you can ask: What does it has to do with evolution? Evolution explains a universal characteristic of life, unity behind diversity, through its core idea of descent with modification from common ancestry. Now consider the fact: all organisms are made of cells. This is a statement of the great unity of life that points to the common origin of all species of living organisms. All cells share some common characteristics, both structural and functional. Yet, cells also differ substantially from one another. This is exactly what theory of evolution would predict: all cells are related to one another through their descent from earlier cells with modification.

What about the genome or DNA in the cell? What do they have to do with evolution? Well, to start with, consider this fact: DNA possesses a simple four-base genetic code that provides the recipe for developing an organism. This genetic code is nearly universal, that is, shared by all organisms from bacteria to plants to animals including human. This means that this genetic code must have been operating since very early in the history of life, and this way links all life on Earth to a common ancestor (or ancestors): prime evidence for evolution. Furthermore, experimental studies have revealed that the whole history of evolution of each species is recorded in the genome of each of its organisms.

Using modern tools and techniques in biotechnology and bioinformatics, such as DNA sequencing, genomes of various species are being deciphered and compared with one another. Take the example of human, our own species. In 2003, the sequence for a reference human genome – the whole genome with about 3 billion base pairs–was determined in a process called DNA sequencing or simply sequencing. About four

And God Said: Let There Be Evolution

years later when the DNA sequence of James Watson was made public; Craig Venter, the founder of Celera, also made his genome sequence public. Scientists then could compare the genomes of three humans. Many more followed. Scientists found that on average, a human sequence differs from the reference sequence by about 0.1 percent, or 1 in 1000 base pairs. This may not sound much, but given that we have about 3 billion base pairs in our genome, it adds up to a difference of about 3 million base pairs. The genome comparison has been extended to other species as well.

This comparison reveals the history of biological evolution from very early times to the present. This also provides the test bed for the existing theory of evolution. For example, one prediction of the theory of evolution is that the more closely related two species are in their evolutionary history, the more similar their genomes will be. An example is human and chimpanzee, or dog and cat. Conversely, two species which are more distantly related in the evolutionary tree, the less similar their genomes will be. An example is human and chicken, or human and bacteria. The trend reflected from the data in Table 4.2 supports these predictions of the theory of evolution. The genetic similarity between chimpanzee and human is from 96 to 98 percent, supporting the claims that these two closely-related species diverged from a common ancestor recently, that is, about six million years ago. Genome comparison and data analysis also indicates that three domains of life: bacteria, archaea, and eukaryotes diverged between 2 and 4 billion years ago.

Another prediction of the theory of evolution is that the genes that evolved a long time ago should still be similar in strikingly different species. Genome comparisons are supporting this prediction. As another example, certain genes in yeast have been found to be very similar to their

140

counterparts in humans. These similarities have their practical applications, discussed in Chapter 11. Also, by comparing the genome of two different species, scientists can pinpoint exactly which genes uniquely define a species from another species. All this helps in building the detailed evolutionary tree of life. Genomic research is still in progress. So far not only it has supported the existing theory of evolution, but has also provided new and more detailed insight into evolution.

Table 4.2 Genetic similarities between various organisms and humans.

Animal	Percentage of Genetic Similarity with Humans	Comments
Human	99.9	All humans share 99.9 % of their genes
Chimpanzee	96-98	Depends on how you measure the similarity
Cat	90	Cats have 90% genes homologous to human genes.
Cow	80	Cows are 80% genetically similar to humans.
Mouse	75	75% of mouse genes have equivalent genes in humans.
Fruit fly	60	Fruit flies share about 60% of DNA with humans.
Chicken	60	About 60% of chicken genes correspond to a similar human gene.

As mentioned earlier, genes are the basic units of heredity. For example, one gene in an organism contains instructions to make one type of protein; and this way all the proteins (that is types of proteins) of the body are produced according to instructions in the corresponding genes. Proteins are the workhorses of life in an organism. In addition to making proteins, DNA also has instructions to make a copy of itself, that is, self-replicate. Using this self-replication and cell division, one complete organism originally develops from one cell, the zygote.

The zygote is produced through the process of sexual reproduction, where variation begins.

4.4 Variations from Sexual Reproduction

Recall that each cell in your body contains 23 pairs of chromosomes, that is, two sets of 23 chromosomes. Your two parents combined have four sets of 23 chromosomes, and you got from them your two sets of 23 chromosomes, one set from each parent. This means that you do not have the full genome of either of your parents and that's why you are not a clone of either of your parents. Half of each of your parent's genomes was selected at random to build your genome during the sexual reproduction process. In an analogy of a card game, think of the reproduction mechanism shuffling the alleles of both parents and dealing them at random to build the individual genotypes of the child. The randomness in this process gives rise to uncertainty and chance, and hence the variations across even the children of the same parents. This is why, unless they are identical twins, no two brothers or sisters look exactly alike. This way, the sexual reproduction process contributes to genetic variations, which in turn cause phenotypic variations within a population from generation to generation. This genetic (micro) variation and the resulting phenotypic variation provide the raw material on which natural selection operates.

> **Caution!** Although it's true that inherited phenotypes arise from genotypes, environmental influences also play their role in shaping a phenotype. In general, a phenotypes arises as a combined effect of inherited genotype and environmental influences:
>
> **Genotype + Environment** ➜ **Phenotype**
>
> However, only that part of a phenotype that is genetically determined is passed down to the offspring.

So, the sexual reproduction process gives rise to a zygote, which has the seeds of variations necessary for evolution.

4.5 From Zygote to Organism

As you know by now, a gamete from a male (sperm) fuses with a gamete from a female (egg) to form a complete cell called a zygote. A multicellular organism such as a human baby with billions or trillions of cells develops from a single cell (zygote). This process happens in several stages. To begin with, there is only one cell in the zygote, and it multiplies into many cells by repeatedly using a process called cell division in which a cell grows to a point where it splits into two. All the cells contain the same set of genes inherited from the parents, as before a cell splits into two, each of the original chromosomes (genome) are converted into two copies through a process called replication, and each new cell gets an identical copy of the original genome. As the number of cells increases, the zygote turns into a ball of cells with an inner group of cells and an outer group of cells. The inner group of cells becomes the *embryo*, whereas the outer group of cells becomes the membranes that nourish and protect the embryo. The cells of the embryo, called embryonic stem cells, are *pluripotent;* so called because at this stage a cell has the ability to turn into a cell for any part of the body such as blood, muscle, heart, liver, or nerve. The process that turns an embryonic cell into a specific type of cells is called *cell differentiation.*

Through cell differentiation, embryonic cells turn into different types of specialized cells, which multiply through the cell division process and make different tissues; tissues make organs; and organs make different parts of the body. In this rapid growth, the main external features of the baby begin to take shape.

How is this relevant to evolution? This process of development from one cell to multicellular organism is fundamentally the same for all multicellular organisms that reproduce sexually. This fact points to a common ancestor of all of these organisms. We will revisit this topic to explore further the relevance of the early development process to evolution in Chapter 6.

To summarize, all of the cells in your body are descended from one fertilized egg, zygote, which contained the genome inherited from your parents. That genome was copied to all other cells in your body. Inside a cell of all organisms, proteins are made according to instructions in the corresponding genes, and therefore the variation flows from gene level to the protein level. Proteins give rise to almost all the functionality of the organism, including its traits. Let's take a look on how it happens.

4.6 From Proteins to Traits

Proteins are the workforce of life. These large, complex molecules (polymers) are made up of smaller subunits (monomers) called amino acids. Nearly all dynamic functions of a living organism depend on proteins. For example, a variety of roles that proteins play in a cell include the following:

- **Biochemical role.** A type of proteins called enzymes catalyze (or speed up) biochemical reactions in our body necessary to keep us alive and healthy.

- **Cell signaling and coordination of an organism's activities.** These tasks, central to the functioning of an organism's body, are carried out by a type of protein called hormones. For example, insulin helps regulate the concentration of sugar in our blood.

- **Mechanical functioning.** Proteins facilitate the mechanics of the body. For example, the proteins called actin and myosin are responsible for the contraction of muscles.

- **Structural role.** Certain proteins are responsible for making spiderwebs, feathers, hair, hooves, and many other body parts.

- **Transportation.** Many proteins are responsible for transportation within the body. For example, a protein called hemoglobin transports oxygen from the lungs to other parts of the body through the blood stream.

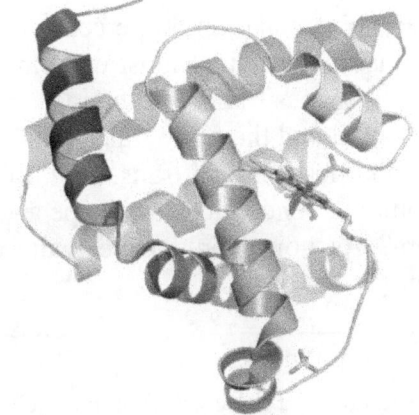

Figure 4.5. The three dimensional structure of a protein molecule called myoglobin. Source: Protein Data Bank.

- **Defense against diseases.** Types of proteins called antibodies combat harmful bacteria and viruses.

Because different proteins have different properties, different organisms with different proteins have different properties or traits; a display of variation or diversity. As an example, a protein called myoglobin was the first protein molecule whose three dimensional structure, shown in Figure 4.5, was discovered. This protein is found in the muscle tissue of almost all mammals. A property of this protein is to store oxygen. Now imagine that an organism has lots of myoglobin proteins in its muscles. What property for the organism would it lead to? Yes, you are right: the organism would be able to hold its breath for a long time because it could use the oxygen stored in the muscle cells. Now, consider the animals that are able to hold their breath for a long time such as diving mammals. You would imagine that those animals such as seals and whales should have high concentrations of myoglobin molecules in their muscle cells that would enable them hold their breaths longer. And this is true.

Proteins make up 50 percent of the mass of most cells, and are instrumental in almost everything a living organism does. Their shapes and structures determine their functions. Structural and therefore functional changes in proteins give rise to modifications in traits, and this way play their role in evolution. Where does the shape and structure of proteins come from? As you should have realized by now: it comes from genes.

Think About It!

Structural and therefore functional changes in proteins give rise to modifications in traits. Can we therefore say that proteins are the raw material (origin) of evolution?

Answer:

The answer is: No. It's the genes and not the proteins that are inherited. Therefore, genes, and not proteins are the raw material or origin of evolution.

The chemical instructions encoded in DNA in the form of genes tell the cell what kind of proteins to make. It happens through a process called central dogma of biology, explained in the next section.

4.7 Central Dogma of Biology: From Genes to Proteins.

Illustrated in Figure 4.6, the central dogma of molecular biology explains how protein is made from genes in a cell in the following steps:

DNA $\longrightarrow$ RNA $\longrightarrow$ Proteins

An enzyme called DNA polymerase replicates the DNA molecule, for example, during cell division. Even though a gene in the DNA encodes instructions for making a specific type protein, it does not produce the protein directly. As a first step, a ribonucleic acid (RNA) molecule is produced from the DNA molecule by a process called transcription, in

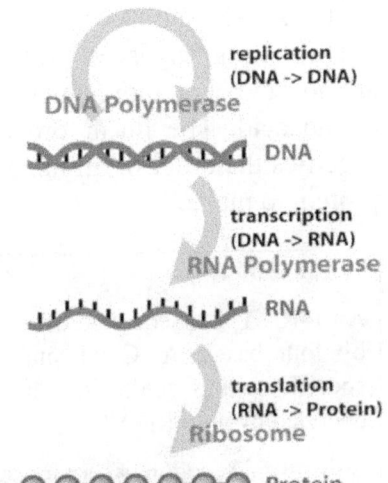

Figure 4.6 Central dogma of life

much the same way that DNA replicates itself. The transcription process is facilitated by an enzyme called RNA polymerase. RNA is very similar to DNA with three important differences: in its structure, it has the sugar called ribose instead of deoxyribose in DNA; it has the base uracil (U) instead of thymine (T) as in DNA; and it is usually single-stranded as opposed to double-stranded like DNA. The fact that it has base U instead of T means RNA has sequences of A, G, C, and U as its bases unlike A, G, C, and T for DNA. Also, RNA comes in various types. One type of RNA, messenger RNA (mRNA), made during the transcription process, has the recipe for protein synthesis. This means that an mRNA molecule encodes a sequence of amino acids, as a protein is a sequence of amino acids. An mRNA molecule, a sequence of bases, is read three bases at a time; and each of these base triplets is called a *codon. Which three bases are in a codon and in which order* specifies a unique codon, which in turn specifies a unique amino acid for the protein. For example ACU, AUC, and AGU are all different codons specifying unique amino acids. A codon can specify only one specific amino acid, but more than one codon can specify the same amino acid. The linear order of codons in an mRNA chain specifies the order of amino acids and this way encodes which protein to make.

Think About It!

An mRNA is a sequence of four types of nucleotides represented by four bases: A, C, U, and G. A unique codon encodes one specific amino acid. How many types of amino acids can be encoded by an mRNA?

Solution. A codon is composed of three bases. Given this, the question can be translated to: how many unique codons can be made out of four bases? That is, how many unique combinations of three can be made out of four items? The answer is: $4^3 = 64$.

However, there are only 20 types of amino acids which are produced in our cells to make proteins. This is because more than one type of codon specifies the same protein.

As a second step of the central dogma of life, mRNA codons are translated into a chain of amino acids called a polypeptide chain, which spontaneously folds into the corresponding protein. The process in this second step is called translation. Translation occurs at a biological structure called a *ribosome*. The mRNA binds to the ribosome and another type of RNA called transfer RNA (tRNA). As the mRNA binds temporarily to the ribosome, each codon of the mRNA links to a transfer RNA (tRNA), which brings the corresponding amino acid molecule to the ribosome. In other words, tRNA molecules bring the amino acid molecules to the ribosome in the order specified by the order of codons in the mRNA. The ribosome joins these amino acids in the order they are delivered by the tRNA into a polypeptide chain, which subsequently folds into a protein molecule. When the information coded in a gene is materialized into a product such as a protein or a trait, the gene is said to be expressed.

Now that you know how the genetic code stored in DNA is materialized into proteins in a cell of an organism, you can again ask the question: what does it has to with evolution? The answer is: the process is fundamentally the same in all organisms pointing to the same origin, that is, common ancestor. Furthermore, the universality of this process has immense potential of useful applications, as discussed in Chapter 11.

In this section, we have discussed some facts in molecular biology and genetics discovered through and supported by experimental research. It's important to understand these facts in order to comprehend evolution at the microscopic scale. Equipped with this knowledge, let's take a step back and take a look at how it was realized that Mendel's genetic factors, that is genes, are found in DNA molecules.

4.8 From Genetic Factors to DNA: From Mendel to Watson and Crick

The role of genes and hence DNA in evolution is so critical that it is not out of place to take a brief look at how Mendel's genetic factors were traced to genes in DNA. At about the same time Gregor Mendel concluded his experiments, studies toward a medical degree of a Swiss student Johannes Miescher (1844-1895) were interrupted as he fell ill with typhus. As a result of this disease, he became partially deaf. So even though he eventually finished his degree (M.D.), he could not become a practicing doctor. Instead, he switched to organic chemistry, where he made a big discovery in 1869 when he was examining the composition of the cell nucleus. Inside the nucleus, Miescher found an acidic substance that contained nitrogen and phosphorous. Later this substance would be called deoxyribonucleic acid, DNA.

Fast forward to 1901 when Hugo de Vries fused Mendelian genetic factors, genes, with the theory of evolution. At that time nobody knew what genes were made of, how they were modified, or how they were inherited. The first real clue to the answers of these questions came in 1944 from the experiments performed by Oswald Avery (1877-1955) at Rockefeller Institute in New York. Working with his colleagues Colin Macleod and Maclyn McCarty, Avery showed that DNA and chromosomes are (or contain) the genetic material. He demonstrated this by transferring the genetic characteristics of one cell to another by simply transferring a chemical in the cell called DNA. Here is how he did it: He started with two sets of bacterial cells: one set that would produce a slimy protective layer of strings of sugar called capsules, which would surround the cell; and the other set which would not produce the capsules. Avery took the DNA from the bacterial cells which would produce capsules and purified it. Then he injected this DNA into the cells that had not produced protective capsules. After receiving the new DNA, however, these cells started producing capsules. This demonstrated that the genetic information to make the capsules was contained in the DNA.

The results and conclusions from the Avery experiments were not widely accepted for years. One of the reasons was the limited knowledge at the time about the structure of DNA, according to which the DNA from different species appeared identical. So, how could it account for the transfer of heritable information which was different for individuals within a species and among different species? Furthermore, there was a prevailing opinion at that time that genes were made of proteins. The root cause of this prevailing belief was that even with the limited knowledge about proteins at that time, it was easy to show that different species had different proteins. So, many scientists believed that Avery's DNA sample was contaminated with proteins, and therefore it was protein and not DNA that actually transferred the genetic material. So prevailing was the belief that even Erwin Schrödinger, a founding father of quantum mechanics, went along with the hypothesis of *genes are proteins* in his book *What is Life?*, which was published in 1944.

In the early 1950s, however, experiments performed by Alfred Hershey and Martha Chase at the Cold Spring Harbor Laboratory in New York (U.S.) proved beyond a doubt that DNA, and not proteins, are the genetic material. They proved it by using bacteriophage, a type of virus that infects bacteria. In general, a virus is a noncellular infectious particle that consists of DNA (or RNA), a protein coat, and in some cases an outer lipid envelope. A virus needs a host cell to make copies of itself. Hershey and Chase used bacteriophages that only consisted of DNA and protein. The question they asked was: When a bacteriophage infects a cell by injecting the genetic material into it to replicate itself, what does it really inject into the cell? Does it inject DNA, protein, or both? They made use of the known facts that proteins contain sulfur but not phosphorous, and that DNA contains phosphorous and not sulfur. In one experiment, they labeled bacteriophage with a radioisotope of sulfur in order to track where protein will go; then infected the bacteria with the phage. So, sulfur labeled only proteins in the virus. Then they dislodged the viruses (bacteriophages) from the bacterial cells by whirling the mixture with a blender. The radioactive sulfur was detected in the solution but not inside the cells. That means proteins were not injected by the bacteriophage into the bacterial cell.

Then they repeated the experiment by labeling bacteriophage with a radioisotope of phosphorous, which would go where DNA goes. This time when they dislodged the virus from the bacteria, they detected the radioactive phosphorous inside the cells. This means that bacteriophage injected DNA into the bacterial cells. These experimental runs combined prove that DNA and not protein is the genetic material.

After the experiments of Hershey and Chase, DNA was accepted among the scientific community as the genetic material. Also in the early 1950s, several biochemists including Herman Branson, Robert Corey, and Linus Pauling started studying the detailed structure of proteins. Now that it was established that DNA is the genetic material, the rise of interest in finding the structure of DNA was imminent. The hope was that once the structure of DNA was discovered, it would help in answering lingering questions like how the genetic information such as the color of your eyes is stored within DNA and how DNA works; for example, how traits of organisms originate from DNA. The topic gained such an importance in the biology community that it started a race among scientists for who would go down in history as having discovered what they called the *secret of life*. It was in this environment that zoologist and biologist James Watson from Indiana University met a physicist (with a background in biology) named Francis Crick in 1951 at Cambridge University. The scientific collaboration of these two young scientists landed right in the middle of the race—that was already on—for finding the structure of DNA. Scientists all over the world were sifting through their and others' data searching for clues to the answers to these questions.

Even though the structure of DNA was unknown, some elements of the structure were, however, known. For instance, it was known that DNA was built of monomers called nucleotides and that each nucleotide consisted of a sugar (deoxyribose), a phosphate group, and one of the four nitrogen-containing chemicals called bases: adenine, thymine, guanine, and cytosine (usually abbreviated as A, T, G, and C). Furthermore, by 1952, a biochemist named Erwin Chargaff (1905-2002) had discovered two very important elements of the DNA structure:

1. The amount of A in any DNA molecule is equal to the amount of T, and this equality also exists between C and G.

   A=T and C=G

2. The proportions of A and C (and hence T and G) is different for different species.

And God Said: Let There Be Evolution

> **Note.** Although strictly speaking, A, T, C, and G are bases, it is a common practice to use A, T, C, and G to refer to the complete nucleotides in which they exist; for example A is used to refer to the nucleotide that contains the base A, and so on.

These facts are known as *Chargaff's rules*. These rules, especially Rule 1, were a clue to how nucleotides might be arranged in a DNA molecule. For example, if A only bonds with T and C only bonds with G (which turned out to be true), Rule 1 immediately follows.

> **Think About It!**
>
> Scientists have determined that about 27 percent of nucleotides in the DNA of a fly is the A nucleotide. Use Chargaff's rule to estimate the percentage of other nucleotides in the fly's DNA.
>
> **Solution:**
>
> According to Chargaff's rule 1:
>
> A=T and C=G
>
> Therefore, Percentage of T = Percentage of A = 27%
>
> Percentage of C = Percentage of G = $\frac{100-27-27}{2}$% = 23 %

Scientists, including biologists, were already using X-ray techniques, developed by physicists, such as X-ray diffraction (or X-ray crystallography) to study biomolecules such as proteins. Among the people who were eager to find the structure of DNA was a great chemist, Linus Pauling, who later became a Nobel Laureate in Chemistry in 1954 for his ground-breaking work on chemical bonds and the structure of crystals and molecules. Pauling had proposed a triple helix (a helix with three strands) model for DNA. One of Pauling's problems was that he thought that the bases (in the nucleotide monomers of the DNA polymer) were outside the helix as opposed to inside. Even though experimental results did not support the triple helix model (at which Watson and Crick had their shot too), it got Watson and Crick thinking: if not a triple helix then what? There was another important character in the story: Rosalind Franklin who by using X-ray crystallography, had generated photographs of some two dimensional patterns of DNA (shown in Figure 4.7a). She was an expert in X-ray crystallography and had already discovered the structure of coal by using this technique. In 1952, when Franklin presented her X-ray diffraction image of DNA in a research presentation, she interpreted the theoretical implications of the picture partly correct and partly incorrect. Watson was among the audience but he did not immediately understand the theoretical implications of Franklin's experimental results either. Crick, given his physics background, would have surely figured out the theoretical implications of Franklin's work, but he was not present at this gathering. So Watson and Crick needed to take another look at Franklin's DNA image.

Although, according to Watson, Franklin was overly secretive about her work, her colleague Maurice Wilkins reviewed one of these images with James Watson. By using their collective diverse background in physics and biology, and their understanding of all the existing knowledge about the DNA, Watson and Crick were able to connect the dots to derive the structure illustrated in Figure 4.7b from the image in Figure 4.7a. This is an example of what it takes to make progress in science: uniting theoretical and experimental skills.

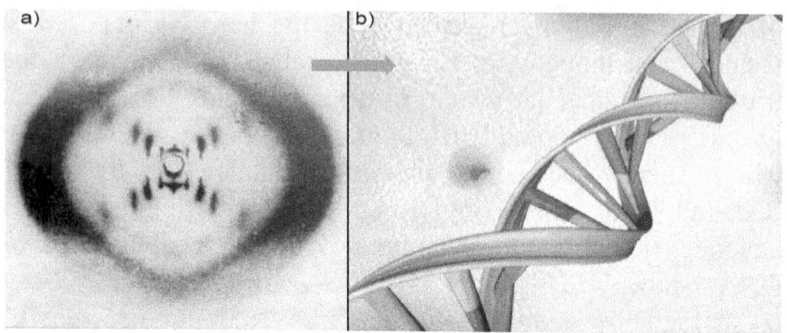

Figure 4.7 By analyzing other scientists' data and results and by connecting the dots, Watson and Crick were able to derive the DNA structure (b) from the X-ray crystallographic pattern of DNA (a).

The discovery of DNA structure started a revolution in biology by igniting a new life into the already existing fields of molecular biology and biotechnology. Nevertheless, there was still a gigantic gap between determining the structure of DNA and understanding the exact role of DNA. It took Crick and other scientists the next quarter century to develop the ideas that now constitute the basics of molecular biology: the genetic code, mRNA, and the translation of mRNA into proteins; parts of the process called the central dogma of molecular biology, discussed earlier in this chapter.

To sum up, the inheritance patterns of Mendel's *genetic factors* were found to be mirrored by the behavior of genes, or DNA which contains genes, or the chromosome in which the DNA is packed. Chromosomes are found in the cell's nucleus. Before the 100th anniversary in 1958 of the publication of *The Origin of Species*, hereditary information had already been traced to genes, which exist in a DNA molecule, which in turn is packaged inside a chromosome.

Before we can fully understand and appreciate the connection of DNA with evolution, we need to understand some details of DNA structure.

4.9 DNA Structure at a Glance

In 1953, James Watson and Francis Crick, based on experimental data, proposed the double helix structure of DNA molecules that had profound implications for our understanding of heredity and variations, and hence evolution. As proposed by Watson and Crick, and illustrated in Figure 4.8, a DNA molecule is a double helix made of two linear strands composed of repeating chains of sugar and phosphate; the two strands curling around each other to make a helical shape. Here are the three points to remember about DNA structure:

1. Each of the two strands is a repetitive chain of sugar and phosphate. This chain is called the backbone.

2. Each sugar in a strand is bonded to one of the four chemical bases: adenine, cytosine, guanine, and thymine; popularly called A, C, G, and T, respectively.

3. A base in one strand is bonded to the corresponding base in the other strand; the two bases are collectively called complementary pair. It is this bond that holds the two strands together in a double helix. These bonds are called hydrogen bonds and arise from the electromagnetic interaction between the bases.

The two strands tied together through complementary pairs of bases are called complementary strands or antiparallel strands. So, the four bases (A, C, G, and T) are at the core of understanding the behavior of DNA. You must know the following three crucial points about these bases:

And God Said: Let There Be Evolution

1. The properties of a DNA molecule are defined by the sequence or order in which A, C, G, and T appear in its strands.

2. As mentioned earlier, base A always pairs with base T and vice versa; and C always pairs with G and vice versa. With this fact in mind, if you know the sequence of bases in one strand of a DNA molecule, you can figure out the corresponding sequence in the other strand, called the complimentary sequence.

3. The process of determining the sequence of bases in the DNA of an organism is called *DNA sequencing*.

As shown in Figure 4.8, a unit in a DNA strand that consists of a sugar, phosphate and a base is called a nucleotide and is often represented by its base. Therefore, it is often said that DNA and therefore genes are made of sequences of four types of nucleotides called A, T, C, and G.

Think About It!

In Section 4.8, you read about Chargaff's rules about DNA structure. Explain how Chargaff's Rule 1 can be derived or naturally follows from the Watson and Crick model of DNA structure.

Solution

In the Watson and Crick model, A bonds only with T and vice versa, and C bonds only with G and vice versa. Therefore, in the DNA molecule: A=T, and C=G, which is Chargaff's Rule 1.

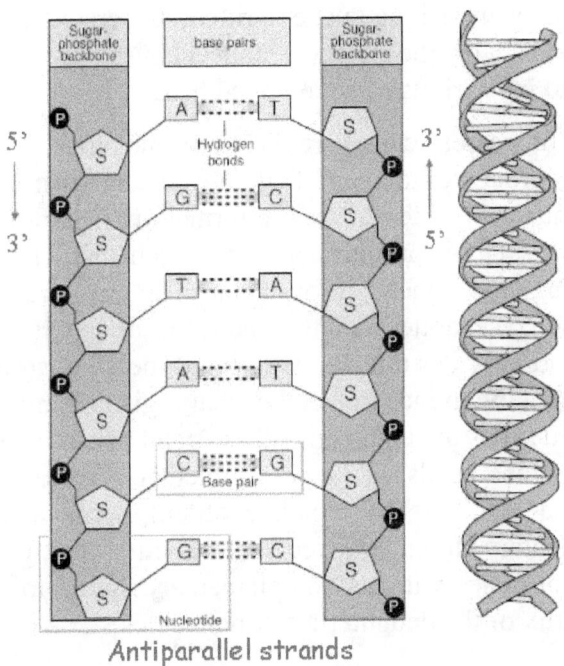

Antiparallel strands

Figure 4.8 Illustration of the double helix structure of the DNA molecule that carries genetic instructions in all living organisms. Image: courtesy of the National Institute of Health.

Because the sequence of A, C, G, and T defines the DNA code, an analogy is often made with the English language while talking about DNA and genes. Just like the 26 letters in the English alphabet, the four chemical bases of the DNA can be considered as four letters (A, C, G, and T) of the genetic language. Just as a sequence of letters in the English (or any) language compose instructions or code (in a computer language), the sequence of four letters (A, C, G, and T) in a DNA strand composes genetic instructions also called genetic code or genetic information. For this reason, DNA molecules are also called information molecules. Realize that the genetic code is a chemical language written, understood, and used spontaneously by the physical and chemical laws of nature; there is no mysterious force involved here.

Two processes called replication and transcription are necessary processes to keep life going. As already mentioned, a DNA sequence carries genetic information. The double helix structure of DNA provides a clear mechanism for copying (replicating) the genetic information. The two strands of the double helix separate, and the sequence of letters (bases) in each strand provides a template to build the other strand called complementary strand. Remember that a given template strand uniquely defines the complementary strand because an A in the template strand needs a T in the complementary strand and vice versa, and a C in the template strand needs a G in the complementary strand and vice versa. Each of the two template strands combines with its complementary strand to make the double helix of the daughter DNA molecules.

Think About It!

Consider the following band of a DNA strand:

ACTGGCTTA

Write down the complementary band that will be prepared from it during replication.

Solution

By using A↔T and C↔G, we obtain the complementary band as:

TGACCGAAT

Now that you have enough knowledge about the DNA molecule, we can explore if DNA studies support or refute the theory of evolution.

4.10 The DNA Connection of Evolution

Genes (or DNA) are the material that is passed from one generation to the next. Therefore if different species are related by evolution, then evolution has something to do with genes. For example, if no significant similarity between the genes of different species is found, one can argue that different species or lines of organisms were created independent of each other and they progressed along their own different lines, and no species arose from another species. However, if to the contrary, we find a great similarity among the genes of different species, we can present it as evidence that different species are related to one another and the details of these similarities may suggest that multiple species emerged from a single ancestor. This would reject the hypothesis of straight line progression in favor of branched tree of life, that is, the theory of evolution.

Think About It!

Species having more similar DNA are more closely related. With this reasonable assumption, consider three species A, B, and C. If the DNA of A is more similar to the DNA of B than that of C, which of the following is the correct prediction of the theory of evolution?

A. The common ancestor of A and B lived more recently than the common ancestor of A and C.

B. The common ancestor of A and C lived more recently than the common ancestor of A and B.

Answer

A

Well, the experimental results are in. As we already discussed in the earlier sections of this chapter, there is a pattern of similarities and differences that support the theory of evolution. For example, there is a 96-98 percent similarity between the DNA of a human and that of a chimp (Table 4.2), pointing to a common ancestor in the recent past. As experimental studies show, the relatedness among different species revealed by genome comparisons goes far beyond in quantitative details than these simple comparisons. For example, one may ask questions like this: If there is about 98 percent genetic similarity between human and chimp, what then accounts for the major morphological and behavioral differences between the two groups? Some of these differences can be accounted for by the differences in the control genes, the genes that control the growth rate. For instance, one of the major differences between human and chimp is that the human brain has four times as many cells. This difference among the number of cells can be determined by just one control gene. A single mutation in a control gene can explain such a big difference in the intellectual capability between chimps and humans. This shows how a little difference among genes at the micro-level can result in a big difference among traits at the macro-level (organism). This however will not come as a surprise to physicists and chemists who know that it takes only one more proton in the nucleus of the atom and an additional electron around it to get a brand new element. For instance, the addition of one proton and one electron would transform gold into mercury. In both cases (living and non-living entities), the reason that a small difference at microscopic level makes a big difference at macroscopic level lays in the fact that macro-entities are constituted from micro building blocks. It's just another reflection of the fact that macroscopic reality has its roots in the microscopic world.

The genome comparisons are being performed at different levels such as at the level of sets of chromosomes, DNA, and different genes inside DNA. These kinds of comparisons offer a detailed insight into the history of life, way beyond what the fossil records alone could offer. So, this genome comparison data not only generally supports the theory of evolution originally offered by Darwin and Wallace, but also by presenting more details, it extends the theory and puts it on the firmer grounds of the microscopic world. These studies enable biologists to build trees (or sub-trees) based on the comparison of different gene sets, and then they can combine trees to build a big picture. As an example, Figure 4.9 presents what is called a phylogenetic tree of life based on the sequencing of genes that encode for the production of ribosomal RNA (rRNA), a major component of ribosomes, a structure that is the site of the protein production in a cell. Based on this and other genomic comparison data, scientists can group all organisms at the highest level into three domains: Bacteria, Archaea, and Eukarya. The figure does not present the whole detailed tree of life; this is just a simple illustration as a big picture. A branch of the tree represents a group of organisms such as a species, and the length of the branch represents the amount of genetic change in the group or lineage. As the tree shows, all three domains of life (and lineages in them) originated from a common ancestor and did not develop in parallel and independent of each other. The tree clearly shows that prokaryotes (bacteria and archaea) evolved before eukaryotes.

Members of the domain Bacteria and the domain Archaea have one thing in common: they consist of prokaryotic cells and therefore are called prokaryotes. A prokaryotic cell is the cell that has no nucleus or any other membrane-enclosed internal structure called organelle. Most prokaryotes are unicellular (one-celled) microscopic organisms. Domain Bacteria consists of the most diverse and widespread, and most ancient prokaryotes as Figure 4.9 shows. Domain Archaea consists of prokaryotes including those that live in extreme environments such as salty lakes and boiling hot springs. Eukaryotes, the members of the domain Eukarya, are organisms made of eukaryotic cells, the cells that contain membrane-bound internal structures called organelles including a nucleus. Eukarya includes both unicellular organisms such as protozoans (a diverse group of unicellular eukaryotic organisms), and multicellular organisms such as plants, fungi, and animals including humans. On the evolutionary scale, archaeans are closer to eukaryotes than to bacteria.

Phylogenetic Tree of Life

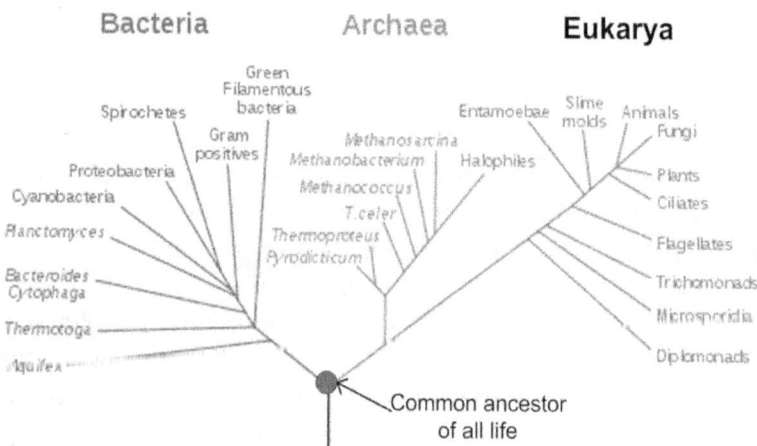

Figure 4.9 From gene comparisons based on sequencing of ribosomal RNA (rRNA) genes, all life can be grouped into three groups called three domains: Bacteria, Archaea, and Eukarya.

So, comparing the genes of different organisms help us to study the details of evolution and build evolutionary trees as never before. But how exactly do the genes in DNA drive evolution?

4.11 How Genes Drive Evolution

As already explained, variations are necessary for evolution to occur and the root causes for variations are genes, the variations in genes, and sexual reproduction. Variations in genes are of two kinds: First, there are alternative copies, alleles, of a gene corresponding to each heritable trait. This gives rise to variations among a trait such as a cleft chin and a chin without cleft. Second, a change can occur in a gene, called a mutation. This change in a gene can modify an existing trait in a few organisms. If such modifications are favorable in the given environment, they can eventually give rise to new dominant traits at population level where natural selection operates by selecting organisms equipped with traits that give them the ability to survive and reproduce in a given environment. The organisms with the ability to survive and reproduce viable offspring succeed in passing their DNA on to the next generation, and are naturally selected for. The organisms that fail to survive or reproduce viable offspring due to their different traits originating from different genes, fail to pass their DNA to the next generation and hence eventually their traits perish, that is, selected out. An organism (or a species) has in its genome all the mutations that led to its success as a persistent living creature (or a species). So, we humans, who made it all the way from the time of the first cell to today, should be very thankful to all mutations in us. However, do not get carried away with mutations; most of mutations are harmful. But what really is a mutation?

As mentioned early on in this chapter, a mutation is a change in the nucleotide sequence of an organism's DNA. Then, why do mutations occur? There is more than one reason for mutations to occur. For example, mistakes can happen during the replication of DNA, for instance, during cell division. A wrong letter (base or nucleotide) can be substituted for the right one, a letter (base or nucleotide) can be entirely missed, or an extra letter can be added to the sequence. These mutations are called *substitution, deletion,* and a*ddition,* respectively. If these mutations involve only a single nucleotide, they are collectively called *point mutations.* Sometimes, however, a whole sequence (string) of letters can be reversed (an example of substitution), missed out (deletion), or duplicated (addition).

Think About It!

Experimental studies suggest that the earliest forms of life on Earth had minimal number of genes, that is, a very small genome. What kind of mutation may be responsible for evolving larger genomes?

Answer

Duplication. Genome replication occurs during cell division. Due to errors during replication, it is possible that some genes and even the entire set of chromosomes can be duplicated. If this duplication happens in the germ (sex) cells, it would be handed down to the offspring.

Another cause for mutations is the exposure to harmful environmental agents such as high-energy radiations, which can break chromosomes apart into pieces. Some pieces can be lost and this will give rise to deletions of A, C, G, or T in the DNA string. Radiation can also knock electrons right out of atoms and generate a trail of free radicals, which can damage DNA. Other causes for mutations include mutagenic chemicals and viruses. Hermann Muller (1890-1967), an American geneticist, was awarded the 1946 Nobel Prize in Physiology or

Medicine, "for the discovery that mutations can be induced by X-rays."

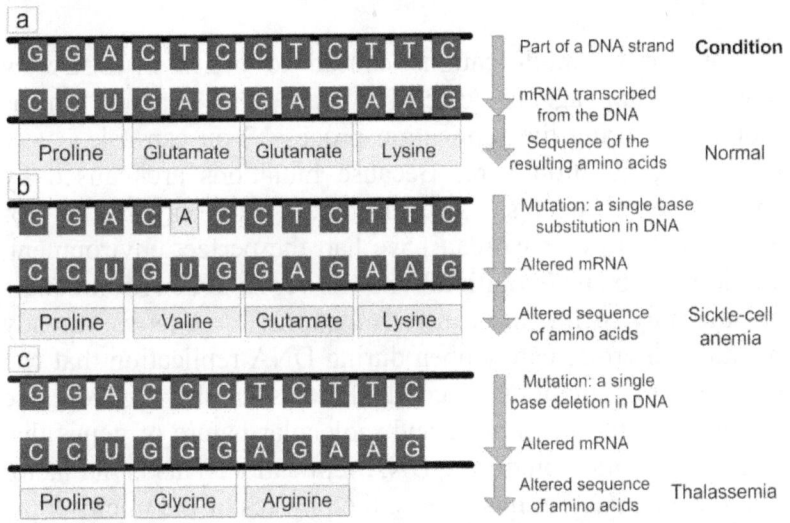

Figure 4.10 Examples of gene mutations and their effect on protein production and the organism

Most mutations are harmful to the organism in which they occur and to the offspring to which they are transferred. Figure 4.10 presents two examples of how mutations occur at the molecular (genetic) level and how their effect flows to the organism level through protein production. Figure 4.10a shows part of the gene that codes for the production of the protein called hemoglobin, which transports oxygen from the lungs to other parts of the body, discussed earlier in this chapter. Figure 4.10b shows the substitution of a single base A for T. As a result, a codon of the mRNA produced during transcription is altered, which in turn results in the production of the valine amino acid instead of the glutamate amino acid for the polypeptide chain of the protein. This insertion of one wrong amino acid into the protein molecule causes a condition called sickle-cell anemia. Similarly, Figure 4.10c illustrates how the deletion of a base, T, changes all the codons to the right and the corresponding amino acids. This results in the condition called *thalassemia.*

Damage or modifications caused to genes in DNA by environmental agents, discussed earlier, is passed on as mutations during the replication process. You can ask a very interesting question here: Because mutations are caused by environmental agents, is environment the only factor for evolution? In other words, if we had the perfect environment, would there be no evolution? The answer is: no, environment is not the only factor that causes mutations. First, as we already mentioned, errors can happen during DNA replication that can give rise to mutations. Second, there is another unavoidable factor built into the atomic and molecular nature of genes that also causes mutation during DNA replication. This is the factor that makes quantum mechanics, the physics of the microscopic world, the underlying science behind evolution, and is therefore explained in a separate chapter, the next chapter.

Note. Free radicals are atoms and molecules with single unpaired electrons, and are responsible for many diseases including aging Their motion is governed by quantum mechanics.

You can determine the flow of genetic variation from one generation to another by comparing the DNA sequence of an organism with its offspring. Scientists have done exactly that and have confirmed that mutations do occur fairly regularly and they are passed down to the subsequent generations.

Caution! It is obvious to realize that mutations in organisms that reproduce sexually will flow from one generation to the next through germ cells (sex cells). A mutation in a non-sex cell in these organisms will not be transferred to the next generation and therefore does not participate in evolution.

You may ask: at what rate do the mutations occur? Different species have different rate of mutations but the typical average for substitutions (replacing one letter with another) is about one in 100 million base pairs per generation. We, human beings, accumulate about 100 mutations per generation. This may not sound like much given that there are more than one million codons in the human body. However, as mentioned earlier, remember even a single mutation can make a big difference, for example, a single mutation at the wrong place can have a significant impact on the organisms as shown in Figure 4.10, or can sometimes even be fatal.

Here are some discovered facts (supported by data) about mutations:

♦ The most common kind of mutation is substitution in which a letter (base or nucleotide) of DNA is replaced with another.

♦ Each letter in each gamete for each generation has one chance in a billion to be replaced by another letter.

♦ Most mutations that happen in a random way are harmful—imagine typos in your email message or in a computer code; you don't expect them to be good—whereas only a tiny fraction are beneficial to the fitness of the organism. Some mutations, called neutral, are neither beneficial nor harmful.

♦ Only harmful or beneficial mutations play a role in evolution. Individuals with beneficial mutations are selected for and they multiply over generations through reproduction, and individuals with harmful mutations are selected out and they along with their harmful (unfit) traits vanish.

And God Said: Let There Be Evolution

Scientists have also observed that certain types of mutations occur more frequently than others. For instance, changes in certain types of large sequences of bases occur more frequently than the average rate of overall single base-pair substitutions. As another example, *homopolymers*, continuous strings of eight or more identical letters, are especially prone to copying errors during the DNA replication. Same is the case with the sequences of two or more bases repeated over and over; DNA regions called *microsatellites*.

Obviously, mutations affect genomes. All of these spontaneous, uneven (in rate) changes within genomes even within the same species add up over generations to a substantial amount of diversity. Here is a point to note: Although insertion and deletion events (mutations) involving large DNA strings (or segments) are not as frequent as change events involving single base pairs, they represent the majority of total number of bases that differ among genomes (because a string change event changes more bases than a single base-pair change event). Another important point to note is that different individuals have a different number of copies of certain genome regions. This may have profound implications that scientists have just begun to explore.

> Note. A mutation event is either a change of just one base pair or a change of a continuous string of multiple base pairs with each pair in the string changed.

Sequence changes among humans cause one or more of the following changes in a significant proportion of all 23,000 human genes:

1. Protein encoding

2. Regulatory information

3. Number of copies of the genes

This alone becomes an underlying source of variations among traits that differ among different individuals.

We saw earlier in this chapter how environment plays a role in evolution by being one of many factors that cause gene mutations. Here is another role that the environment plays: A neutral mutation in one generation can turn into a useful one in another generation under a changed environment. For example, the gene that protects a given type of bacteria from antibiotics may have been useless in the earlier environment in which there were no antibiotics. However, its presence enabled this type to survive under the changed environment in which antibiotics are being used.

Think About It!

Explain how the theory of evolution predicts that most mutations should be at least somewhat harmful to organisms.

Solution

As a result of their long history of evolution, which includes many generations of selecting favorable traits and corresponding genes, most organisms have achieved a very good fit between their traits and the environment. Therefore, any deviation from the existing genome, that is mutation, is most likely to be harmful.

To summarize, mutations are random (spontaneous) changes within genomes that add up to a significant amount of diversity even within the same species. Taking mutations into account, evolution, as illustrated in Figure 4.11, is therefore in general a three step process:

1. In each generation of a population, mutations result in new genetic variants at the molecular level of DNA.

2. New genetic variants may result in new trait variants at organism level.

3. Natural selection works as a filter over generations: the interaction of the population with the environment increases the frequency of beneficial variants (favored by the environment) and decreases the frequency of harmful variants (not favored by the environment). As a result the gene pool of the population changes.

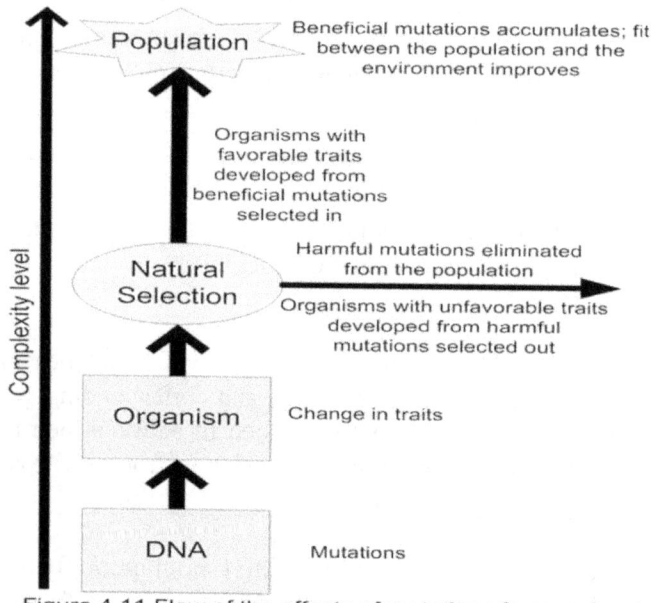

Figure 4.11 Flow of the effects of mutations from molecular to population level.

Note. A gene pool of a population is the sum total of all the alleles of all the genes in all organisms of the population.

Step 3 above means the organisms (or species) with good variants (better fitness) survive to pass the mutation down to the next generation, whereas organisms with harmful variants (poor fitness) do not survive to pass their genes along. In this way, the environment filters out the lethal changes and absorbs in the beneficial changes to the genome of the population. This way, mutations at molecular level are expressed in terms of the evolution of traits and species at the macroscopic level.

> **Note.** Random changes in DNA, called mutations, can give rise to changes in an organism's traits and thereby provide a constant source of variations within a population and species over time, given that DNA is inherited.

Natural selection working over variants originating from positive (beneficial) mutations is called positive natural selection, whereas natural selection working over variants resulting from negative (harmful) mutations is called negative natural selection. Natural selection corresponding to neutral mutations is called neutral natural selection. Harmful mutations are largely filtered through natural selection because they decrease the reproductive success of organisms. Many mutations occur in somatic (non-sex cells) and therefore are lost when the organism dies. Therefore, a very small fraction of all the mutations that occur in a population become dominant among the population.

Because neutral mutations have no appreciable effect (beneficial or harmful) on the fitness of the population, their frequency increases over generations. These mutations–silently and unnoticeably spread through the population–subsequently change the genetic makeup (genome) of the population. Notice that this is happening without interference from natural selection in this case. The frequency of some genes or alleles (different versions of the same genes) can also decrease over generations. So, the frequency distribution of alleles in the genome of a population changes over generations. This is a way of looking at evolution right where it originates, microevolution.

4.12 Microevolution

From research in physics, we know that most of the physical reality that we experience every day at macroscopic level has its root in the microscopic world. For example, you and every physical thing you see around you are made of atoms, and most of the physical phenomena that you see around you are really happening at the microscopic (atomic and molecular) level. Physicists have learned that the microscopic behavior of physical entities often cannot be explained in terms of their behavior at macroscopic level, but nothing happens at the macroscopic level that cannot be explained in terms of microscopic processes. So, in this spirit, we can ask the question: What are the smallest units of evolution? From the previous sections of this chapter, by now, you should know the answer: population and genes. Population is the smallest group of organisms that evolves; which at its most fundamental level means the allele frequencies of the population change over generations. Evolution at this (allele or gene) level is called microevolution. This evolution is the origin of macroevolution, the evolution that Darwin, Wallace, and other scientists during their times observed in the macroworld at the level of populations and species of organisms. The changes of microevolution accumulate and crystallize over a longer period of time in terms of new species, visible to us as macroevolution.

To explore microevolution, you need to understand two concepts: population and its gene pool. A population is a group of individual organisms of the same species living in the same geographical region at the same time and therefore can interbreed to produce viable and fertile offspring. You have already learned in this chapter that the genetic makeup of an individual organism is defined by its genome. Well, the genetic makeup of a population is determined by its *gene pool*, which is the sum total of all the alleles of all the genes in all individual organisms of the population.

In order to understand microevolution in quantitative terms, you need to understand the allele frequency assigned to each allele in a population. Let's work out an example to understand what an allele frequency is. Consider the gene that determines if you have a cleft chin or non-cleft chin. This gene has two alleles: one dominant allele for a cleft chin (*D*) and the other recessive allele for non-cleft chin (*d*). For this gene you received one allele from each of your parents. So during the sexual reproduction process there were three possibilities for the inheritance this chin gene. If you received *d* from both of your parents, your genotype is *dd*; you have a non-cleft chin as your phenotype; and you are said to be homozygous for the allele *d* or for non-cleft chin. If you received *D* from one parent and *d* from the other, your genotype is *Dd*; your phenotype is cleft chin; and you are heterozygous for this trait. If on the other hand you received *D* from both of your parents, you have a cleft chin; your genotype is *DD*; and you are homozygous for *D* or for the trait.

Now consider a population of 1000 individuals in which 700 individuals are homozygous for non-cleft chin (*dd*), 200 are heterozygous for cleft chin (*Dd*), and 100 are homozygous for cleft chin (*DD*). Because each individual has two alleles for this gene, the total number of alleles for this gene in the gene pool of the population is $1000 \times 2 = 2000$. The total number of dominant cleft-chin alleles, *D*, is 400 ($100 \times 2 = 200$ for *DD* individuals, plus $200 \times 1 = 200$ for *Dd* individuals). Therefore the allele frequency for *D* in the gene pool of this population is $400/2000 = 0.20 = 20\%$. Hence, the allele frequency for the *d* allele is 80 percent because the sum of all allele frequencies for

And God Said: Let There Be Evolution

a gene must be 100 percent, and in this case there are only two allele types (*D* and *d*).

Here is the rule of thumb for microevolution: when the allele frequencies of a population's gene pool change, we say that evolution has occurred. This change in the gene pool of the population exhibits itself in terms of the genetically inherited changes in the individuals of the population, who obviously are the members of the population's gene pool. This quantitative approach to measuring evolution at the microscopic level was developed independently by Godfrey Hardy, an English mathematician, and Wilhelm Weinberg, a German physician, early in the twentieth century. They developed their mathematical model of evolution based on the theory of probability. A simple derivation of their equation called Hardy-Weinberg equation is presented in the box following for mathematical-minded readers. Other readers can ignore the content in the box.

Godfrey Hardy (1877-1947)

Wilhelm Weinberg (1862-1937)

There is Nothing Unscientific About Chance in Evolution: Derivation of the Hardy-Weinberg Equation

The proper name for what biologists call randomness or chance is probability, a concept that is well understood in science. Mendelian genetics uses probability to predict the genotypes and the corresponding phenotypes from fertilization. Let's continue with the process of sexual reproduction in which a child gets the cleft chin gene. As discussed earlier in this section, there are three possible genotypes for this gene: *DD*, *Dd*, and *dd*. Assume that the probability that the child receives *d* from one parent is q. Therefore the probability that the child receives *d* from both parents (for genotype *dd*) = q × q = q^2.

Similarly, assume that the probability that the child receives *D* from one parent is p. Because the child either receives *D* or *d* form a parent, therefore p+q=1.

The probability that the child receives *D* from both parents (for genotype *DD*) = p × p = p^2.

Now, the probability that the child receives *D* from the father and *d* from the mother = p×q=pq

Similarly, the probability that the child receives *d* from the father and *D* from the mother = q×p=qp = pq

Therefore total probability for the child to receive the genotype *Dd* = pq + pq = 2pq

Because the total probability is always equal to 1, by adding the probabilities of acquiring *Dd*, *DD*, *dd*, we obtain:

$$p^2 + q^2 + 2pq = 1 \qquad (4.1)$$

Equation 4.1 is called the Hardy-Weinberg equation as it was first derived in 1908 by Godfrey Hardy and Wilhelm Weinberg. This equation is used to test if evolution is occurring in a population. It provides the same kind of tool for population genetics that Punnett square diagrams, discussed in Chapter 3, provide for Mendelian genetics, which works at individual level.

According to the Hardy-Weinberg model, a population is not evolving only if the measured frequency of one homozygotic genotype (such as *DD*) is equal to p^2, the measured frequency of the other homozygotic genotype (such as *dd*) is q^2, and the measured frequency of the heterozygotic genotype (*Dd*) is 2pq. In other words, a population is not evolving if it satisfies the Hardy-Weinberg equation. Recall that we have used Mendelian genetics rules to derive the Hardy-Weinberg equation. This means that the population must satisfy the Hardy-Weinberg equation if the only *change process* allowed in the population gene pool is Mendelian segregation and recombination. An evolving population violates one or more of these conditions specified above. The sources that cause these violations are discussed outside this box.

The mathematical rigor of the Hardy-Weinberg model is out of the scope of this book. However, let's take this opportunity to explore an important scientific approach. In mathematics, philosophy, and science, there are two logical approaches to prove a doctrine: You derive it from facts or existing principles. Alternately, you begin by assuming that the doctrine is false and then logically run into some kind of false result or a contradiction, hence your initial assumption was false, and therefore the doctrine must be true. The latter is the approach that Hardy and Weinberg took toward evolution as explained in the following:

Let's assume all the evolutionary scientists along with Darwin are wrong; and evolution does not occur. Now, we will explore where this assumption will lead us or what the requirements are to make this assumption true? In Chapter 3, you learned how during fertilization alleles of a sex cell separate into games (sperms in a male and eggs in a female) and recombine to form a zygote in processes called Mendelian segregation and recombination. If mating between males and females happens randomly, then the fertilizations in a population are like shuffling of a deck of cards, and therefore it would change neither the allele frequency nor the genotype frequency. As long as we assume mating in the population is random, fertilization will not bring about evolution. But we would also have to stop all other processes that might change the allele frequency or the genotype frequency of the population. This is exactly what the Hardy-Weinberg theorem states:

If only Mendelian segregation and recombination of alleles are at work in a population, the allele frequencies and the genotype frequencies of the population will remain constant from generation to generation, and therefore the evolution will not occur.

Such a non-evolving population is said to be in Hardy-Weinberg equilibrium, that is, it is not evolving. Remember, for this to happen, only Mendelian segregation and recombination of alleles is allowed, that is, no other process that may change the genetic makeup of the population is allowed. This means, in order to be in Hardy-Weinberg equilibrium, that is not to evolve, the population must meet the following requirements:

1. **Absolute mating randomness**. This means all members of the population breed and choose their breeding partners absolutely randomly. There are no preferences. Also everyone produces the same number of offspring. Under these conditions there is no change in allele frequencies of the gene pool due to sexual reproduction. Obviously this condition is practically never met, and hence there will be change in allele frequencies leading to evolution.

2. **No mutation**. Mutations modify the gene pool by actions such as altering alleles, deleting genes, and duplicating genes. No mutation means no changes (in allele

frequencies) from the mutations to the gene pool. In practice, mutations do happen and therefore evolution is inevitable.

3. **No gene flow.** No genes are flowing outside or into the population. That means no immigration into or emigration from the population is happening. This condition, in practice, is not commonly met either, hence evolution happens.

4. **No genetic drift.** Genetic drift is the process in which events based on chance or probability cause changes in allele frequencies of the population's gene pool from one generation to the next. Genetic drift has a noticeable effect only in small populations. For example if in a population of one hundred people, there are only ten blue-eyed individuals and eight of them are killed in an accident before getting married and produced no offspring; it is a huge change in the blue-eye allele frequency of the population. The effect of this accident will go unnoticed if it happened in a population of millions or even thousands of individuals where there are many blue-eyed individuals.

5. **No natural selection.** Natural selection does not affect the gene pool directly. It operates on individual organisms, and if an individual is selected out, all of its genes are expelled out of the gene pool of the population. But because natural selection is directional, that is, it favors some genes (corresponding to favorable traits) against others, it changes the allele frequencies of the gene pool by passing alleles to the next generation in different proportions than they are in the current generation. Therefore, for a population not to evolve, natural selection must not be operating. If natural selection is operating, and therefore gene pool is changing, then the population is evolving.

Think About It!

A catastrophe struck a population of 50,000 organisms of a population, and only 100 of them survived. The genetic pool of the survived part of the population is different from that of the original population. What is the source of the evolution that this population has gone through?

Answer

Genetic drift because there is chance involved in who survives the catastrophe.

If any of these conditions are violated, evolution will happen. We know that most of these conditions are commonly violated, therefore evolution does happen. For example, it's scientifically proven that mutations do happen, we know that gene flow due to immigration and emigration does happen, and absolute randomness in sexual reproduction is hardly ever met.

Do not be misled into the conclusion that the Hardy-Weinberg theorem is just a theoretical intellectual exercise. Scientists actually measure the allele and genotype frequencies of real populations and then compare them with the predictions of the Hardy-Weinberg equation in order to monitor if evolution is happening. If the data from the population violates the predictions of the Hardy-Weinberg theorem, then the population is said to be evolving.

Mutations along with sexual reproduction basically produce all of the genetic variations necessary for evolution to occur. That said, mutation rates are very low in plants and animals, about one mutation in one hundred thousand genes. So, mutations do not drastically change the allele frequency directly over a few generations. However, for organisms with short lifespan, such as bacteria, mutations can directly change the allele frequencies of a population and therefore cause

181

genetic variation rather quickly. In plants and animals, natural selection, genetic drift, and gene flow are the three major mechanisms that cause most of the evolutionary changes.

Note that we have just pronounced that natural selection is not the only mechanism for evolution. However note that gene flow and genetic drift may cause *adaptive evolution*, that is, evolution that improves the fit of the organisms with their environment; or they may cause the evolution of new traits which may neither be useful nor harmful for the population in adaptation to the existing environment. In a changed environment, however, the new traits may contribute to the survival or the demise of the population under the filter of natural selection, depending on if the new traits are adaptive or not to the new environment. Evolution under natural selection is also called adaptive evolution, because this evolution consistently results in a better match between the population and its environment. Gene flow and genetic drift may cause non-adaptive or adaptive evolution, but only natural selection consistently causes adaptive evolution.

Think About It!

Consider genetic drift occurring in a small population. What kind of change can it cause in the small population that will be unlikely or less likely in a very large population?

Answer

Some alleles, by some chance event, can completely disappear from the genetic pool of the small population resulting in the reduction of genetic variants or diversity.

In adaptive evolution, the following three things are essentially involved:

1. Trait variations in the population exist. Recall that natural selection operates on variations; no variations means no selection.

2. Trait variations have their roots in genetic variations; otherwise they cannot flow from generation to generation to participate in evolution.

3. Genetic variations that play role in evolution are not neutral; some of them are harmful (unfit) and some are useful (fit) for adaptation. In other words, the genetic variants and traits rising from them affect the individual's ability to survive and reproduce to leave behind their viable offspring.

Equipped with the knowledge of mutations and variations discussed in this chapter, let us now see where we are in answering the question that haunted Darwin, Wallace, and other biologists of their times.

4.13 From Mutations to Trait Variations: How Are We Doing, Sir?

As discussed in Chapter 3 (Section 3.2), Sir John Herschel wanted to know the origin (or the *why*) of the variants among the traits before he could appreciate the Darwin-Wallace idea of natural selection acting on these different traits. Well, now we know that spontaneous changes in the DNA (mutations) are the *why* of variations that Herschel was looking for. Throughout this chapter, we have discussed how these mutations actually result in trait differences (or trait variants): genotype change reflects in phenotype change. As explained in this chapter, scientists have been studying and comparing genomes of different organisms. They have even gone beyond just comparing the genomes to pinpoint specific reasons for variations. Here are a few examples of what scientists have already figured out:

❖ The genomes of Mendel's tall and short pea plants differ by a single G to A substitution in a gene that codes for the enzyme (a type of protein) called *gibberellin oxidase*. This substitution changes a single amino acid in the enzyme responsible for reducing the enzyme activity, which in turn causes a 95 percent drop in the production of a hormone that is responsible for stimulating growth in the stems of pea plants; hence short pea plants.

❖ Mendel's wrinkled seed trait is caused by a mutation (or change event) that inserts an 800 base-pair-long sequence in a gene that codes for a starch-related enzyme. This affects the enzyme production, which in turn reduces the starch synthesis, changing the sugar-water proportions, subsequently resulting in wrinkly but sweeter peas.

❖ A single base-pair change in a dog's genome has been found responsible for improved racing performance and increased muscle size in whippet dogs. This base-pair change inactivates a signal that otherwise suppresses muscle growth.

Exploring how mutations translate into differences among traits is becoming an active field of research with implications far beyond the realm of evolutionary studies.

Think About It!

What are the two microscopic phenomena that create genetic variations in populations? Do not use the term genes.

Answer: Sexual reproduction and mutation.

4.14 Summary

For about a century now, physicists have known that the everyday reality that we experience in the macroworld has its roots in the microworld of molecules, atoms, and subatomic particles. So, it should not come as a surprise that biologists are discovering the roots of evolution in the microworld of molecules of DNA and genes. The genius of Gregor Mendel rests in the fact that he hypothesized the existence of entities that he called genetic factors based on the empirical facts he discovered through his experiments. He and other scientists of his time knew nothing about DNA or genes, rest aside their structures. It has taken longer than a century of scientific work to experimentally discover the genetic factors, that we now call genes, and understand how they cause traits and trait variations.

Research in molecular biology and genetics has not only proved that genes and gene mutations are the origin of traits and trait variations, but has also demonstrated to the detail of molecules and atoms how exactly traits and trait variations originate from gene and gene mutations. It is a well proven fact that some traits originate from genes, and that individuals within the same population of a species have variations in their traits, that is, they vary genetically in their ability to survive and reproduce in a given environment. These facts, combined with the fact that environment does change, make natural selection and adaptive evolution inevitable. Darwin and Wallace only knew evolution in the macroscopic world; so they presented only natural selection as the mechanism of evolution. In the process of discovering the roots of evolution in the microworld, scientists have discovered that natural selection is not the only mechanism of evolution. Although other mechanisms such as gene flow and genetic drift also bring about evolutionary changes, natural selection is the only mechanism that consistently causes adaptive evolution and thereby improves the match of the population with its environment. These processes and facts are scientifically so well understood that to deny evolution is equivalent to denying experimentally-verified facts and principles of molecular biology and genetics.

In a nutshell

✓ The genetic variation that makes evolution possible is originally produced by mutation and sexual reproduction.

✓ Scientific research has demonstrated that Mendel's discrete genetic factors are genes, which exist inside a DNA molecule.

✓ When a gene is activated to form a product, say a protein, the gene is said to have expressed itself.

✓ Each cell of an organism contains the same set of genes, called its genome. However, not all the genes are used in all the cells; different types of cells (such as blood cells and skin cells) use different subset of genes. In general, not all the genes express themselves in all the cells.

✓ The cell converts the information coded in a gene to the gene product, the protein, by using a process called the central dogma of molecular biology. One gene contains instructions to make one type of protein.

✓ Proteins, produced by the corresponding genes, play a role in giving rise to specific traits. Different proteins and therefore different genes are responsible for different traits.

✓ The origin of the modifications in heritable traits and the rise of new heritable traits is gene mutations.

✓ Heritable traits are modified by gene mutations and not directly by the environment. However, environment is one of many causes of gene mutations, for example, radiation causes gene mutation. Also environment does provide grounds for natural selection to operate: the organisms with traits that enable them to adapt to the environment are selected for.

✓ The processes such as mutation, genetic drift, gene flow, and natural selection prevent all natural populations from reaching a genetic equilibrium and therefore evolution does happen in nature.

✓ The unavoidable built-in cause for mutations is the atomic and molecular nature of genes predicted by quantum physics.

✓ The rise of traits and trait variations from genes and gene mutations has been well understood to the detailed level of molecules and atoms. Natural selection operates on these variations to cause evolution.

✓ A bonus for discovering the causes of variations is the discovery of mechanisms of evolution in addition to natural selection such as gene flow and genetic drift. However, natural selection is the only mechanism that consistently improves the fit between the population and its environment, that is, brings about adaptive evolution.

✓ Macroevolution originates from microevolution, which is measured as the change in gene pool of a population in terms of change in allele frequency. In other words, as evolution has its roots in genes, the theory of evolution has found its roots in genetics.

4.15 Behold

As discussed earlier, Herschel's strong argument against the theory of evolution introduced by Darwin and Wallace was that useful new traits and species could never arise from simple random variations. Also his question was: what causes these variations? Herschel's objections persisted because Darwin could not explain the cause or origin of these random variations and how they are passed down to subsequent generations. So in 1859 and in the following years, the origin of variations was still a mystery. Herschel obviously did not know that he was diametrically opposed to the reality that would be discovered more than a century later when he argued that the appearance of new traits would always require:

" mind, plan, design, to the plain and obvious exclusion of the haphazard view of the subject and the casual concourse of atoms."

Not only has it been demonstrated that genes are responsible for traits and trait variations, the study of genes and genomes have also proved that "the casual concourse of atoms" can in fact generate useful new traits. Here are some examples of new traits as a result of simple rearrangements of atoms and molecules in the DNA sequence of organisms from peas to humans: sweeter peas, ability to run faster, bigger muscles, and improved ability to digest certain foods.

In other words, progress in molecular biology and genetics has taken the mystery of the origin of variations out of the picture and as a result has opened the door to endless applications and opportunities, discussed in Chapter 11.

Fine, mutations are the origin of variations that drive evolution. Biologists often refer to mutations as random mutations. So, in about one and half century after the theory of evolution was originally introduced by Darwin and Wallace, we have traveled from random variations at trait level to random mutations of genes. The term random comes from the fact that these mutations occur spontaneously, using probability at some points, regardless of whether they would be helpful or harmful to the organism. One approach that some opponents of evolution take is to mock the term *random* by reducing it to *chance* or *accident* to discredit evolution. We scientists take a different approach. We keep asking questions and exploring the issue in ever more detail. We are questioning this randomness now with questions such as: What does randomness really mean here? Why do certain types of mutations occur more frequently than others?

We have already discussed the probability component of randomness in Chapter 3 and in this chapter. However, to fully address this question, we must recognize the elephant in the room: quantum physics. Because with genes, biology and studies of evolution have entered the realm of atoms and molecules, and therefore quantum physics must be counted in. With that in mind one thing is sure, that the randomness we are witnessing in gene mutations is scientific randomness such as uncertainty in quantum mechanics, at least in part. We explore the connection of evolution with quantum mechanics, a well-established theory of physics, in the next chapter.

And God Said: Let There Be Evolution

FIVE

Quantum Evolution: The Physics Connection

And God Said: Let There Be Evolution

Every advance in knowledge brings us face to face with the mystery of our own being.

Max Planck

All science is either physics or stamp collecting.

Ernest Rutherford

Quantum physics thus reveals a basic oneness of the universe.

Erwin Schrodinger

5.1 Once Upon a Time

Erwin Schrödinger (1887-1961), one of the founders of quantum mechanics, was born on August 12, 1887, in Vienna, Austria. He received his doctorate in physics in 1910 at the University of Vienna. During his stay at University of Zurich from 1920 to 1926, he developed the wave formulation of quantum mechanics including his famous equation, known as Schrödinger equation, which captures the main concepts and principles of quantum mechanics. In 1933, he shared a Nobel Prize with Paul Dirac "for the discovery of new productive forms of atomic theory."

Erwin Schrödinger (1887-1961)

Schrödinger worked in several fields including physics, mathematics, biology, history, philosophy, language, and literature. He was at Dublin's Institute in Advanced Study from 1939 until 1956 before he returned to Vienna. It was there at Dublin where he became interested in biology. His book *What Is Life?*, published in 1944, introduced biologists to a way of looking at life from the perspective of physics. It was this book that started James Watson and Francis Crick (a physicist himself) on their journey of the search for the secrets of genes. This led to the discovery of DNA's double helix structure in 1953 for which Watson and Crick were awarded the 1962 Nobel Prize in Physiology or Medicine, which they shared with Maurice Wilkins.

Schrödinger died of tuberculosis (TB) in Vienna on January 4, 1961 at the age of 73.

5.2 In the Beginning

We explored in Chapter 3 how the results from Mendel's experiments rejected the hypothesis of blendable variants in favor of discrete variants which are passed down from parents to offspring in form of discrete genetic factors, which were later recognized as genes. For many reasons described in Chapters 3 and 4, scientists failed to immediately connect the dots between Mendelian genetics and Darwin-Wallace theory of evolution. Along with other scientists, Hugo de Vries, a Dutch botanist and one of the pioneers of genetics, contributed to identifying the connection between Mendelian genetics and the theory of evolution in the beginning of the twentieth century. Hugo de Vries, one of the three botanists who re-discovered Mendel's ignored work, presented his mutation theory in 1901, proposing that evolution occurred due to mutations in Mendelian genes.

Following the re-discovery of Mendelian genetics, T. H. Morgan (1866-1945), an American embryologist and geneticist, performed experiments to see if mutations could produce new species. What he discovered through his experiments is that rather than immediately creating new species in a single step, mutations increased the genetic variation in the population. This helped establish the link between Mendelian genetics and the chromosomal theory of inheritance. Morgan demonstrated that genes (Mendelian genetic factors) are carried on chromosomes and are the mechanical basis of heredity. The 1933 Nobel Prize in

Physiology or Medicine was awarded to Morgan "for his discoveries concerning the role played by the chromosome in heredity." It was the first Nobel Prize for the field of genetics.

Another geneticist, R. A. Fisher (1890-1962), showed that the apparently continuous variation measured by biometricians could be the result of the action of many discrete genetic loci (genes). Overall, Fisher's work demonstrated how Mendelian genetics was completely consistent with the theory of evolution driven by natural selection.

Recall from Chapter 4 that there are three types of genetic mutations: harmful which decreases the fitness of the organism, beneficial which increases the fitness of the organism, and neutral which has no effect on fitness. Harmful and beneficial mutations play their roles in evolution. Individuals with beneficial mutations are selected in due to their increased fitness through the traits corresponding to the beneficial mutations, and they multiply over generations through reproduction. Individuals with harmful mutations are selected out and they, along with their traits, vanish.

Although in the last chapter we explored some causes of mutations such as harmful environmental agents which cause damage to DNA, in this chapter you will see that these are not the only causes. So, we are going to carry over the question about the causes of the gene mutations.

In general, from what you have read so far in this book, you can ask the following questions:

1. You learned in Chapter 4 that mutations are the source of variations that drive evolution; but what's the origin of these mutations? In Chapter 4, we explored some causes of mutations such as errors during replication, but the question

here is more fundamental: how and why do these errors occur?

2. Why do certain types of mutations occur more frequently than others?

3. What is the relationship between mutations and natural selection? In Chapter 4, we discussed the obvious aspect of this relationship: how mutations give rise to variations which are the raw material for natural selection. But we can ask more questions such as this one: are mutations generated in response to the need for adaptation due to change in the environment, or do they occur independent of such a need?

We explore the answers to these questions in this chapter. Toward the end of this chapter we put together all the components of the theory of evolution operating at different levels of complexities of life from population to genes into one under the name of standard theory of evolution.

Exploring the answers to the three questions stated above requires help from physics. So, let's take a quick and easy glance at the Universe from the eyes of physics followed by a discussion on what physical laws have to do with evolution.

5.3 Physics: The Most Fundamental Science

Recall from Chapter 4 that evolution has its roots in genes and mutations of genes, that genes are strings of base pairs (bp) of A-T and C-G, and that these strings are parts of a DNA molecule. Talking about size, a DNA molecule has a diameter of about 2.5 nanometers (nm), and one bp is about 0.34 nm long. The bonds that keep a gene together and functional operate at the level of protons and electrons, subatomic particles. Obviously, the behavior of matter at this level is governed by the laws of quantum physics, the physics of the microworld. Figure 5.1 and Table 5.1 present some examples of material systems on the size scale from macro to micro including nano and beyond nano.

> **Note.** When physicists use the term micro, they usually mean all sizes non-macro including micro (10^{-6}), nano (10^{-9}), pico (10^{-12}), and smaller. Roughly speaking, macro includes everything that you can see with your naked eyes, and micro includes smaller particles and structures, for example microbes (unicellular organisms), cells, molecules, atoms, and subatomic particles; which you cannot see without the aid of tools such as light microscopes, electron microscopes, and particle detectors.

The physics used to describe the entities and phenomena of the macroworld is called classical physics or Newtonian physics and is an approximation of quantum physics. Newtonian physics works extremely well in the macroworld, but its validity breaks down in the microworld, where quantum physics becomes necessary to understand entities and phenomena. Physics, in general, is the most fundamental science used to explain the behavior of all physical entities from subatomic particles to molecules, to everything on our planet, to galaxies, and to the expanding or shrinking Universe. Principles discovered in physics apply across all natural sciences including astronomy (if you don't consider it a part of physics), chemistry, earth sciences, and biology; and also make up the foundations of all engineering and technology.

And God Said: Let There Be Evolution

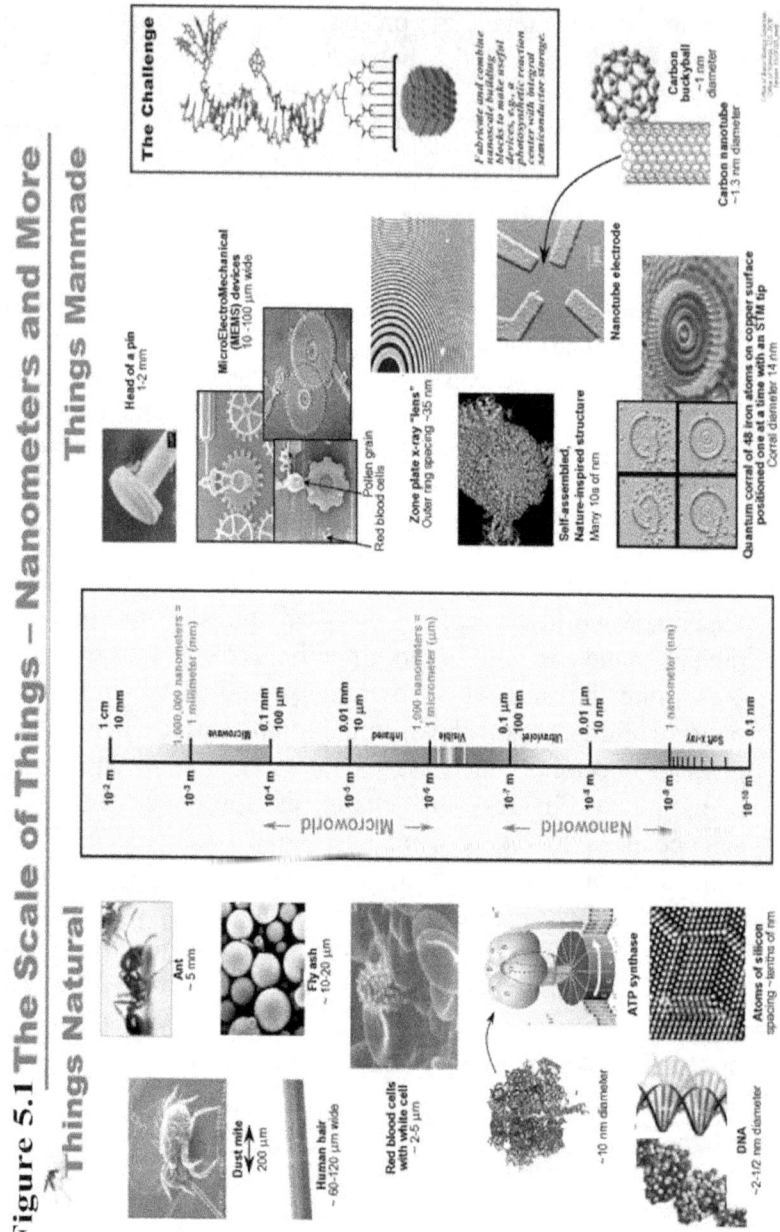

Figure 5.1 The Scale of Things – Nanometers and More

198

Table 5.1 Examples of material systems on the size scale.

Size	Scientific notation	Example (In this neighborhood)	Observation tools/techniques
1.7 m	1.7×10^{0} m	Human height	Human eye
1 cm	1×10^{-2} m	Wedding ring	Human eye
1 mm	1×10^{-3} m	Thickness of a CD	Human eye, optical microscope
100 μm	1×10^{-4} m	Plant cell	Human eye, optical microscope, electron microscope
10 μm	1×10^{-5} m	Animal cell	Human eye, optical microscope, electron microscope
1 μm	1×10^{-6} m	Bacterial cell	Optical microscope, electron microscope
100 nm	1×10^{-7} m	Virus	Electron microscope
10 nm	1×10^{-8} m	Virus, protein molecule	Electron microscope
1 nm	1×10^{-9} m	Protein molecule, aspirin molecule	Electron microscope
100 pm	1×10^{-10} m	Water molecule	Electron microscope
10 pm	1×10^{-11} m	Atom	Indirect observation by tools such as cyclotrons
0.001 pm	1×10^{-15} m	Proton	Indirect observation by tools such as cyclotrons
<1 pm	$<1 \times 10^{-12}$ m	Other subatomic particles	Indirect observation by tools such as particle accelerators and particle colliders

And God Said: Let There Be Evolution

The totality of physics research so far shows that the physical universe is governed by the following four fundamental forces of nature to which all other forces can be traced:

- **Gravitational force.** This is the force that keeps us on the Earth, and keeps the Moon revolving around the Earth, and keeps Earth revolving around the Sun. It is more important at macro-level.

- **Electromagnetic force.** This is the force that helps us hold, lift and move things. It also holds the molecules and atoms together that form all the materials in this world. This is the prevalent force governing the functioning of organisms from molecular to organism level.

- **Weak nuclear force.** This is the force that governs the radioactive decay of certain substances.

- **Strong nuclear force.** This is the force that holds the nucleus of an atom together.

Electromagnetic force is largely responsible for keeping the cellular machinery in our bodies functioning. In terms of the order in the Universe, gravitational force holds the planetary bodies together in a system such as solar system or a galaxy; electromagnetic force helps keep together the living organisms and non-living bodies on the planets and their basic building blocks, atoms and molecules; strong nuclear force helps to keep the nuclei (and their constituent protons and neutrons) of atoms together; and weak nuclear force transforms one type of nucleus (and hence atom) into another type of nucleus (atom).

As an example, there are different forms of electromagnetic force that are responsible for biochemical reactions in our bodies, for giving the double helix shape to a DNA molecule, and for providing different shapes and therefore different functionalities to different proteins, the workhorses in our bodies.

Physics describes the functioning of the Universe and everything in it. If you ask me to describe the Universe in one word, that word would be *energy*. I could have equivalently used the word *matter* instead of energy because matter and energy are interconvertible. Most of the Universe and its phenomena can be described as an interplay between matter and energy. For example, all matter is made of atoms and an atom is held together by the electromagnetic energy states of the electrons around its nucleus. Most of the phenomena at the atomic level can be described by theses electrons changing their energy states. Our solar system is held together by the gravitational energy states held by the planets due to which they keep revolving around the Sun. The Sun itself burns and converts matter into energy that supports life on Earth. Even on Earth, living organisms exchange energy with their environment. I can go on but you get the idea.

So, it should not come as a surprise then, that the dynamics of the Universe and the systems in it at a fundamental level is governed by a few energy-related fundamental laws or principles of physics. These laws are in effect both at the macroscopic and microscopic levels and run through many other laws and theories. Here are the most important ones:

Principle of Conservation of Energy. One of the fundamental principles of physics valid on both microscopic and macroscopic scale is the principle of conservation of energy, which states that energy is a quantity that can be converted from one form to another but can neither be created nor destroyed. For example, when you are driving your car the energy stored in the fuel is partly converted to the kinetic (motion) energy of the car and partly to heat (thermal) energy. Heat is one of many forms of energy, and matter (or mass) is another.

An equivalent statement of this law is that the energy in a process is conserved, that is, energy gained by a system under study is equal to the energy lost to its environment. In other words, the energy of the Universe in any process remains conserved (unchanged in amount).

> **System.** A system, in our context, is a physical entity (which may be composed of multiple components) under study, on which a process is occurring; and the rest of the Universe is called its environment or surroundings.

Principle of Conservation of Matter and Energy. Once you realize that matter is a form of energy, you can rewrite the energy conservation principle in form of the principle of conservation of mass and energy, which states that the total amount of mass and energy is conserved; that mass and energy are interconvertible. This means that if a mass is lost, it's accounted for by the equivalent amount of energy produced by it and vice versa. Power is generated at a nuclear power plant based on this principle. Once you realize that energy is a form of matter, you can say matter can neither be created nor destroyed.

> **Note.** Mass, m, and energy, E, are equivalent and interconvertible according to Einstein's famous equation:
> $$E = mc^2$$
> where c is the speed of light through a vacuum, a constant; m is the relativistic mass; and E is the relativistic energy. For example, the Sun is powered by the conversion of mass into energy, which follows this equation.

The principle of conservation of energy that we have already discussed states that the energy (mass form of energy included) of the Universe is constant; it can neither be created nor destroyed, but can only be transferred or transformed. The study of energy transfer and energy transformation is called thermodynamics, and energy conservation principle is called the first law of thermodynamics. There is another law of thermodynamics, an equally important law: the second law of thermodynamics.

Second Law of Thermodynamics. Even though energy cannot be created, it can be transferred and transformed. This is how living organisms get their energy. For example, our electric energy is obtained by transforming the energy of fossil fuel such as coal, crude oil, or natural gas; the mechanical energy of falling water; or the energy of the atomic nuclei. As another example, plants transform the energy from sunlight into the carbohydrates that we eat, and the energy from these carbohydrates is converted by our cells into the energy molecule called adenosine triphosphate (ATP), which then cells can use to perform functions. However, it turns out that not all the energy can be transferred and cycled in a usable form. During these energy transfers and transformations, some energy is lost as unusable energy, which adds to the disorder (entropy) of the Universe. The *second law of thermodynamics* states that in each energy transfer, the entropy of the Universe either remains the same or increases, but it never decreases.

Entropy. A degree of randomness and disorder in a system.

For a process or event to occur spontaneously, it must increase the entropy of the Universe. The term *spontaneous* in this book (and in science) means that the process occurs automatically without the aid of any external effort, that is, without any input of energy. However, it does not mean that the process would occur quickly: it may occur instantly such as an explosion, it may take years, or centuries, or any time interval in between.

Principle of Stability. Systems in the Universe tend to be in a state of equilibrium or stability. This is usually not stated as a specific principle, but this is a natural tendency that underlies many specific equilibrium principles in physics and chemistry. Changes in a system occur in a direction to accomplish this (stability) by decreasing energy or increasing entropy or both. Note that the term stability is used here in the sense of equilibrium and not necessarily in the sense of order because an increase in entropy brings an increase in disorder.

So what does all this have to do with the evolution of life? Well to start with, life is work and energy is the capacity to perform work or facilitate change. A cell, the fundamental unit of life, is a microscopic factory in which thousands of biochemical reactions are occurring, which involve energy transfer and energy transformation. All the fundamental laws we have discussed in this section apply during these energy transfers. These laws also have their hand in the evolution of cells and organisms.

5.4 Biological Evolution Obeying Physical Laws

A physical law makes a physical entity function in a certain way or makes a natural phenomenon occur in a certain way. In other words, physical laws impose constraints on physical entities and natural phenomena. Because living entities are also made of matter, they are not free of physical laws. Their development and evolution obey the relevant physical laws of nature. So, obviously, biological evolution is not free of physical laws either.

In Chapter 4, you learned how mutations, the root cause of variations at the macro (organism) level, occur randomly. We already have tremendous variations at organism level: millions of species and within each species no two organisms are exactly alike except identical twins. Still, mathematically speaking, total randomness should produce a lot more variation. For example, where are the mammals with wheel-like feet? Why don't we have organisms in all perceivable shapes and forms at each level of complexity? The answer to such questions is that physical laws impose bounds on the variations of organisms during the course of evolution. For instance, the range of animal forms is limited by physical laws that govern factors such as diffusion rates, gravity, heat transfer or exchange, motility (movement), resistance, strength, and viscosity.

Viscosity. The resistance offered by a fluid to flow.

Diffusion. Spontaneous spreading or flow of a substance down its concentration gradient, that is, from the space where it is more concentrated to the space where it is less concentrated.

Here is an example of how physical laws affect evolution: Physical laws govern the relationship between body plan, body size, fraction of total body mass used for muscles, and motility of the organism. An increase in body size often offers limitations to motility, which requires muscles. For instance, as body size increases, the fraction of total body mass used for muscles also increases. The science here is accurate enough that given the fraction of body mass in the leg muscles, we can estimate the force that the muscles can generate and hence the maximum running speed for a given body plan. In accordance with physical laws, increase in size also influences the body plans of organisms. Physical laws suggest that the thickness of the skeleton to support the body should increase with an increase in body size. This would affect the body plan. By looking at the actual relationships between these different physical parameters of existing species, it is clear that evolution has operated (favored or disfavored organisms) in accordance with or within the limits of these physical laws.

Here is another example of how physical laws constrain evolution. Water has a higher density and viscosity than air, and therefore offers more resistance for travel or motility. The effect of the physical laws corresponding to the relationship between viscosity and motility (travel) is evident in that speedy birds, fishes, and mammals that swim in water have a streamlined body contour, that is, a body shape tapered at both ends. Examples are penguins, tuna fish, and seals. Such evolution in independent evolutionary lineages due to adaptation by different evolutionary lineages to similar environments in similar ways, including physical laws, is called convergent evolution. Whether you are a bird, a fish, or a mammal; if you are in the water and if you have a streamlined body, natural selection favored you because streamlined bodies facilitate speedy movement through water. This type of convergent evolution demonstrates the power of the limits imposed by physical laws on evolution.

There are many more examples of physical laws and parameters determined by those laws imposing limitations on the evolution of body plans, but these few examples presented here are enough to make this important point. All physical laws directly or indirectly relate to the four fundamental forces described in Section 5.3. Physicists however believe that these forces are different facets of a single unified force and they are in search of that force.

Biologists have their own single unified *force* (speaking metaphorically) called evolution.

5.5 Evolution: The Unifying Force of Life

The whole history of physics, the most fundamental of all sciences, can be looked upon as a history of discovering the following three facts:

1. As already discussed in this chapter, the physical universe is governed by four fundamental forces of nature: gravitational force that keeps us on Earth and keeps the Earth circling around the Sun, electromagnetic force that is largely responsible for keeping the cellular machinery in our bodies functioning, weak nuclear force that is responsible for the decay of a radioactive substance, and strong nuclear force that keeps the nucleus of an atom together.

2. All these four forces, which look different when seen from the energy levels that we live in, are actually different manifestations of a single unified force at a higher energy, for example, at the energy level that existed in the very beginning of this Universe.

3. Nature does its most fundamental design work at the microscopic level, that is, at the atomic and molecular level. Therefore, the physical reality of the macroscopic world that we experience in our daily lives has its roots in the microscopic world of atoms and molecules.

The theories in physics that demonstrate the unification of the four forces are called unification theories. In a similar fashion, the theory of evolution originally introduced by Charles Darwin and Alfred Wallace, and first put forward by Charles Darwin in his book *On the Origin of Species,* has been proven over time to be the unification theory in biology that ties together apparently separate biological facts into a single unifying framework. It has withstood scientific scrutiny up to date. By its very nature, if evolution has taken place, it must have its signature on all forms of life, micro and macro. This prediction of the theory of evolution, which has been proven by facts to be true, makes evolution the unifying force of life.

The discovery and study by physicists of the four fundamental forces of nature, gave us–with the help of scientists, technologists, and corporations–the electronics age and now the information age. This completed the Copernican revolution by not only establishing but benefiting from the notion that nature is a lawful system of matter that can be explained without appealing to supernatural powers. In much the same way, Darwin's theory of evolution and molecular biologists are completing this revolution in the area of living matter: life. Both of these revolutions in physics and biology combined are giving rise to the coming age of nanotechnology –the molecular age. And in this age, living meets the nonliving, unified by the laws of quantum physics.

Through the study of evolution, biologists are learning a lesson that physicists began learning toward the end of nineteenth century: The macroscopic reality that we experience every day while living in the macroworld has its roots in the microscopic world of atoms and molecules, because macroscopic things are made of microscopic objects. In the course of this learning and realization, physicists including Einstein, Planck, Schrödinger, and Louis de Broglie developed quantum physics, also called quantum theory or quantum mechanics, to explain the behavior of microscopic entities such as subatomic particles like protons and electrons, atoms, and molecules; which could not be explained with Newtonian physics, also called classical physics or classical mechanics.

5.6 Here Comes Quantum Mechanics

Classical physics, also called Newtonian physics, is successfully used to explain the entities and phenomena in the Universe at macroscopic scale ranging from the motion of a ball in a soccer game, to the motion of planets, to the motion of galaxies, to the motion of the expanding and shrinking Universe. Quantum mechanics was developed to explain the results of experiments performed toward the end of nineteenth century to study the behavior of entities and phenomena at microscopic scale, as classical physics failed miserably to explain these results.

In the classical world, physical entities are classified into two groups: particles and waves. Particles have some properties, such as mass, that waves don't have, and waves have some properties, such as wavelength, that particles don't have. Particles are more localized. As a matter of fact, any physical entity (or object) of any size, for example ranging from a marble to the Sun, can be considered as a point particle by identifying a suitable point in the entity to determine its position in space. By having some information about a particle, we can measure its position and momentum, and predict its position and momentum at any time in the future without any uncertainty. Unlike a point particle, a wave is a more spread out entity, a traveling disturbance that transports energy in space from one place to another. The amount of energy carried by a wave is continuous, that is, it could acquire any value. So, the distinction between particles and waves, certainty of determination of physical quantities (characteristics), and continuity of the values of physical characteristics were some underlying features of classical physics which were challenged by the experiments performed on phenomena at microscopic level. The validity of quantum mechanics is also proved through its enormous applications including the electron microscope, used to study the structure of proteins and other biomolecules.

Quantum mechanics saved physics by tearing down the wall between particles and waves. In quantum mechanics, a physical entity is neither just a particle nor just a wave. Instead, it behaves as a particle or as a wave depending on what it is put through. This is called the *wave-particle duality* of nature. Because a particle can behave as a wave, it's not as much localized any more as a point particle, and its position can be predicted only in terms of probabilities. For example, there is 25% probability (or chance) that the particle will be found at point 1, 30% chance that it will be found at point 2, and so on. For example, the position of an electron in an atom is represented by the probability distribution. The uncertainty in quantum mechanics is represented by the famous Heisenberg's *uncertainty principle* which states that that there are a number of pairs of physical observables (also called quantities or characteristics) such that both observables in a pair cannot be simultaneously determined precisely. One such pair is the position and momentum of a particle. Another pair is time and the energy measured at that time. This uncertainty is the built-in uncertainty of nature and has nothing to do with the precision of the experimental apparatus.

Because a particle can behave as a wave and its location is determined only in terms of probabilities, it can show up in unlikely places where its appearance is prohibited in classical physics. This is called *quantum mechanical tunneling* or just tunneling. Furthermore, when a particle is bound in a region such as an electron in an atom, it can only have certain discrete values of its observables such as momentum and energy. This is called *quantization.*

So, is classical physics wrong? The answer is no. Quantum physics is more accurate physics and classical physics is a good approximation of quantum physics. This approximation works well on macroscopic scale but breaks down at microscopic scale. Why? Because macroscopic objects are made of microscopic objects; the macroscopic reality that we experience has its roots in the microscopic world. Quantum mechanics is developed to explain entities and phenomena on microscopic scale, so it is more accurate science. Classical physics, the approximation of quantum physics, works on macroscopic scale because quantum effects, which are significant in the microscopic world, are negligible in the macroscopic world. For example, a few microns (micrometers) or even nanometers could be very significant on a micro scale, but absolutely negligible on a macro scale, that is, the world of centimeters and meters. Furthermore, the quantum effects of individual micro objects may cancel each other out in a macro object (or system) made out of those micro-objects.

Now, equipped with this knowledge, let us return to the three questions raised in the beginning of this chapter. First, what is the origin of gene mutations?

5.7 Natural Selection, Adaptive Change, Mutation, and the Quantum Connection

As we discussed in Chapter 4, evolution has its roots in genetic mutations; but where do the mutations come from? For example, we can begin this enquiry by asking questions like these: Are these mutations generated by the need of an organism to adapt to a change in the environment? Or are they independent of any such need at organism level? In this section, we will explore this issue.

Also in the process of doing so, we distinguish among different but related concepts and roles of mutation, adaptive change, and natural selection. This will also present an example of how the Darwin-Wallace evolution theory is different from Lamarck's hypothesis, which we discussed in Chapter 1 and found that it was not supported by evidence. During this exploration, we will also explore an experiment that not only demonstrates the validity of Darwin-Wallace theory and the invalidity of Lamarck's hypothesis, but also historically made the first connection between evolution and quantum physics.

Our exploration begins with a bacterium, which is a unicellular organism that divides into two cells typically within 20 minutes. Due to their short lifespan and simple structure, bacteria make good subjects of experimental studies. It is well known that some bacteria develop antibiotic resistance, and therefore antibiotic drugs don't work on them anymore. Similarly, bacteria also develop resistance to viral infections.

Now, here are two very relevant and simple questions:

1. Is the resistance in bacteria caused by adaptive change or gene mutation?

2. If the resistance is caused by gene mutation, is that mutation random or directed?

Directed mutation is mutation that arises only in response to adaptive change or natural selection, whereas random mutation is mutation that occurs randomly and spontaneously independent of adaptive change or natural selection. In other words, the probability of these mutations occurring is independent of whether they would be helpful or harmful to the individuals in which they are occurring. So, if the resistance is caused by directed mutation, then it is really caused by adaptive change.

After his research fellowship to work on genetics at Caltech ran out in 1939, a physicist named Max Delbrück (1906-1981) continued pursuing his research interests in genetics while teaching physics at Vanderbilt University. In 1942-1943, he addressed the two questions stated above in an experiment that he performed in collaboration with Salvador Luria (1912-1991) of Indiana University. This Nobel-prize winning research is known as the Luria-Delbrück experiment.

Luria and Delbrück inoculated a small number of bacteria into separate culture tubes (test tubes) and let the bacteria grow for a certain period. Subsequently, they transferred equal volumes of these separate cultures onto plates containing phage (virus). They knew that if the virus resistance in the bacteria was caused by a spontaneous activation in response to the virus and not due to heritable genes or gene mutations independent of the virus, they would expect that each plate contain roughly the same number of resistant colonies. A resistant colony is colony of bacteria that have developed resistance to the virus. However, this was not what Delbrück and Luria found at the end of their experiment. The number of resistant colonies on each plate varied drastically.

And God Said: Let There Be Evolution

Luria and Delbrück successfully explained how these experimental results prove the occurrence of a constant rate of random mutations in each generation of bacteria growing in the initial culture tubes. These mutations were found to be independent of the presence of a virus and therefore independent of adaptive change or natural selection; hence were not directed by adaptive change. Furthermore, Luria and Delbrück used a quantitative approach to demonstrate how the experimental results agreed with the hypothesis that bacterial resistance to viral infection is caused by random mutations which are independent of adaptive change, and how the data disagreed with the hypothesis that bacterial resistance is caused by adaptive change.

Think About It!

Does the Luria-Delbrück experiment support or refute Lamarck's evolutionary hypothesis?

Answer

The Luria-Delbrück experiment refutes Lamarck's evolutionary hypothesis, which would predict that the bacteria develops resistance mutations in response to the viral infection and not independent of it.

Later the results of the Luria-Delbrück experiment were confirmed by an experiment performed by H.B. Newcombe. He incubated bacteria on a Petri dish for a few hours and then plated the bacteria onto two new Petri dishes which were treated with phage. The difference between the two dishes was that the bacteria on first dish were left unspread, and the bacteria on the second dish were then respread, that is, bacterial cells were moved around allowing single cells in some colonies to form their own new colonies. Newcombe knew that if colonies contained resistant bacteria before entering into contact with the virus, he would expect that some of these bacteria would form new resistant colonies on the respread dish (because they had more room there). Therefore he would find a higher number of surviving bacteria on the respread dish. After both dishes were incubated for growth, at the end of the experiment, Newcombe found that there was actually as much as 50 times greater number of bacterial colonies on the respread dish than on the unspread dish. This demonstrated that bacterial mutations to viral resistance had randomly occurred during the first incubation, obviously independent of the virus and hence independent of adaptive change or natural selection. In other words, mutations occurred before natural selection was applied. In other words, mutations were not directed.

> **Note**
>
> James Watson, the co-winner of the Nobel Prize for discovering the structure of DNA molecules, was the first graduate student of Salvador Luria at Indiana University. Although Luria did his Nobel work before Watson, Watson got his Nobel Prize seven years earlier than Luria.

These experiments together clearly showed that bacterial resistance is based on mutations and not adaptive change, and that gene mutations are not directed. In other words, these experiments demonstrated that mutations and adaptive change (or natural selection) are two decoupled steps in evolution. First mutations occur independent of adaptive change or natural selection and foster variations among the population, and then natural selection operates and eliminates those organisms who cannot adapt to the environment or the change in it.

And God Said: Let There Be Evolution

Max Ludwig Henning Delbrück (1906-1981)

Max Ludwig Henning Delbrück was born on September 4, 1906 in Berlin, Germany. He studied astrophysics and shifted to theoretical physics during the latter part of his graduate studies. After receiving his Ph.D. in physics in 1930, he traveled through England, Denmark, and Switzerland for his postdoctoral research.

During the years of his postdoctoral research, he had the opportunity to associate with Niels Bohr and Wolfgang Pauli, two inventors of quantum mechanics. Delbrück got interested in biology under the influence of Bohr who speculated that wave-particle duality (at the heart of quantum mechanics) might have wide applications encompassing other fields of science including biology. In 1935, Delbrück wrote a paper with Timofeeff and Zimmer on mutagenesis. A popularization of this paper in Schrödinger's book *What is Life?*, and the book itself had a strong influence on the development of molecular biology in the late 1940s.

To further pursue his interests in biology, Delbrück moved to the United States in 1937. While working at Caltech in 1939, he co-authored a paper called *The Growth of Bacteriophage* with E.L. Ellis in which they demonstrated that viruses reproduce in "one step," rather than exponentially as cellular organisms do. After his research fellowship at Caltech ended in 1939, Delbrück remained in the U.S. during World War II, teaching physics at Vanderbilt University in Nashville while pursuing his genetic research interests. In 1942-1943, he and Salvador Luria of Indiana University demonstrated that bacterial resistance to viral infection is caused by random mutation and not adaptive change, in an experiment famously known as the Luria-Delbrück experiment. Delbrück and Luria won the 1969 Nobel Prize in Physiology or Medicine in part for this work. Delbrück returned to Caltech in 1947 as a professor of biology where he remained for the rest of his life. He was one of the pioneers and the most influential scientist in the movement of using physical sciences in biology.

Max Delbrück died on March 9, 1981 at the age of 74 in Pasadena, California.

216

So, these (along with other) experiments bring us to a point where we can say that mutations in genes occur randomly, and are not directed by the need to adapt to the environment.

Let's do a simple exercise of imagination. Imagine there was no such thing as mutations? Imagine living in a perfect world where the DNA replication process made no mistakes, environmental agents (such as radiations) never caused any changes to the genes, and so on. In such a world, there would be no gene mutations. What are the implications of the absence of random mutations? The local implication for the discussed experiment is that if random mutations did not exist, no bacteria would survive the virus and there would be nothing for natural selection to act on. On a global scale, if we switched off random mutations, there would be very little room for modifications in existing traits, the rise of the new traits, and the rise of new species. A drastic change in the environment could eliminate whole populations; again very little for natural selection to select from. Absence of random mutations is an invitation to doomsday. We wouldn't be here without random mutations. So, you can see how random mutations provide the ground for natural selection to operate on, and in this way allow the party of life on our planet to continue.

At this point, however, Herschel might say: you have just moved the random change or the chance game from the macroworld of organisms to the microworld of genes. But what causes these random mutations anyway? We scientists never stop asking questions. However, we ask questions to make progress and not to stop it. You may wonder: didn't we say in Chapter 4 that environmental agents and errors during the DNA replication cause mutations? Yes, but as mentioned in the beginning of this chapter, our question now is even more basic. For instance, what causes the errors during DNA replication, what causes these mutations (errors) to occur randomly, and what is the mechanism for these random mutations to occur?

In that spirit, what causes random mutations? This is where quantum mechanics comes in, whose discovery started to solve physics problems around the same time Hugo de Vries was connecting the dots between Mendelian genetics and the evolution theory of Darwin and Wallace.

The Luria-Delbrück experiment, which largely won them (Luria and Delbrück) the 1969 Nobel Prize in Physiology or Medicine, opened the door for discovering the connection between evolution and quantum mechanics.

5.8 The Quantum Basis for Evolution

As discussed in Chapter 4, harmful environmental agents and errors during DNA replication are the fundamental causes of mutations. Stating this in more general terms, the causes of mutations fall into two categories: external to genes such as environmental and internal causes due to the atomic and molecular nature of genes. When genes were discovered by Mendel in terms of discrete genetic factors, it was not obvious that they were molecules. However, once their molecular and atomic nature were predicted, it became obvious to physicists like Schrödinger that their behavior would be governed by the laws of quantum mechanics, the physics of the microscopic world of atoms and molecules.

In the previous section, we showed how the Luria-Delbrück experiment showed that mutations occur independent of adaptive change. Encouraged by this experiment and other experimental knowledge existing at that time, Schrödinger showed in his famous book *What is Life?* how quantum mechanics predicts that mutations are bound to happen in the form of isomers, different atomic configurations of gene molecules. Isomers arise as a result of energy fluctuations due to the uncertainty principle in quantum mechanics discussed earlier in this chapter. These mutations are unavoidable because they are due to the intrinsic quantum nature of the DNA code or genes. Although this book of Schrödinger's inspired Watson and Crick to discover DNA structure and also inspired the beginning of the whole field of molecular biology, the quantum nature of mutations pointed out by Schrödinger in his 1944 book was largely ignored. A couple of decades later, the quantum nature of mutations was further explored by Swedish geneticist Per-Olov Löwdin (1916-2000).

The discovery of DNA structure had already revealed that the genetic code in DNA is made of the sequence of A-T and C-G pairs called base pairs. It was also known that these pairs were formed due to a hydrogen bond between A and T, and C and G. A hydrogen bond is simply a chemical bond between a proton of one member of the pair and an electron of the other member of the pair. Protons and electrons are fundamental particles and their behavior is undoubtedly governed by quantum mechanics. Their positions are only given by probability distribution, and the uncertainty in their position enables them to tunnel to unlikely places. This gives rise to different structures of the bases and genes and therefore mutations. The replicating enzyme will replicate both: probable (standard) structures and unlikely structures. The replication of unlikely structures would be very natural for a quantum physicist but are considered incorrect (replication errors) by biologists who are thinking classically as opposed to quantum mechanically. According to quantum mechanical calculations, *incorrect* (or *erroneous*) bases should amount to 0.01 % of the standard bases. However, the DNA replication machinery has a proofreading mechanism in which an enzyme corrects most of the errors made during the first run. This leaves only about one incorrect base per billion correct base pairs. These incorrect bases that escape the correction machinery of the DNA replication process are the source of unavoidable mutations and their source is the laws of quantum mechanics.

This is one of many ways quantum physics makes the basis for evolution. So, we have answered the first of the three questions raised in the beginning of this chapter: how and why the random errors occur that cause mutations. Although a lot more work need to be done, there is already enough evidence that quantum principles are behind these so-called random errors. The second question still remains: Why do certain types of mutations occur more frequently than others?

5.9   Occurrence   Rates   of   Random Mutations

Why do certain types of mutations occur more frequently than others? The short answer is: it depends on the structural properties of the four bases in the DNA: A, C, G, and T; and the natural tendency for stability, that is, the chemical stability of the molecules and the entities composed of the molecules. As mentioned earlier, stability is the general (and tested) law of physics that states that each system in this universe tends to reach and stay in its most stable (or equilibrium) state. This state has different names under different contexts such as the least energy state, the maximum entropy state, and the state of equilibrium.   Stability is an underlying law behind many biochemical phenomena in our body when it comes to asking why something happens this way and not the other way.

A system acquires stability through two possible channels: decreasing the energy content of the system and increasing the randomness or dispersal of the system, called entropy. In other words, stability is acquired through the interplay between decrease in energy and increase in entropy. So, the tendency for stability does not always lead to increased order because stability may be acquired by increasing entropy and hence increasing chaos.

Nevertheless, you will realize the importance of the stability principle once you conceptualize that we owe our lives to the stability of our genes throughout our lives and throughout our species or the species we originated from. This stability or permanence plays an important role in evolution. By stability here it does not mean no change or no mutations. It means the intact structure and functionality of the DNA being transferred from generation to generation for billions of years.

Are the rates of mutations determined by the need of adaptation or by the laws of quantum mechanics? If you believe that mutations occur through quantum mechanical processes, the answer is obvious. But what do the experimental results suggest? Let's explore.

5.10 Adaptive at the Surface, Quantum at the Bottom

In order to understand if mutation rates are adaptive (directed) or determined by the laws of quantum mechanics, you need to understand how measurements are made in quantum mechanics. Not only in science but also in many non-scientific fields, the importance of measurement cannot be overstated. However, according to quantum mechanics, when we measure something in the microscopic world, we run into uncertainty, and we must take that uncertainty into account. If we look at microscopic measurements with the macroscopic eye, we will be deceived. This is what we learned from the physics experiments conducted toward the end of nineteenth century, and this is also what we learned from a biology experiment run by John Cairns a century later. In the following, we discuss this experiment and its implications about the role of quantum mechanics in mutations.

The Cairns Experiment. Let's consider this experiment that John Cairns, a biochemist of Harvard School of Public Health at Boston, performed in 1988 on *E. coli,* a type of bacteria shown in Figure 5.2, found in the human digestive tract. Cairns took a specific strain of the bacteria that had a mutation, say *lac⁻*, which would disallow the bacteria to eat and digest the milk sugar called lactose. The *lac⁻*

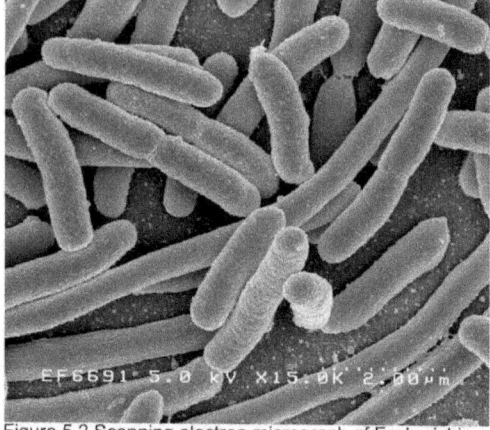

Figure 5.2 Scanning electron micrograph of Escherichia coli, grown in culture and adhered to a cover slip. Image: courtesy of The National Institute of Health.

mutation does so by disabling the bacteria from generating beta-galactosidase, an enzyme necessary for digesting lactose. Cairns made sure that the only food available to this sample of bacteria was lactose, which this sample of bacteria should not be able to eat. So, most of the bacteria should die (and not multiply) except a few that will have a mutation, say lac⁺, which would facilitate the production of the beta-galactosidase enzyme to digest lactose. From the viewpoint of directed mutation, the bacteria would be under stress to have lac⁺ mutations, that is, the mutation will be created in response to the environment that contained lactose as the only food. From quantum mechanical viewpoint, the rare bacteria with lac$^+$ mutation among a sample of bacteria with lac$^-$ mutation is expected to exist independent of the kind of food available. This existence is expected from quantum mechanical tunneling discussed earlier in this chapter. Even if one bacterium in the sample has this advantageous mutation (lac$^+$), this bacterium would eat lactose and multiply, the rest would starve to death.

At the same time, Cairns took a second sample of *E. coli* bacteria with the lac$^-$ mutation and let it feed on yeast instead of lactose. We expect that due to quantum mechanical fluctuations (or tunneling), some bacteria in this sample will contain lac$^+$ mutations as well. However, in this case, *E. coli* can eat even in the absence of beta-galactosidase. So, the bacteria have no stress on them; both lac$^-$ and lac$^+$ will grow. Bacteria grew in both cases as expected.

However, Cairns went a step further. He measured the rate of lac$^+$ mutations in both samples. He found that the rate of mutations in the first sample where *E. coli* were forced to starve on lactose had a much higher rate of lac$^+$ mutations. It almost appeared as if *E. coli* sensed that it was starving (it was under stress) and responded to the situation by developing more lac$^+$ mutations. Cairns et al published the results of their experiment in 1988. Some classical (as opposed to quantum) biologists who see natural selection (and no other principle) operating everywhere at all levels interpreted the results as evidence for adaptive mutations (or directed mutations). Following the 1988 paper of Cairns et al, hundreds of publications have appeared supporting and opposing the hypothesis of adaptive mutations while avoiding quantum mechanical explanation. This situation is very much the same as faced by physicists at the end of twentieth century, when they were unsuccessfully trying to use classical physics to explain the results of experiments performed at atomic and molecular levels. This situation is also similar to the situation discussed in Chapter 1, when Cuvier presented the hypothesis of catastrophism in order to interpret the patterns observed in the fossil record in favor of creationism, whereas the patterns were actually presenting evidence for evolution.

The results from one of Cairns experiments, presented in Figure 5.3, demonstrate that the quantum mechanical interpretation as opposed to directed mutation interpretation is the correct one. In this experiment, Cairns added lactose immediately after spreading the bacteria on the first set of agar plates, 24 hours later on the second set of plates, and 72 hours later on the third set of plates. The graph in

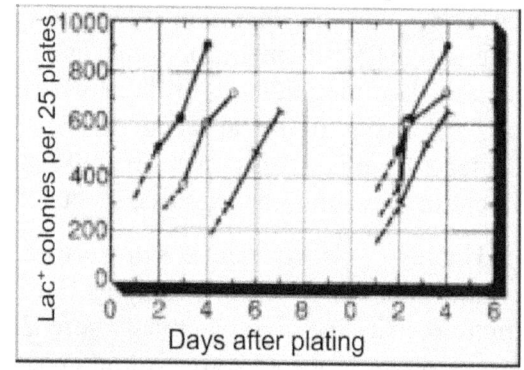

Figure 5.3 Results from one of Cairns experiments

Figure 5.3 shows the number of colonies that appeared. On the left, the number of colonies in real time are shown, and on the right the number of colonies shown are relative to the time at which the lactose was added. As these results demonstrate, delaying the addition of lactose delays the appearance of the colonies, but it does not affect the total number of mutants. It demonstrates that the rare mutants were already there before adding lactose; the quantum mechanical view. If the view of directed mutations was true, one would expect more mutants in case of early lactose than delayed lactose because more bacteria would go through the stress to produce mutants (lac$^+$) in case of early lactose.

224

Recall that the Luria-Delbrück experiment, discussed in this chapter, had already shown that mutations occur independent of adaptive change. In Cairns experiment, we are dealing with the rates of mutations. Why does the stress (of starvation) put on the bacteria sample increase the rate of advantageous mutations? This question stems from the classical way of looking at the situation. The quantum way of looking at the problem will generate the following question: why do the bacteria with lac^- mutations give higher values for the measurements of lac^+ mutations in the presence of lactose only? This (or similar) approach to enquiry has been applied by a few, including Johnjoe McFadden and Vasily Ogryzko. What follows is the quantum mechanical explanation of the difference in the mutation rates in the two samples of bacteria in Cairns experiment.

Quantum Measurements of Mutations. Before we can explore the quantum mechanical explanation of results from the Cairns experiment, we need to understand the quantum mechanical view of measurements. When we make a measurement on any system, we interact with it in order to make the measurement, and therefore we change it. Those changes are negligibly small in the macroscopic world, but significant in the microscopic world, where everything is so small.

As an example, consider a base A in one strand of a DNA sequence corresponding to a T in the second strand. Now focus on the proton in A that makes bond with an electron in T. Also assume that there is a G next to T in the second strand. Normally, A always bonds with T. However, as explained earlier, in quantum mechanics, the proton can behave as a wave and therefore as a wave can bond with both electrons, the electron of T and the electron of the next door G. In other

225

words, the probability of bonding with the electron of T is very high and the probability of bonding with G is very low. Now assume that we make the measurement to determine if the proton of A is bonded with the electron of T or with the electron of G. In our measurement, we change the system enough to get a single result, that is, we will either see the A-T bond or the A-G bond. Because the probability of A bonding with T is much higher, most of the time we will see A-T bond but as a rare event, we will see an A-G bond and this will be regarded as a mutation. During our measurement, we are acting as environment on the proton of the base A. When an enzyme is reading the DNA strand to replicate it or to make protein, the enzyme is the environment and is "making measurements" on the DNA strands and on the protons and electrons involved in bonds.

This process in which the environment (in this case enzyme) interacts with a system to extract one value out of many possible values is called *quantum decoherence*. So, during quantum measurements if the proton spends more time with G then there will be more mutations. So the mutation rate depends on the environment making the measurements.

Now going back to Cairns experiment, the presence of lactose in the environment adds to the complexity of the environment that is making measurements on the DNA of the bacteria. The environment that includes lactose pushes A's proton to the G side more than the environment that does not contain lactose. This results in more mutations in the presence of lactose, and that's what Cairns observed.

Bottom line: The rate of mutation may depend on the quantum environment of the genes. This environment acts on the wave nature of the particles to generate events in the face of quantum mechanics uncertainty.

> **Think About It!**
>
> After reading the section *Quantum Measurements of Mutations*, answer the question: why do different types of mutations occur at different rates?
>
> **Answer**
>
> The reason, at least in part, is that different types of mutations have different quantum environments, which gives rise to different mutation rates.

So, uncertainty in the microscopic world is inescapable.

5.11 The Game of Chance

As mentioned earlier, skeptics of evolution often mock the theory of evolution for its aspect of chance or random change, what we call random mutations. They would say, for example, we cannot be just a product of some random chance or accident. Let me begin responding to this by saying it just off the bat, yes we are the production of random chance in many ways. Like it or not, it's a reality and we live with it. First, it is a scientifically proven fact that what you call chance plays a role in determining which of the two alleles (versions of gene) from each parent are chosen during the formation of a sperm and an egg, as discussed in Chapter 4. Then there are more chances that a sperm will not meet the egg than it will. So, chance enters our life right from its conception and then lives with us for the rest of our lives. For example, we take risk in everyday activities from stepping outside our beds to climbing down the stairs (we might fall), to driving (we might run into an accident), to making investments, and so on. Any project manager would tell us that the root of risk is uncertainty, that is, chance. So, chance is a fact in all aspects of life and all disciplines of knowledge; it should not be such a mockable concept. In science, chance is usually referred to as probability, and probability arises from uncertainty. However in science, uncertainty and therefore probability is a measurable mathematical quantity. It's not a plain wishy-washy *may be* or *I don't know* thing. The terms random, chance, or uncertainty when used in science have a very specific and certain meaning. For example, random genetic mutations mean that these mutations occur spontaneously and the probability of their occurrence is independent of whether they are helpful or harmful to the organism in which they occur. There is however more to it.

Heisenberg's uncertainty principle, an inescapable law of the microscopic world, states that certain uncertainty is inherent to the microscopic nature of things, including genes. As we have shown in this chapter, the chance behind the variations of traits in Darwin-Wallace theory of evolution have been traced down to randomness in mutations, and ultimately to Heisenberg's uncertainty principle. Heisenberg's uncertainty principle was discovered in 1927, 70 years after Darwin and Wallace presented the theory of evolution. So, we can mock the theory for leaving the cloud of chance hanging on it unexplained, or we can appreciate it for being so ahead of its time.

To summarize, randomness in evolution, that is, in genetic mutation means the following three things:

1. Genetic mutations occur spontaneously.

2. The probability of a genetic mutation to occur in an organism is independent of whether it will be helpful or harmful to the organism. In other words, mutations are not directed.

3. There is enough evidence to suggest that the probabilistic nature of genetic mutations is an expression of scientific physical law such as the Heisenberg's Principle of Uncertainty.

However, you are quite justified to ask a question at this point: how does it really matter if quantum mechanics is at work at some step in evolution or not?

5.12 Quantum Mechanics or Not: What's the Difference?

By definition, scientists are out to find the truth about how things work. They are often driven by curiosity, and yes they every now and then get some intellectual satisfaction in the process. But these are only means to an end of finding the truth, and the implications of scientific findings often go beyond labs and scientific papers. Historically, the human species and society has benefited from scientific findings enormously. Look at anything around you from electricity to computers, from cellular phones to satellites, from your car to a plane; all these things we take for granted are based on scientific discoveries.

So, a scientific theory is not just an intellectual exercise, it may have profound implications and applications. For example, fundamental theories behind biological processes may find their applications in medicine. Let me illustrate this with a trivial example. We showed in this chapter that genetic mutations are caused by the quantum nature of genes. Once we have proved that quantum mechanics is behind mutations, we can use quantum mechanics to control the processes that involve mutations. For instance, due to the use of antibiotic drugs, some harmful bacteria have developed resistance to drugs. The drug kills all the bacteria except those few with mutation to resist the drug. These survivors multiply and take over because they are resistant to the drug. The drug provides the environment in which the resistant bacteria (although few in the beginning) are selected and they eventually prevail. Now, quantum mechanics predicts and experiments support that before using the drug, drug-resistant mutations are very rare. The exact probability of the existence of a drug-resistant mutation depends on the bacteria but it could typically go as low as one in a billion (10^9). A patient may harbor a billion or more bacteria in the body during active disease. After the drug kills all non-resistant bacteria, the one with resistance will multiply. Once you have figured that out, now you can control the process. For example, use three different drugs simultaneously which would require three different mutations simultaneously to resist all three drugs. Now let's go back to the theory. The probability that any of the bacteria will have three independent mutations developed simultaneously is the multiplication of individual probability of having each of these three mutations. Let's assume for simplicity that the probability of having each of these mutations is 1 in a billion. Then the probability of having all three mutations is 1 in 10^{27}. This means 1 in 10^{27} will be drug-resistant bacteria. Even the whole world does not have these many, that is 10^{27}, bacteria for a given disease. This means this is a safe solution.

So if you believe in the theory, you can develop this kind of antibiotic treatment and treat patients without the danger of creating platoons of drug-resistant bacteria. Actually such a treatment has already been developed and it's called multi-drug therapy (MDT). Cancer cells also grow uncontrollably into a tumor due to mutations. So, understanding the theory behind mutations more thoroughly may help develop a cure for cancer.

Note. The MDT will work and has worked if applied correctly. Here is an example of incorrect application: If the patient skips doses or does not take all the drugs in the set of drugs prescribed, a mutant can be selected. Actually it has happened in the case of an MDT treatment of TB. Understanding the theory behind the process in more detail will give an even better handle on the process and hence the development of a better treatment.

In Chapter 11, we will explore a great many more examples of applications based on the theory of evolution.

5.13 Standard Theory of Evolution: Putting It All Together

As we have seen so far in this book, the theory of evolution has been put on firm grounds by experimental facts. As we explored in this chapter, it's already on its way to get well rooted in the well-established theory of physics, quantum mechanics. You should have also noticed that what Darwin and Wallace presented was just the beginning; the theory has expanded after that. So far in this book, we have explored components of the theory of evolution at different levels of complexities from populations to organisms to genes. In this section, we put together all of these components well supported by data and call it the standard theory of evolution. From the perspective of the standard theory of evolution, we can look at the history of life as a continuous cycle of evolution, illustrated in Figure 5.4.

And God Said: Let There Be Evolution

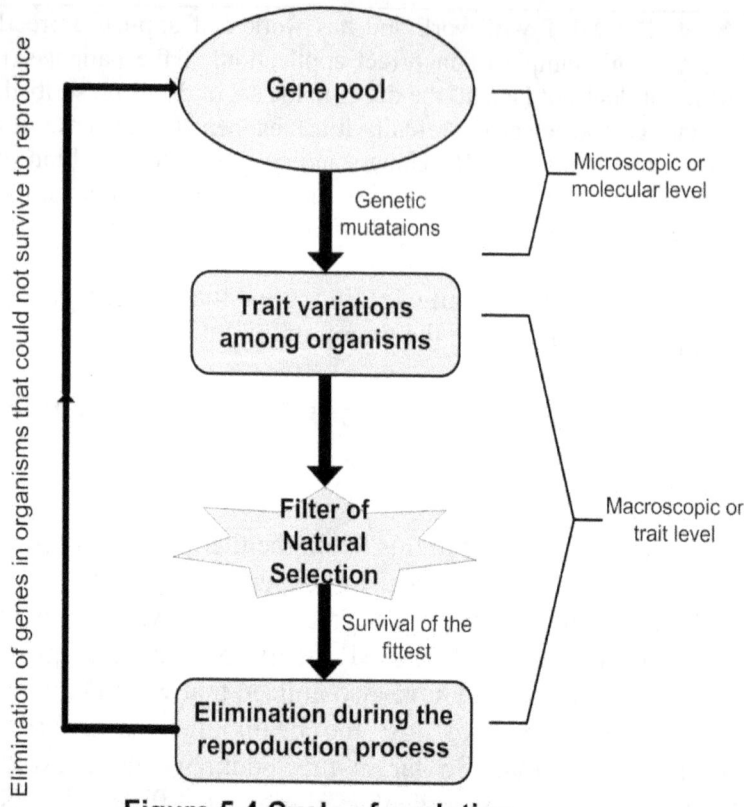

Figure 5.4 Cycle of evolution

Different steps and aspects of this cycle of evolution are described in the following:

1. Genetic mutations occur independent of the need for adaptation to change or natural selection.

2. Although some mutations may occur due to external causes such as harmful environmental agents, evidence suggests that naturally occurring genetic mutations have their roots in the intrinsic quantum nature of genes.

3. The quantum environment of genes plays a role in determining mutation rates. This gives rise to different mutation rates for different types of mutations or in different regions of the DNA.

4. Genetic mutations occurring at the molecular (microscopic) level result in trait variations in organisms both at microscopic level (bacteria) and at macroscopic level.

5. Natural selection operates on variations by selecting organisms with trait variants that help them to adapt to environmental changes and eliminating those who do not have these variants and hence cannot adapt.

6. The elimination of organisms obviously eliminates some genes from the gene pool of the population because the eliminated organisms are not reproducing.

7. The genes corresponding to the selected traits are passed on to the next generation through reproduction.

8. This endless re-iterative process (or cycle) selects one kind of genes and mutations against others and gives rise to new traits and eventually to new species from old ones.

Think About It!

What is the relationship between mutations and natural selection?

Answer

Mutations and natural selection are decoupled processes in the sense that mutations are not influenced by natural selection; they occur independent of natural selection and provide variations (trait variants) for natural selection to operate on.

Note that natural selection is not evolution; it's a step in the process of evolution, or a mechanism for evolution. In a nutshell, evolution observed at the organism level is caused by genetic mutations at microscopic scale, that is, microscopic level. Remember, natural selection is an organism-level phenomenon which operates on trait variations generated by gene mutations at microscopic level, which evidently have their roots in the laws of quantum physics. Gene mutations and natural selection are decoupled (independent) processes. Of course for microbes, organism level is also the micro-level.

Think About It!

Biologists often refer to natural selection operating on variations. Does this mean that natural selection is an external mechanism imposed on a population?

Answer

No. This terminology is just shorthand for saying that natural selection is a process within the population that inevitably brings a change in the genetic pool and therefore the set of traits of the population over generation to favor adaptation to the changing environment.

5.14 Summary

The evolution process occurs in three steps:

1. Gene mutations occur at particle, atomic, and molecular levels independent of the need to adapt and independent of natural selection.

2. Alleles (versions of genes) and gene mutations give rise to variations of traits. Genotypes (plus environment) give rise to phenotypes.

3. Natural selection operates at organism level on the trait variations (phenotypes) to select the traits (and corresponding organisms) that adapt to the environment and to eliminate those that do not. The genotypes of the selected organisms are passed down to the next generation.

The causes for gene mutations fall into two categories: external and internal. Harmful environmental agents such as X-rays are examples of external causes, whereas the intrinsic quantum nature of genes and molecules and particles around them is an example of an internal cause. The molecules, atoms, and particles around a gene are called the gene's quantum environment. It is this environment that gives rise to different rates for different types of mutations.

In a nutshell

✓ Natural selection is not evolution; it's a step in the process of evolution.

✓ There is enough evidence to suggest that evolution has its roots in the laws of quantum physics, which govern the generation of mutations in genes. Rooted in physical laws, evolution is as certain as gravity.

✓ In addition to external causes such as harmful environmental agents, gene mutations also occur due to the inherent quantum nature of genes, which cause so-called replication errors.

✓ Overall, gene mutations occur independent of adaptive change or natural selection and contribute to trait variations for natural selection to operate on.

✓ The rate of mutation depends on the quantum environment of the genes.

✓ The chance behind the trait variations in the Darwin-Wallace theory of evolution has its roots in the randomness of gene mutations, which in turn has its roots in the laws of quantum physics.

✓ The randomness of gene mutations means that mutations occur spontaneously and the probability of their occurrence is independent of whether they would be helpful or harmful to the organisms in which they occur. So, this randomness is very much a part of the scientific law and may have its roots in Heisenberg's uncertainty principle of quantum mechanics.

✓ Bottom line: The theory of evolution originally introduced by Darwin and Wallace is finding its roots in quantum mechanics, a well-established theory of physics.

5.15 Behold

As we discussed in this chapter, through the study of evolution and molecular biology in general, biologists are learning a lesson that physicists began learning toward the end of nineteenth century: The macroscopic reality that we experience every day has its roots in the microscopic world of atoms and molecules because macroscopic things are made of microscopic objects. This is true for both living and non-living entities. The study of microscopic entities such as genes, genomes, cells, and microbes has already proven that "the casual concourse of atoms" can in fact generate useful new traits of organisms. Here are some proven examples of new traits that arose as a result of simple rearrangements of atoms and molecules in the DNA sequence of organisms from peas to humans: sweeter peas, ability to run faster, bigger muscles, and improved ability to digest certain foods.

We have already seen in this book how extraordinary diversity of life has an underlying unity in terms of cells and genes: all living organisms are made of cells, their traits are developed from their genes, and their genes have the same universal genetic code. Evolution sheds light on this unity behind diversity by functioning as a mechanism to derive diversity from unity. In the next chapter, we explore evolution as a unifying thread of life by visiting evidence of evolution from different areas of life and from different complexity levels of life.

And God Said: Let There Be Evolution

SIX

Evolution:

A Unifying Thread of Life

And God Said: Let There Be Evolution

It is not the strongest of the species that survives, nor the most intelligent that survives. It is the one that is the most adaptable to change.

Charles Darwin

Nothing in biology makes sense, except in the light of evolution.

Theodosius Dobzh

6.1 Once Upon a Time

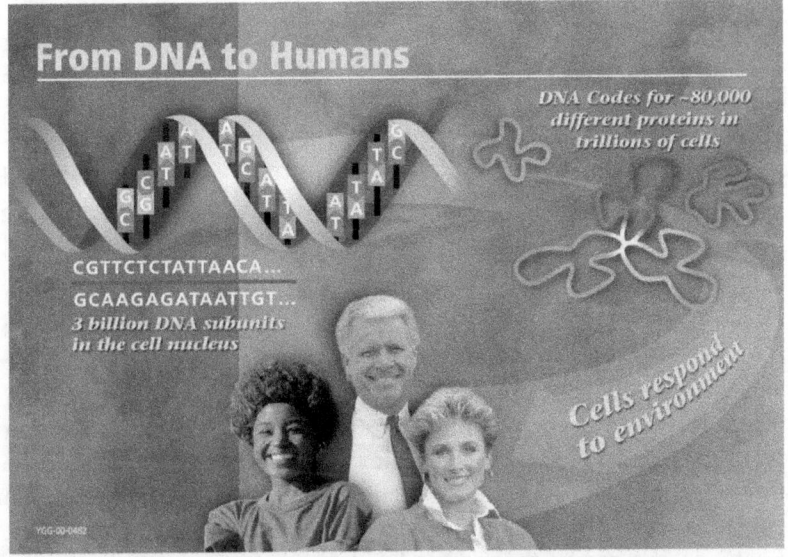

Figure 6.1 From DNA to humans. Source: Genome Management Information System, Oak Ridge National Laboratory.

Once upon a time, life originated on Earth in terms of the first cell (or cells) and evolved into all forms of life we see today around us. Biology is the study of life, and life appears to present the biggest contradiction of all: unity behind diversity. Endless different forms of life consist of the same building blocks, cells, and develop from the same blueprint, DNA. Not only that, processes inside cells are also basically the same among different species, such as how the

And God Said: Let There Be Evolution

DNA code is translated into proteins, and how raw material (food) is converted into energy. Can please someone explain to us this unity behind diversity? Nothing but evolution does that. Evolution explains the 3.5 billion-year-long history of life on Earth from the first DNA to humans: a wonderful display of unity behind diversity.

As explained in Chapter 4, each DNA molecule contains several genes, which are the basic physical and functional units of heredity. The human genome (set of genes in a human cell) is known to comprise about 25,000 genes. A given gene is a specific sequence of nucleotide bases representing the information required for constructing proteins, which are the workhorses of life. All living organisms are composed largely of proteins, which are made in the cells according to the instructions coded in the genes. Proteins provide the structural components of cells and tissues, and facilitate all the dynamic functions of an organism. A type of protein called enzymes serve as catalysts for biochemical reactions essential to maintain life in an organism.

Proteins, responsible for almost all the dynamic functions of an organism, are large and complex molecules made up of long chains (polymers) of subunits (monomers) called amino acids. The shape and structure of a protein, given to it by different forms of electromagnetic force working at atomic level between different parts of the molecule, determines its functionality. There are thousands of types of proteins comprised of only 20 types of amino acids: an example of unity behind diversity at the molecular level. Each specific sequence of three DNA bases (codon) acts as a code for the cell's protein-synthesizing machinery to add a specific amino acid to the polymer of protein under construction. For example, the base sequence ATG codes for the amino acid methionine. Because three bases code for one amino acid, the protein coded by an average-sized gene (3000 bp) will be comprised of 1000 amino acids. The genes in a DNA molecule consist of a series of codons that specify which amino acids should be combined to make a protein polymer.

So, here is the 3.5 billion year history of life on Earth in one sentence: Four kinds of bases (A,T,C, and G) lead to 20 kinds of amino acids, which lead to many thousands kinds of proteins, which in turn lead to countless organisms with varying traits, which can be grouped into billions of species. Evolution is the mechanism for this grandeur display of unity behind diversity.

If you wanted to squeeze the 3.5 billion year history of life on Earth into a single day of 24 hours on the scale of time, it took about 20 hours for multicellular life to evolve, 1.6 hours for vertebrates to occupy the land, another 1.6 hours for flowers to evolve; and listen to this: only in the last 2.88 seconds we humans arrived. So, we are the late comers to this grandeur party of life on Earth.

6.2 In the Beginning

In the beginning, there was a cell which evolved into all forms of life that we see around us. The central success of the theory of evolution lies in the explanation that it offers for the tremendous diversity among living organisms superimposed on underlying unity. Evolution connects the phenomena involved in the growth of an organism from alleles to gametes, to a zygote (cell), to an organism, to a species, and to multiple (different) species; the phenomena on the size scale from unicellular organisms to multi-cellular organisms of different sizes; and phenomena across different species. Mutations, that is heritable changes in DNA, result in variations in traits of organisms related to body form, functioning, and behavior. Diversity is the sum total of variations within and among species of organisms that have accumulated since the origin of life.

The concept of unity behind diversity even goes beyond life. Indeed, the sum total of all the great discoveries and breakthroughs in the history of science (bio and non-bio) have revealed a secret over and over again: there is an underlying unity behind apparent diversity. This is a key point to understand not only the things and phenomena around us but also their diversity. For example, all life is made of the same basic building block of life: the cell. This applies to both living and non-living entities. Look at the countless diverse things in the non-living world from sand in the desert to water in an ocean, from a rock on a beach to moons and planets and galaxies, from your wine glass to your Mercedes car, and so on. Think of the plains, the mountains, the clouds, the night sky, and everything in them. It's not hard to imagine how boring the world and the Universe would be without this diversity. Well, it turns out that there would be no world or universe without this tremendous diversity. For instance, it is due to the diversity built into life that organisms were able to survive through changing environments that would otherwise have eliminated them long ago. Yet, there is a great unity behind this diversity. This unity goes even beyond life. Anything and everything is made of some type of atoms out of a total of the 94 naturally occurring kinds of them, such as hydrogen, iron, and sodium. Yes, living entities are also made out of the same set of atoms as non-living entities. All atoms are made of an even much smaller set of fundamental particles that we can count on our finger tips.

The unity behind diversity can be explained in terms of a few fundamental particles and a few forces (also called interactions) among them. All other particles and systems are made of these fundamental particles, and all other forces are different facets of these few fundamental forces. As explained in Chapter 5, there are four fundamental forces: weak nuclear force, strong nuclear force, electromagnetic force, and gravitational force. Physicists are even discovering evidence that these four forces are also different low-energy facets of a single force that was present in the beginning of the Universe at very high energy.

Now look at the diversity of life from a bacterium to an elephant, from a tree to a human, from a chicken to a lion, and so on. Out of about 100 billion different kinds (species) of organisms that ever lived on Earth, about 100 million are still living with us today. Even within the same species, there is great diversity. Look at us, the 6.5 billion human beings with varying heights, eye color, hair color and other traits. No two of us are quite exactly the same except perhaps identical twins. In spite of this tremendous diversity among a mind-boggling number of species and within each of those species, there is an underlying unity: all organisms are made of cells. Any cell, the smallest unit of life, is made of four kinds of molecules: carbohydrates, lipids, nucleic acids (DNA and RNA), and proteins. All these molecules in turn are made out of the same set of atoms that non-living entities are made of. This big picture of unity behind diversity itself suggests that all species point back to a single or a few common ancestors that had these common traits and passed them down to their descendants, which grew into diversity over a long time and many generations.

Evolution is a unifying thread of life as it connects all the different species to one another and all the organisms within a species too. Accordingly, the theory of evolution is a unifying theory that weaves together all the major concepts and principles in biology. As you have seen in the previous chapters, evolution begins its work at the level of fundamental particles (for example, protons and electrons in the hydrogen bonds of genes) to create gene mutations, the seeds for variations of traits; and caries on its work to the level of organisms and species where *natural selection* operates.

And God Said: Let There Be Evolution

In a nutshell, the unity of life behind incredible diversity can be attributed to descent (with modification) of all organisms from a common ancestor or a few ancestors. Because the descendants of that ancestral organism lived in various habitats, different groups living in different environments accumulated different modifications that helped them to adapt to their specific environmental conditions. Over long periods of time, descent with modification resulted in rich diversity of life among species and also among organisms within each species.

All through this book, we have presented evidence including observations and experimental results that support evolution and the theory of evolution originally introduced by Darwin and Wallace. In this chapter, we present a few more examples from a pool of an overwhelming amount of evidence for evolution. The examples cover several lines of evidence in the diverse areas of life reflecting the fact that evolution is a common thread that runs through all life. If I had to eliminate just one concept in order to make a course in general biology illegible with all its nicely connected concepts falling apart, it would be evolution. It's like turning the light bulb off.

In Chapter 1, we discussed how during Darwin's times the prevailing belief of *perfect creation* was being challenged by new scientific observations, which were raising some serious questions in various fields including biogeography, comparative morphology, embryology, geology, and paleontology (study of fossils). We discussed how the doctrine of perfect creation failed to answer the questions raised by observations in these fields. In this chapter, we also explore how the theory of evolution successfully answers those questions.

Let's begin with the fossil record.

6.3 The Fossil Record: Telescope into the Past

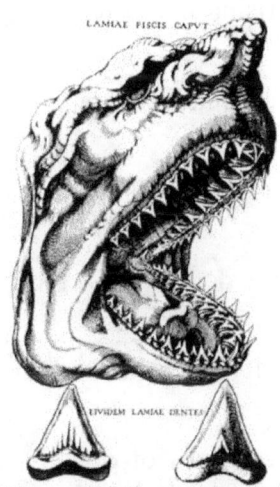

Figure 6.2 Comparison of the teeth from a shark head with a fossil tooth. Source: Nicholas Steno's 1667 paper.

In 1667, Nicholas Steno (1638-1686), a Danish pioneer in anatomy and zoology, shook the world of science by reporting the similarity between shark teeth and the rocks commonly known as *tongue stones*. This was our introduction to the fact that fossils are a record of the past life on Earth. Figure 6.2 presents Nicholas Steno's anatomical drawing of an extant shark head (top), shark tooth (bottom right), and fossil shark tooth (bottom left). Steno, in his 1667 paper, concluded that the fossil teeth indeed came from the mouths of once-living sharks.

Two centuries later, Mary Ann Mantell (1799-1847), a British fossil collector and paleontologist, picked up a tooth, which her husband Gideon thought to be of a large iguana (lizard); but it was later identified to be the tooth of a dinosaur, *Iguanodon*. This discovery demonstrated that fossils may represent forms of life that are no longer with us today.

> Note. The history of Earth and life on Earth is written on the rocks in terms of fossils. These rocks and fossils are like the pages of an encrypted history book scattered all over the globe. The information collected from the study of fossils provides a compelling evidence for evolution.

And God Said: Let There Be Evolution

For centuries now, fossil studies have been a very well-established and reliable technique to look into the history of life on Earth. Furthermore, the fossil record is also used to cross-test the evolutionary hypotheses arising from other kinds of evidence.

Overall, the study of the fossil record demonstrates the following evolution-related points:

1. Organisms in the past differed from organisms in the present.

2. Many species that existed in the past do not exist today; they have become extinct.

3. Evolutionary changes have occurred over time in many groups (such as species) of organisms.

4. Over longer time scales, major new groups of organisms have appeared (evolved).

Illustrated in Figure 6.3 is an example of how the modern horse, the *Equus,* evolved. This illustration is supported by an almost complete fossil record found in North American sedimentary deposits from the early Eocene Epoch (*56 – 34 million years ago*) to the present. This makes it one of the best examples of evolutionary history, called phylogeny. The Eocene Epoch is a major division of the geologic timescale (in the Cenozoic Era) that lasted from about 56 to 34 million years ago (55.8 ± 0.2 to 33.9 ± 0.1 million years). The evolutionary journey from the *Hyracotherium* to *Equus*—the modern horse—took at least 12 genera, which includes several hundred species. Figure 6.3 shows the gradual evolution of the horse lineage from its ancestral to its modern form by presenting a dominant genus from each geological period.

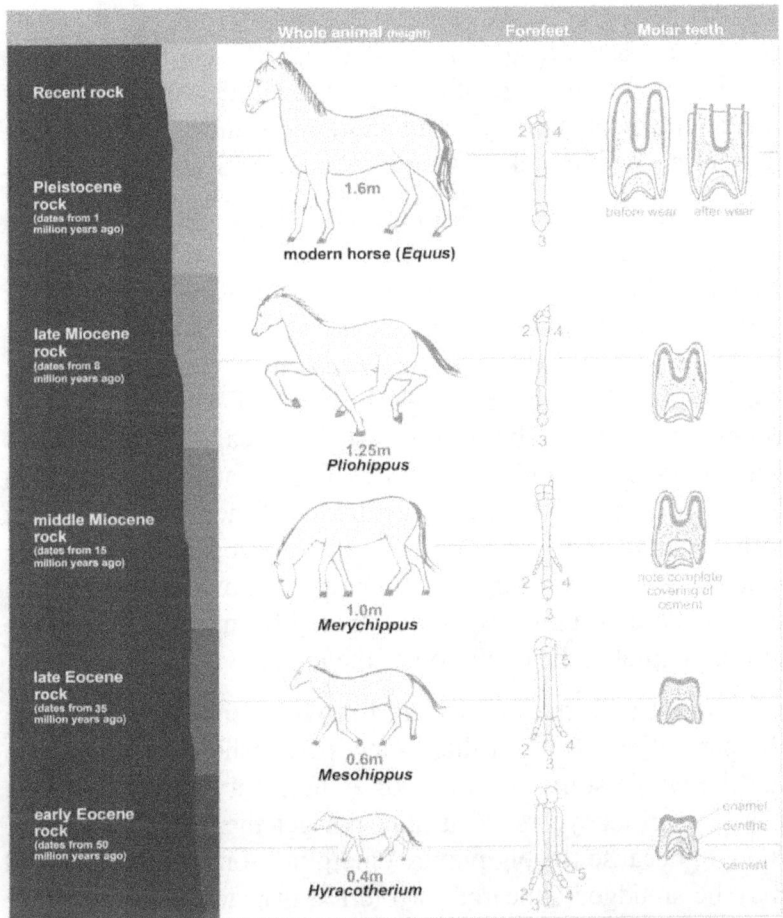

Figure 6.3 Evolution of the horse, showing species in the evolution branch reconstructed from the fossils obtained from successive rock strata. The foot diagrams are all front views of the left forefoot. The third metacarpal is shaded throughout. The teeth are shown in longitudinal section.

This evolutionary sequence of the horse begins with a rather small animal called *Hyracotherium*, commonly referred to as *Eohippus*. Before spreading across to Asia and Europe, the *Hyracotherium* lived in North America about 54 million years ago. The fossil records show that the *Hyracotherium* was a small animal of the size of a fox. It was lightly built as its body was adapted for running. With short and slender limbs, its feet were elongated to make the digits almost vertical.

Pattern of plant fossils found in different strata show that *Hyracotherium* originally lived in marshy, wooded country, which gradually became drier. So, the adaptation and hence the survival and reproduction advantage went to those individuals in the population who had their head in an elevated position, which provided them a good view of the surrounding countryside. To be able to run faster to escape from predators was another advantageous trait. These factors favored an increase in body size and the evolution of the splayed-out foot to the hoofed foot. The original splayed-out foot useful in the marshy, wooded country for support; was now unnecessary on the dry, hard ground. The change in diet from soft vegetation to grass, facilitated the evolution of the teeth.

In a nutshell, the fossil record provides snapshots of life in the past. When you put these pieces (snapshots) together, you get the whole story (a movie) of evolutionary change over the past 3.5 billion years. You may not get the complete picture though, because the big picture emerging from the fossil record may be smudged, smeared, or blurred in places and may have bits missing here and there, as not all forms of life left behind fossils. However, overall, the fossil record clearly shows that life is billions of years old and has been changing over time. Here is the bottom line: The picture emerged from the fossil record fits well with the theory of evolution.

To be specific, here are the major elements of the big picture emerged from the fossil record: Life began 3.8 billion years ago, insects diversified 290 million years ago, and the human and chimpanzee lineages diverged only five million years ago. It's fair to ask this question at this point: How have scientists figured out the dates of long past evolutionary events? Here are some of the methods and evidence that scientists use to assign dates to events:

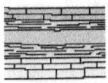
1. Stratigraphy provides a sequence of events from which relative dates can be extrapolated.

2. Radiometric dating relies on half-life decay of radioactive elements to allow scientists to date rocks and materials directly.

3. Molecular clocks allow scientists to use the amount of genetic divergence between groups of organisms to extrapolate backwards to estimate dates.

Note. If you are curious enough to learn more in a fun way, you can easily find an exhibition displaying fossils in your nearby museum, such as a natural history museum or a science and tech museum.

Radiometric dating and molecular clocks are discussed in the next section.

6.4 Scientific Techniques to Analyze Fossil Data

As compared to Cuvier's or Darwin's times, we now have more sophisticated modern scientific techniques based on scientific discoveries at our disposal to analyze the fossil record. Two of these techniques are discussed here as examples:

And God Said: Let There Be Evolution

Radiometric Dating: Dating with the History of Life. With help from discoveries in physics and chemistry in the area of radioactivity, fossil dating has become more accurate over time. Now, based on radioactivity, scientists use a technique called radiometric dating to estimate how long ago a given rock formed. This enables scientists to infer the age of fossils contained within the rock. Radioactivity, also called radioactive decay, is a phenomenon of spontaneous emission of radiation from atomic nuclei. This phenomenon is exhibited by the atoms of certain elements, called radioactive atoms, such as carbon-14 ($^{14}_{6}C$) and uranium ($^{238}_{92}U$). It turns out that radioactive atoms are inherently unstable and as a result radioactive *parent atoms* over time decay into stable *daughter atoms*.

The time in which half the parent atoms of a given sample of a specific radioactive element decay into daughter atoms is called the *half-life* of that element. For example, the half-life of $^{14}_{6}C$ is 5,715 years, and the half-life of $^{238}_{92}U$ is 4.5 billion years. This means that half of a given sample of $^{14}_{6}C$ decays into $^{12}_{6}C$ in 5,715 years. The key point to note here is that the half-life of an element is a constant and is not influenced by environmental conditions such as temperature, pressure, and chemical bonding; and therefore serves as a perfect *nuclear clock* to determine the age of different entities.

Let us work through an example to illustrate how radiometric dating works. Carbon dioxide in our atmosphere mostly contains carbon-12, but also a small portion of carbon-14. Through a process called photosynthesis, carbon-14 makes it into plants along with carbon-12. Carbon-14 becomes incorporated within animals when they eat plants. Because living plants have a constant intake of carbon, they maintain a constant ratio of carbon-14 to carbon-12 inside of them, which is nearly identical to the ratio in the atmosphere. Now consider an organism that dies and leaves behind fossils. The organism does not intake any carbon after death because it no longer digests carbon compounds (food). However, the carbon-14 inside of it continues decaying into carbon-12, and the ratio of carbon-14 to carbon-12 starts decreasing due to this radioactive decay. In other words, the *nuclear clock* begins ticking from the time of death onwards. The ratio of carbon-14 to carbon-12 diminishes to half in 5715 years, to a quarter in 11430 (5715×2) years, to one eighth in 17145 (5715×3) years, and so on. So you can see that by working backwards, we can determine the age of the fossil. First, the present carbon-14 to carbon-12 ratio is determined. Then by comparing this ratio to the original ratio (the atmospheric ratio), and by using the half-life, we determine the age of the fossil. This technique using carbon is called *radiocarbon dating*. Due to a relatively small half-life of carbon-14, radiocarbon dating cannot be used to date entities that are older than 50,000 years because after this amount of time, radioactivity becomes too low to be measured accurately. To date entities older than 50,000 years other radioactive elements can be used. For example, the half-life of uranium-238 is 4.5 billion years, longer than or about the same as the history of life on Earth.

> Note. A careful reader will recognize an assumption in our example: the ratio of carbon-14 to carbon-12 in the atmosphere has been constant over the age of the fossil. This is a reasonable assumption. Nevertheless, the fluctuations in the carbon-14 to carbon-12 ratio over time are usually accounted for by using other kinds of data.

As mentioned earlier, scientists use molecular clocks to extrapolate backwards from the current amount of genetic divergence between organisms to estimate dates. What are molecular clocks? Let's explore.

Molecular Clocks. Remember, if a species evolved from another species, it has the same DNA as the parent species but in an evolved form. It has been discovered that some genes or other regions of the DNA evolve (change) over the course of millions of years at constant rates or reliably known rates. Once the rate of change of a DNA region is determined, it can be used as a yardstick to measure the absolute time that it took for an evolutionary change to occur. This technique of measuring time by using the rate of mutation is called a molecular clock. For example, the gene that codes for the protein alpha-globin, a component of hemoglobin, undergoes base changes at a rate of 0.56 changes per base pair per billion years, considering only the changes that affect the structure of the protein. Given that this rate is reliable, the gene could be used as a molecular clock. Suppose we find two species differing by one base pair in their alpha-globin gene. This means that these two species separated from their common ancestor 1/1.79 (*or* $\frac{1}{0.56}$) billion years ago.

Here is an illustrated yet simple example of how the concept of a molecular clock becomes a powerful tool to determine the date of a lineage-splitting event. As shown in Figure 6.4, consider a species O with a DNA string ATGCAATTTATCG. Now assume that we know two species (species 1 and species 2) and each of these two species at present differ from species O by two bases, as illustrated in Figure 6.4. Also note the two changed bases that are different from each other in the two species. We also know that the DNA string of species O (that is, ATGCAATTTATCG) changes at a rate of one base per 20 million years. This means that these two species diverged from a common ancestor 40 million years ago: 20 x 2 = 40. As illustrated in the figure, after 20 million years, two lineages diverged from this ancestor due to a different mutation in each: a G at the end of the string got replaced with a T in one, and a T at the fifth place from the end of the string got replaced with a G in the other lineage. So at this point in time the two lineages differed from each other by two bases. After another 20 million years, that is, 40 million years away from the common ancestor, the two descendant lineages diverged farther by another base mutation in each. Now the two lineages are farther apart by four bases. This means that the two species (1 and 2) differ from each other by a total of 80 million years of evolution (20 x 4 = 80). Note that here we are talking about work performed by evolution on species and not just merely counting the years. This kind of analysis can be applied to all lineages during the 3.5 billion years of evolution of life on Earth.

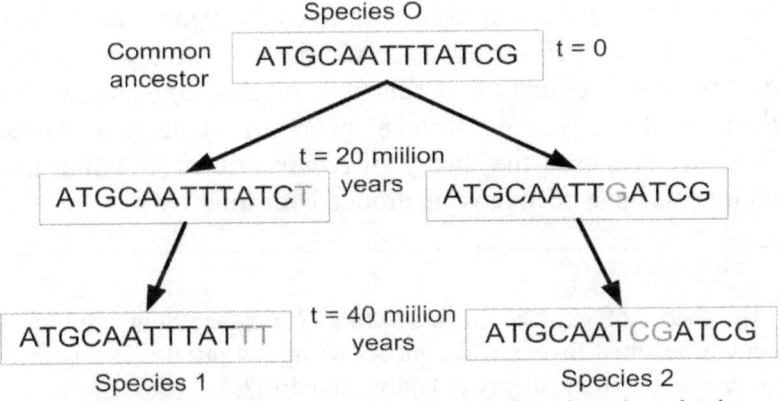

Figure 6.4 Illustration of the use of molecular clocks.

Compared to the size of the Universe, our planet Earth is just a dot; but compared to the size of a typical organism, it is a big wide world. The study of geographic distribution of species of organisms on one hand represents another dimension in diversity and on the other hand adds to the body of evidence for evolution.

6.5 Biogeography

Biogeography is the field that studies the geographic distribution of species of organisms. About 70 percent of the Earth's surface is occupied by oceans and the rest by land masses: continents and islands. Along with other factors, geographic distribution is influenced by *continental drift*, which is the slow movement of Earth's continents over time, and the formation of islands.

6.5.1 Role of Continental Drift in Biogeography and Evolution

Initially, scientists believed that the positions of Earth's landmasses (continents) have been fixed throughout the history of life on Earth. However, scientific theories are always tested against observations, experimental results, and alternative theories. In 1912, a German geophysicist named Alfred Wegener proposed the theory of *continental drift* stating that continents have been moving around over time.

Think About It!

The shapes of some continents such as Africa and South America could be fitted together like pieces of jigsaw puzzle. Is this fact consistent with the theory of continental drift?

Answer

Yes. The fact that the shapes of these continents can fit together strongly suggests that these continents were actually together some time in the past. Alfred Wegener's idea of continental drift was initially inspired by this observation.

Based on data from different sources, scientists have estimated that about 250 million years ago these movements (drifts) united all of Earth's land masses into a single large continent called *Pangaea* (meaning "all lands" in Greek). As illustrated in Figure 6.5a, this supercontinent began to break apart about 225-200 million years ago (mya). By 20 mya, as scientists estimate, the continents that we know today were within a few hundred kilometers of their current locations.

Data has validated the inter-consistency of the theory of evolution and the theory of continental drift: the two fit together like the two pieces of a puzzle. For example, combining the theory of evolution with our understanding of continental drift, scientists have predicted where fossils of different groups (species) of organisms might be found. These predictions have often been upheld by data. For instance, scientists have derived evolutionary trees of horses from the anatomical data. Combining the information from these evolutionary trees with the ages of the fossils of the ancestors of the horses, scientists have estimated that the present day horse species originated about 5 million years ago in what we call today North America. According to the theory of continental drift, 5 million years ago when this species originated, North America was quite close to South America, but unlike today, it was not yet connected to South America. So, horses from North America would not be able to migrate to South America. All of this predicts that the oldest horse fossils will only be found in North America. This and all other predictions like this one about other groups (species) of organisms have been upheld by data from observations and discoveries, and therefore provide evidence for evolution.

And God Said: Let There Be Evolution

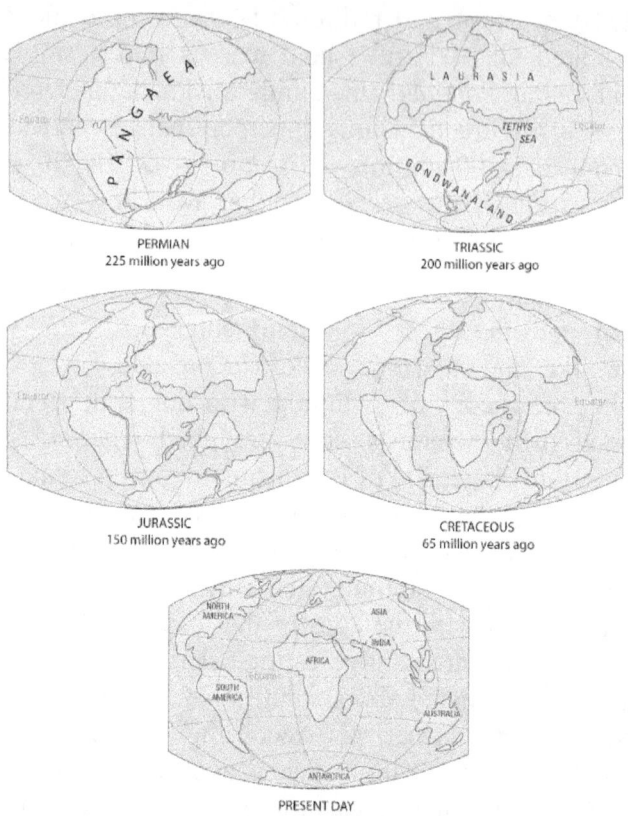

PERMIAN
225 million years ago

TRIASSIC
200 million years ago

JURASSIC
150 million years ago

CRETACEOUS
65 million years ago

PRESENT DAY

Figure 6.5a Illustration: According to the continental drift theory, the supercontinent Pangaea began to break up about 225-200 million years ago, eventually fragmenting into the continents as we know them today.

Antonio Snider-Pellegrini and Wegener noted that if the continents were joined in the past, the locations of certain fossils of plants and animals found on today's widely separated continents would form certain definite patterns, as shown in Figure 6.5b. The figure shows the present day locations of certain plants and animals originally determined from fossils. These patterns are shown by the bands of colors in Fig 6.5b

258

and are supported by data. This presents compelling evidence for continental drift. Similar animal and plant fossils have been found around the shores of different continents, consistent with the hypothesis that they were once joined. As an example, the fossils of Mesosaurus, a freshwater reptile (that looks much like a small crocodile), have been found both in Brazil and South Africa. Further evidence is the discovery of fossils of the land reptile Lystrosaurus from rocks of the same age from multiple locations in Africa, Antarctica and South America. Furthermore, evidence is also presented by living animals: the same animals being found on two continents. Some earthworm families (such as Acanthodrilidae, Ocnerodrilidae, and Octochaetidae) are found in both Africa and South America. In a nutshell, there is an overwhelming amount of data validating the theory of evolution by crosschecking with continental drift.

Think About It!
Why is the flora of southern South America so similar to that of southern Africa?

Answer
Because once upon a time these continents were connected.

Another major piece of evidence in support of the theory of continental drift came from the discovery of the widespread distribution of Permo-Carboniferous glacial sediments in Africa, America, Antarctica, Arabia, Australia, Madagascar, and South America. The pattern of the oriented glacial striations and deposits called tillites implied the continuity of glaciers, which in turn suggested the existence of the supercontinent of Gondwana (also known as Gondwanaland) about 300 million years ago. The pattern of striations has also suggested glacial flow away from the equator and toward the poles, in modern coordinates. This data has supported the idea

And God Said: Let There Be Evolution

that the southern continents in the past have been in dramatically different locations, as well as contiguous with each other, which obviously implies continental drift. The pattern of fossil evidence suggests that Gondwana drifted and merged with other landmasses to form Pangaea. The point here is that continents are always on the move; they collide, split, and unify. Since Earth's outer layer solidified 4.5 billion years ago, a single supercontinent has formed at least five times. The motion of the continents together with the dynamics and impact of water, wind, and asteroids, influenced life's evolution on Earth. For example, as continents collided and shorelines disappeared, some lineages of organisms disappeared with them, and some new lineages opened up, changing the directions of evolution of life on Earth.

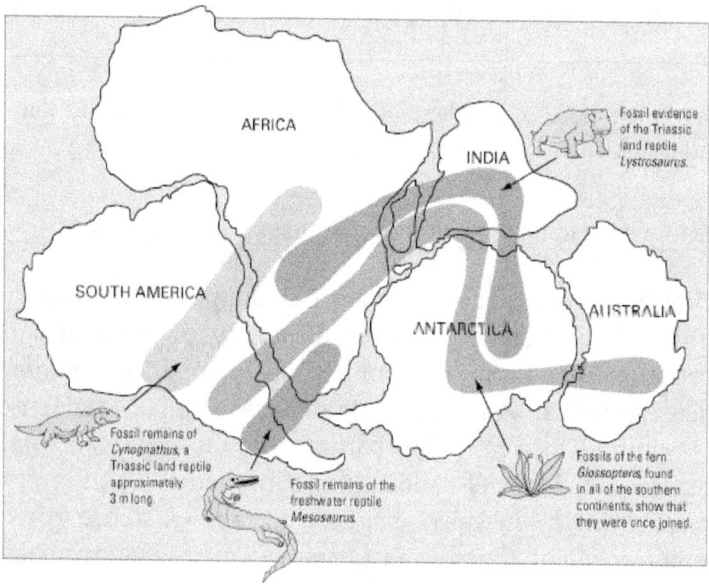

Figure 6.5b Fossil patterns across continents.

If you still have any doubt that the continents really move, here is the ultimate evidence for it. Have you or someone you know ever used the GPS (global positioning system) car navigation system to get to a destination? The GPS, a constellation of 24 satellites, is used to make precise geodetic position measurements, which are in turn used for navigation. Satellite signals recorded by the GPS receivers on the ground are used to determine daily position measurements. Data analysis of these signals demonstrates that the continents move apart two to four inches every year.

Think About It!

As you know from this section, there is compelling evidence that continents were once connected. Using this fact with the information from GPS, should you be able to see Europe from America if the Young Earth doctrine of creationism is true?

Answer

According to GPS, continents move at most four inches a year. According to the Young Earth doctrine, Earth is at most 10 thousand years old. This implies that Europe and America should be at most 40,000 (10,000 multiplied by 4) inches away, which is less than a mile. Therefore, yes, you should be able to see some European cities from somewhere in America. But can you?

There is an important dual characteristic of the continental distribution of life:

1. Different continents or areas in the world with similar ecology (climate and terrain) harbor different animals and plants.

2. The animals and plants in different continents with similar ecology, although fundamentally different, look and behave alike.

And God Said: Let There Be Evolution

It would be reasonable, based on continental drift, to expect to find the same species in two similar habitats in different parts of the world. However, this is not the case. For example, we find native cacti in the deserts in North and South America; whereas the deserts in Africa, Asia, and Australia harbor euphorbs, which are very different. Even though cacti and euphorbs are different species, they share some similar characteristics. For instance, both are water-retaining plants adapted to arid climates or soil conditions; and therefore both belong to a group called succulents. Similarly, the similar geographic areas of Africa and South America harbor different species of animals: Africa has Cercopithecidae (commonly known as Old World monkeys), apes, elephants, giraffes, hornbills, and leopards; and Platyrrhini (commonly known as New World monkeys), cougars, jaguars, llamas, sloths, and toucans are found in South America.

Think About It!

Fossil records show that dinosaurs originated about 200-250 million years ago. Using this fact, would the continental drift theory predict that dinosaur fossils should be found on only a few continents or many continents?

Answer

According to the continental drift theory, the whole landmass formed only one big continent called Pangaea during the time dinosaurs were originated. Because Pangaea later split into many continents, we expect to find dinosaur fossils on many continents.

In the light of creationism, you may wonder: if all species were created perfectly from scratch, why would the creator put different species on different continents with similar ecology? And then why do these different species of animals look and behave alike? Creationism provides no viable answer for such questions. Theory of evolution, however, provides an explanation for this pattern in terms of *convergent evolution*, which is a process through which unrelated lineages acquire the same or similar biological traits for reasons such as occupying similar ecological niches. These species on different continents are different because their common ancestor in the evolutionary tree is in the distant past. However, these different species have acquired the same or similar traits because they went through the same evolutionary or selective pressure as they have been living and reproducing in similar habitats.

In a nutshell, the patterns of distribution of species across continents offer compelling evidence for evolution. Another biogeographical area of interest here is the distribution of life across islands.

6.5.2 Role of Islands in Biogeography and Evolution

An impressive addition to the diversity of tests for the theory of evolution comes from the disparate data on the distribution of species across islands. First, note that there are two types of islands, continental islands and oceanic islands, as defined in the following:

Continental islands. A continental island is an island that was originally part of a continent but later separated by any of a number of events such as moving continental plates, or rising sea levels that flooded the land connecting it to the continent. Examples of continental islands include Great Britain and Ireland off Europe, Kangaroo Island off Australia, Long Island off North America, Madagascar off Africa, Sri Lanka off Asia, and Trinidad off South America.

Oceanic islands. An oceanic island is an island that was never part of a continent, but arose from the deep ocean floor due to events such as volcano eruption or continental plate movement. Examples of oceanic islands include: Saint Peter and Paul Rocks in the Atlantic Ocean; and the Galápagos Islands, Hawaiian Islands, and Juan Fernández Islands in the Pacific Ocean.

So, what does this have to do with evolution? Because islands have been isolated since their formation, they make unique living test labs for evolution. Recall from Chapter 1 that data collected by Darwin from the Galápagos Islands played an important role in the formulation of the theory of evolution. Here we explore two distinctive patterns of life on oceanic islands: unbalanced distributions of native species and adaptive radiation.

Unbalanced Distributions of Species. Let's look into an evolution-related distribution pattern of life observed on islands. Oceanic islands are missing many kinds (not necessarily all) from the set of species that are found on continental islands and continents. In other words, the distributions of native plants and animals on oceanic islands are asymmetric or unbalanced compared to the biotas (the total collection of organisms) found on continents or continental islands. When compared to the biotas on continents and continental islands, there is a pattern of the same missing groups of organisms on oceanic islands across the Indian, the Pacific, and the South Atlantic Ocean. For example, oceanic islands are consistently missing amphibians, native terrestrial mammals (you do occasionally find bats and seals), and fresh water fish across the board. You may say, well, these islands are not suitable for these missing organisms to survive and reproduce. However, it has been shown, by introducing these organisms manually to these islands, that these organisms thrive there. Furthermore, there is another observation: Species similar to those on an oceanic island are often found on the nearest mainland, continental island or continent.

So, why have these groups of organisms been missing on these islands? If all species were created from scratch, as creationism claims, why did the creator put different sets of species on oceanic islands from those on continents or continental islands? For example, why did the creator decide to bar amphibians, fish, mammals, and reptiles from oceanic islands but not from continental islands? Furthermore, why were species on oceanic islands created to resemble those on the nearest mainland (continent) or the nearest continental island? Creationism has no real answers for these questions either. As discussed in Chapter 1, this asymmetry of populations can be very well understood in the context of the theory of evolution. The species on the oceanic islands originally arrived from the nearest mainland (continent) or the nearest continental island; and only some selective species could make it from the mainland to the oceanic islands.

There is another distinctive feature of life on oceanic islands called *adaptive radiation* that needs to be explored.

Adaptive radiation. Although oceanic islands are missing many basic groups of animals, the ones that they harbor are living in profusion; in groups of very closely-related species called *adaptive radiations*. An example is a spectacular radiation of 800 species of the fruit fly family *Drosophila*, which makes nearly half the world's total of 2000 species of Drosophila; and they are endemic to the Hawaiian Islands. Another illustrative example from Hawaii is the Silversword alliance, which is a group of over thirty closely-related species of the sunflower family Asteraceae, found only on those islands. These are two of many such examples.

And God Said: Let There Be Evolution

How would you explain these radiations of life endemic to oceanic islands? Why did the Creator put radiations of species on oceanic islands but not on continental islands or continents? Creationism again has no explanation. However, this pattern can be well understood in terms of evolution. What follows is an explanation of adaptive radiation on oceanic islands in the framework of the theory of evolution.

Both on oceanic islands and on continents and continental islands, there are different environmental niches, such as different types of food. The unique thing about an oceanic island is that it was colonized in the distant past by one or a few species usually from a nearby mainland, a continental island or a continent. Different environmental niches on oceanic islands gave rise to different results from those on the mainland because there were more empty habitats and less competition and predators on the oceanic islands. This resulted in clusters of closely-related species that fill a variety of ecological niches, the niches that are often filled by species very different from one another on continental islands and

1. Geospiza magnirostris   2. Geospiza fortis
3. Geospiza parvula   4. Certhidea olivacea

Finches from Galapagos Archipelago

Figure 6.6 Four of the 14 finch species found on the Galápagos Archipelago: evolved by an adaptive radiation that diversified their beak shapes to adapt them to different food sources

continents. This evolutionary diversification of a species or a single ancestral lineage into various closely- or narrowly-related forms with each of them adaptively specialized to a specific environmental niche is called *adaptive radiation*.

As mentioned earlier, oceanic islands provided fertile grounds for adaptive radiation by offering numerous unoccupied niches with minimal competition for resources. As a result, the speciation process went wild producing clusters such as the finches of the Galapagos (Figure 6.6), Hawaiian honeycreepers, members of the sunflower family on the Juan Fernandez Islands and the wood weevils on St. Helena. Such clusters are called adaptive radiation because they can be best explained by a single species colonizing an island and then diversifying to fill available ecological niches.

Table 6.1 summarizes the geographical pattern of life on islands explained by evolution, which we have just discussed.

Table 6.1 Geographical patterns of life explained by evolution.

Life Pattern	Description	Evolutionary Explanation
Dualism of discontinuous distribution of species across continents.	Different continents with similar ecology harbor different species, which look and behave alike.	**Convergent evolution.** Different species living and regenerating in similar habitats experience similar evolutionary or selective pressure to acquire same or similar biological traits.
Unbalanced distribution of species across oceanic islands.	Oceanic islands are missing many kinds from the set of species that are found on continental islands and continents.	The species on oceanic islands originally arrived from the nearest mainland (continental island or continent); and only some selective species could make it from the mainland to the oceanic islands.
Adaptive radiation of species on oceanic islands.	The species on oceanic islands live in profusion, that is, in groups of very closely-related species; this is called *radiation*.	A very few species arrived at oceanic islands. A species that occupies different environmental niches without much competition evolves to give rise to a cluster of closely-related species; this is called *radiation*.

We can draw the following three major conclusions from biogeographical studies:

1. Creationism. The doctrine of perfect creation fails to explain the observed patterns of the distribution of life across continents, continental islands, and oceanic islands.

2. Environment. The environment, in form of wind, change, need, and opportunity, played its role in determining which species migrated to the oceanic islands.

3. Evolution. The patterns in the diversity of life across continents, continental islands, and oceanic islands can be well understood in terms of the theory of evolution combined with the theory of continental drift.

Islands were colonized by selective plants and animals that were able to reach them either out of need or accidently. For example, think of migrating birds, plant seeds carried by the migratory birds or by the wind, insects being blown out over the sea by the wind, floating from a continent by the sea, and so on.

So far in this chapter, we have presented evidence for the theory of evolution that involves looking back into the past from the present. In light of the scientific method, discussed in Chapter 2, a scientific theory on the history and development of life on Earth needs to explain past, present, and future. Just explaining past and present will not make a theory a successful theory; it must be able to look into the future and make testable predictions. Theory of evolution has that ability.

6.6 Experimentation into the Future

As evolution is assumed to occur slowly over generations and ages, it's not easily perceptible to us in terms of a human lifetime. Therefore, biologists use organisms with short lifetimes to conduct their experiments in order to scrutinize the theory of evolution by observing many generations. It is about testing the predictions of the theory, experimentation into the future. With regard to this, we consider three types of investigations or experiments: testing evolution in outdoor labs, testing evolution in indoor labs, and evolution of antibiotic pathogens.

6.6.1 Testing Evolution in the Outdoor Labs

To test the theory of evolution, nature presents the largest lab on the planet, the outdoor lab. After all, this is where evolution took place. Scientists do make use of this lab. For example, take the case of Peter Raymond Grant and Barbara Rosemary Grant, a married couple, both evolutionary biologists, at Princeton University. They are known for their multi-decade study concerning Darwin's finches on an isolated volcanic island named Daphne Major (Figure 6.7), one of the Galápagos Islands (recall from Chapter 1 that this is one of the locations

Figure 6.7 Daphne Major, one of the Galapagos Islands.

explored by Darwin). Since 1970, for four decades, the Grants have spent six months of the year almost each year on the island capturing and tagging finches, taking their blood samples, and releasing them. They also keep record of the environment changes through which the finches go through. This way, by observing and studying more than 20,000 finches, they have conclusively shown that the average beak and body size of these birds change (evolve) in new generations when they are shifted from one climate to another. Not only that, they have also recorded evidence for the beginning of the emergence of new species as a result of the response of the finch populations to these environmental changes.

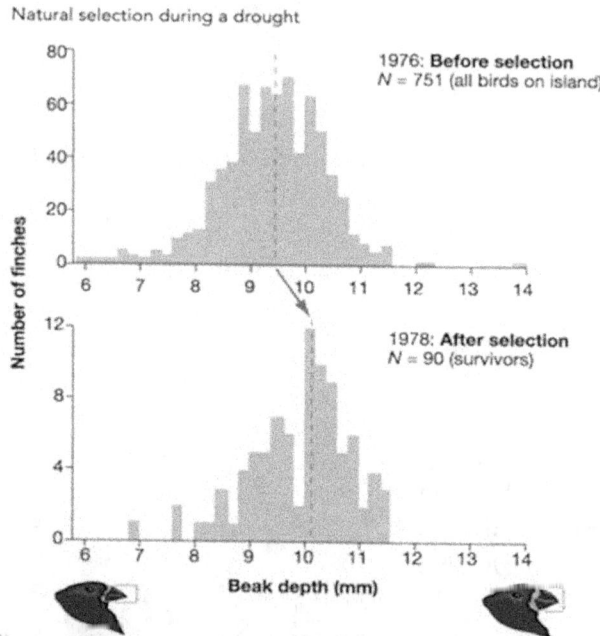

Figure 6.8 Natural selection operating during a drought. Source: Peter Grant and Rosemary Grant.

For instance, Figure 6.8 presents one set of data from their studies of the medium ground finch, which live on seeds. Usually there are different sizes of seeds with different degree of hardness available, with each size suitable for finches of specific beak size (depth). The finches crack open seeds by applying the force of their beak. In 1977, the island had only 2mm of rain instead of 130 mm that it normally had. Due to this drought, the only primary food source left was the seeds of *Tribulus cistoides,* which were hard to open and hence required large, deep beaks. Theory of evolution predicts that in this environment, natural selection will work in favor of the finches that had large (deeper) beaks, and other finches would die of starvation. As a result, the average beak depth of the offspring in the subsequent years is predicted to be greater. This is exactly what the Grants observed, as the drought wiped out 84 percent of the medium ground finch population. Figure 6.8 shows the increase in average beak depth of the survivor population. The average beak depth of the offspring in the subsequent years was observed to be 0.5 mm greater.

So, biologists like the Grants have been using the outdoors such as the Galápagos as a giant laboratory. Usually, evolution in nature (outdoor lab) is a slow process. Indoor laboratories have also been used to test evolution.

6.6.2 Testing Evolution in the Indoor Labs

In order to understand the history of life on Earth, you must realize that evolution occurs within the dynamics of four dimensional space-time (three dimensions of space and one dimension of time), which includes all the water and land and their moving positions along with other geographical changes. One difficulty in grasping evolution is that the timescale over which evolution occurred is much bigger than the human lifespan. However, now, indoor laboratories have been used to test evolution on organisms with short lifespans such as fruit flies and bacteria, which can reproduce themselves relatively frequently. For example, bacteria can typically reproduce in as little a time as half an hour. We discuss two examples here of testing evolution in the indoor lab on bacteria and fruit flies:

And God Said: Let There Be Evolution

Testing Evolution on Bacteria: Antibiotic Resistance. As an example, let's walk through the following simple experiment on bacteria:

1. Consider a population of bacteria composed of two different types: type A and type B. Assume that both types are initially present in equal numbers.

2. Assume that type A bacteria only produce type A offspring and type B only produce type B.

3. The environment of the population is changed by introducing an antibiotic to which type B is resistant but type A is not. The only difference between the two types is that type B has a mutation that enables it to be resistant to antibiotics, and type A does not have this mutation.

4. Theory of evolution predicts that type B is more fit than type A in the new environment, and therefore can adapt better. As a result, type B produces more offspring than type A.

5. Over a number of generations, type B will survive, whereas type A will vanish.

Some kinds of experiments along these lines, proving the predictions of the theory of evolution, are common in a first year biology course these days. In general, from the data collected in the laboratory, scientists can develop models about concepts such as selection and fitness. In context of evolution, *fitness* of a species in a given environment is the probability for that species to survive and reproduce viable offspring in the given environment. All the plants, animals, and bacteria that we find in nature today are the outcome of evolution through various processes, such as natural selection, repeated countless times over a long period of time.

272

Testing Evolution on Fruit Flies: Speciation. As discussed in Chapter 1, the process of splitting one species into two new species is called speciation. Recall that splitting one species into two (speciation) is the core concept of evolution as opposed to simultaneous and independent creation of species permanently, a core concept of creationism. Speciation through natural evolution, however, is usually a slow process that happens over generations. You would agree that it would be hard to get an eye-witness account of a natural speciation event which happened in the distant past. However, by using the organisms with short lifecycles, the speciation aspect of evolution can be tested in a lab.

As a matter of fact, the first steps of speciation have already been produced in the laboratory in several experiments using geographic isolation. For example, in late 1980s, William Rice and G.W. Salt conducted a speciation experiment on fruit flies, *Drosophila melanogaster.* They bred fruit flies over 35 generations by using a maze with different choices of habitats such as light/dark and wet/dry. They placed each generation into the maze, and separated the two groups of flies that came out of two of the eight exits of the maze. The flies of each of the two groups were then set apart to breed with each other only in their group. This happened over 35 generations. Consider what had happened over these 35 generations: The two groups remained reproductively separate because of their strong habitat preferences; they mated only within the areas they preferred, and so did not mate with flies that preferred the other areas. After 35 generations, they mixed the two groups and their offspring. However, Rice and Salt observed that flies from these two groups would not breed with each other even when doing so was the only opportunity offered to them to reproduce. According to the definition of species, this is at least the beginning of the split of one species into two.

273

And God Said: Let There Be Evolution

As another example, Diane Dodd conducted an experiment during late 1980s as well to study the effects of geographic isolation and selection on fruit flies. As illustrated in Figure 6.9, Dodd divided the fruit flies of a single population into two populations and let them live in different cages. This simulated geographic isolation. She put one population on maltose-based food, and the other population on starch-based food. After only eight generations, she mixed the flies from the two populations to examine their mating preferences. Dodd observed that flies

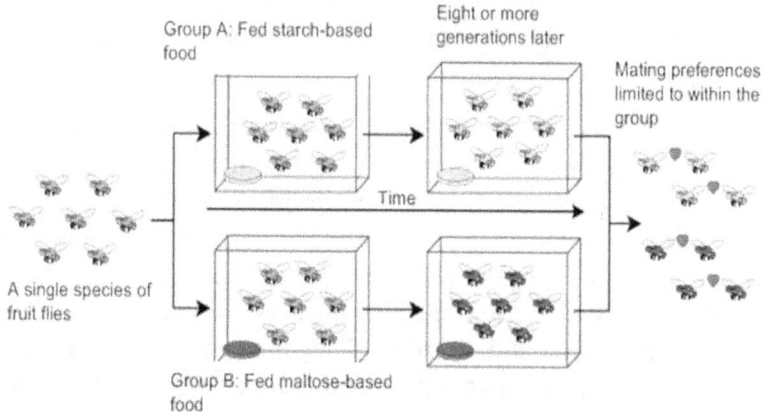

Figure 6.9 An illustration of Diane Dodd's experiment demonstrating speciation in fruit flies, *Drosophila psudoobscura.*

from these two different populations would not mate with each other; they would only mate within their own groups. These results were reproduced by similar experiments conducted on different flies and by using different foods.

These fruit fly experiments suggest that isolating populations to different environments (e.g., with different food sources or habitat environments) can lead to the beginning of reproductive isolation leading to speciation. These results are consistent with the hypothesis that geographic isolation is an important step of some speciation events.

In this section, we discussed how some aspects of the theory of evolution can be tested by running an indoor lab experiment on bacteria developing resistance to antibiotics. As a matter of fact, evolution of antibiotic-resistant pathogens has been a major piece of evidence in support of evolution.

6.6.3 Evolution of Antibiotic-Resistant Pathogens

Antibiotic resistance is the ability developed by microorganisms such as bacteria to survive exposure to an antibiotic such as a drug that kills bacteria.

The first antibiotic was penicillin, discovered in 1929 by Alexander Fleming. In the past 75 years, antibiotics have been critical in the fight against infectious diseases caused by microorganisms (also called microbes), such as bacteria. As an example of widespread use of antibiotic drugs, in 1998, in the United States alone, 80 million prescriptions of antibiotics for human use were filled, which amounts to the sale of 12,500 tons of antibiotic drugs in one year. On the positive side, antimicrobial chemotherapy (killing microbes such as pathogens with chemicals such as antibiotic drugs) was a major cause for the dramatic rise of average life expectancy in the twentieth century. On the negative side, the use of antibiotic drugs has caused the evolution of disease-causing microbes that have become resistant to antibiotic drugs. There is a danger that this may turn the clock back on us; that currently treatable diseases may again become untreatable just like in the days before antibiotics were developed. Already there are a number of diseases that have become hard to treat with antibiotics due to the rise of drug-resistant bacteria. Some examples are childhood ear infections, wound infections, gonorrhea, pneumonia, septicemia and tuberculosis; and the list is growing. At present, about 70 percent of the bacteria that cause infections in hospitals are resistant to at least one of the drugs that are commonly used for treatment. The challenge is to detect antibiotic resistance problems as they emerge, and take actions immediately to at least contain them if not solve them.

The evolution-related question however is: how do these microbes develop the ability to resist antibiotics and become remarkably resilient? The one word answer is: evolution. The development of antibiotic-resistant pathogens is the theory of evolution in action right in front of our eyes. To start with, almost all the pathogens in a patient's body are prone to the antibiotic drug; only 1 in 10^8 (one in a hundred million) has a mutation in their genes to combat the drug. The antibiotic action against the pathogens acts as an environmental pressure. It kills (selects out) most of the pathogens that are prone to the drug. The rare pathogens that have a mutation for survival will live on (are selected in) to reproduce. They then pass their mutation, and therefore the corresponding trait to resist the drug to their offspring, which will eventually result in a fully-resistant colony.

Pathogen. A pathogen is an organism or a virus that causes disease. A virus is not an organism, not even a cell. It's a non-cellular infectious particle that consists of DNA or RNA packed in a protein coat. It can only replicate itself in a host cell.

Another line of evidence for evolution comes from analyzing similarities among different organisms.

6.7 Homologies

Theory of evolution predicts that organisms related through evolution will share similarities that are derived from their common ancestors. Similar characteristics resultant from common ancestry are known as *homologies*. Homologies can be revealed in many different ways such as comparing the anatomies of different living organisms, studying embryological development of organisms, studying vestigial structures within individual organisms, and by examining similarities and differences at cellular level. Remember, by definition, homologous characteristics (or traits) are the characteristics in different organisms that are similar because they were inherited from a common ancestor that also had those characteristics. These different ways of exploring homologies are discussed below in this section.

6.7.1 Comparing Anatomies of Different Living Organisms

There are numerous examples of homologies in anatomies that can serve as evidence for evolution. Let us consider one of them: the forelimbs of tetrapods (vertebrates with limbs). As shown in Figure 6.10, birds, frogs, lizards, and rabbits all have different forelimbs, which support their different lifestyles. However, if you look closely, you will note that those different forelimbs all share the same underlying structure composed of the same set of bones: the humerus, the radius, and the ulna. The prediction of the theory of evolution in this situation would be that there must be some animal species in the past with this underlying structure from which these existing species evolved. As a matter of fact, this prediction has been found to be true, as the fossils of an extinct transitional animal named *Eusthenopteron* has the same set of bones. This demonstrates that *Eusthenopteron* is the common ancestor of birds, frogs, lizards, and rabbits. It had evolved four limbs, and its descendants inherited those features and modified them in order to adapt to their environmental conditions. Therefore, the presence of four limbs is an example of homology, and evidence for evolution.

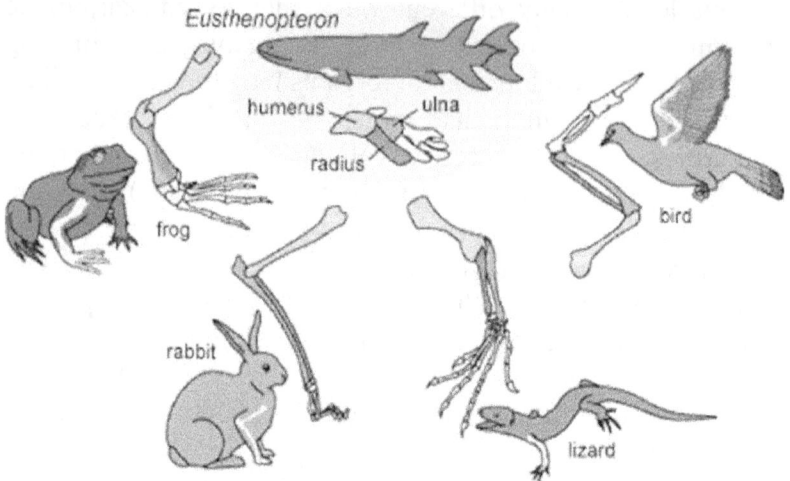

Figure 6.10 Anatomic homologies among the forelimb structures of birds, frogs, lizards, and rabbits pointing to common ancestry.

Another type of evidence for evolution comes from embryology, the study of embryos, unborn (or unhatched) animal young (including human) in their early stages of development.

6.7.2 The Palimpsestic Life of Embryos: Living the History of Evolution

There could be no proof of evolution greater than the fact that all living animals, including us, live their evolutionary history, in some sense, in the beginning of their lives. Literally speaking, a *palimpsest* is a manuscript on which two or more successive texts have been written, the previous one being erased to make room for the next. It's a well-established fact that the embryos of different animals live the life of a palimpsest. For example, studies comparing the early embryos of different vertebrates have revealed remarkable resemblances in the patterns of their early development. All different looking vertebrates in this world begin their life as an embryo that looks like a fish embryo. These look-alike embryos of different species develop into their respective different forms of life as amphibians, birds, fish, mammals, and reptiles. This is an amazing dance of development from unity to diversity that can only be explained with the theory of evolution as shown further on in this section.

To be specific, at some stage in their development, all vertebrate embryos have a tail located behind (posterior to) the anus. Also at some stage in their development, all vertebrate embryos develop structures called throat (pharyngeal) pouches, which subsequently develop into different structures to serve different functions in different species such as gills in fishes, and parts of the ears and throat in human and other mammals. These similarities in development, starting out the same and gradually developing different traits, show that the basic plan for the beginning of these organisms of different species remains the same. It also suggests that these species are related and have common ancestors. The similarities between embryos of a few vertebrate species are illustrated in Figure 6.11.

Here is a fascinating fact: As the early embryos of different vertebrate species develop further, they go through a series of changes that resemble an evolutionary sequence. For example, consider the most striking fish-like feature in the early embryos, which is a series of branchial arches called pouches that exist on each side of the embryo's spot that would later turn into its head. Each of these arches contains tissues that later develop into blood vessels, bones and bars of bones (cartilage), muscles, or nerves. More important: These arches develop into different structures in different species. For example, they almost directly develop into jaw and gill structures in adult sharks and fish. This process includes developing nerves to control the movement of the gills, cartilage to support the gills, and blood vessels.

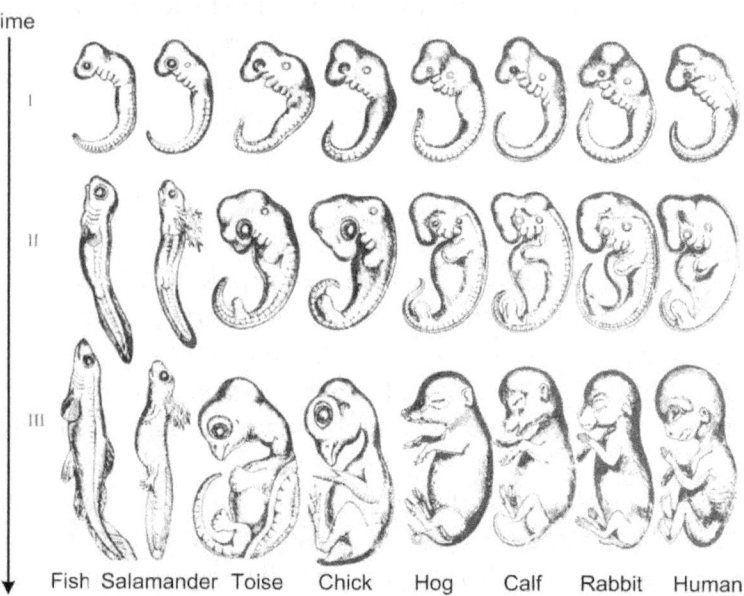

Figure 6.11 Illustrations of the development of different vertebrate embryos. Source: Romanes, G.J. (1892). Darwin and After Darwin, Open Court, Chicago.

As mentioned earlier, shark and fish develop their final product, gills, more or less directly from arches. However, other vertebrate species take twists and turns consistent with the evolutionary tree to build their different final products. For instance, consider the development of blood vessels. In fish and sharks, the vessels develop directly from the original pattern of arches, whereas in other vertebrates the vessels go though changes such as some of them disappear, others move around, and so on. Six pairs of arches is the basic plan that each vertebrate species begins its development with. Fish develop into adult stage by keeping this basic plan pretty much intact. In general, in amphibians embryonic blood vessels turn directly into adult vessels, whereas in human embryonic development, the first, second, and fifth arches disappear by the fourth week. By seventh week, the third, fourth, and sixth arches rearrange themselves and develop into blood vessels (circulatory system), which looks very similar to that of reptiles at this stage. In reptiles, this system directly matures into adulthood, whereas in humans the system continues to change and eventually develops into a mammalian circulatory system equipped with carotid, dorsal, and pulmonary arteries. This happens in an order in which these species evolved on Earth.

A careful reader must have noticed the underlying process of this development: a layer corresponding to a descendant species develops on top of the previous layer belonging to the ancestor species. This is because the descendant has inherited the development program in terms of genome from its ancestor and has made additions to it in terms of gene mutations. In other words, the new species runs its development program on top of the development program of the ancestor species. To a computer programmer this sounds pretty much like object-oriented programming. The features that evolved later in the history of life (appear lower in the inverted evolutionary tree) are programmed to develop later in the embryo. This is very impressive evidence for the theory of evolution and against the doctrine of perfect creation, which states that all species were created from scratch in their perfect form as they are today.

Note. From studies in molecular biology, we now know that the stages of development in which embryos look identical or similar are governed by the DNA molecules they all have in common. An explanation for this from the theory of evolution is that they have inherited these DNA molecules from their common ancestor(s).

Because the new is made on top of the old, imperfect creation is a natural prediction of the theory of evolution. For example, if the theory of evolution is true, we must be able to find some remains of the old in the new. Let's see if we do.

6.7.3 Vestigial Structures: Living Evidence of Imperfect Creation

As opposed to creationism (or perfect creation discussed in Chapter 1), imperfection is the trademark of evolution. This is because the development of an organism or a species is based on and constrained by its entire evolutionary history. In other words, evolution builds a new species from a given ancestral species instead of starting from scratch. Natural selection (a mechanism of evolution) does not result in perfection but only modifications (or improvements in the given environment) over what existed previously. Therefore, theory of evolution predicts the possibility of leftover structures which have no or marginal use in the current species but were useful in the ancestral species. Such structures (or traits) are called *vestigial structures* (or vestigial traits). There are numerous vestigial structures that have actually been identified in different species. For example, the vermiform appendix (Figure 6.12) is a vestige (remnant) of an organ called cecum that would have been fully functional and used to digest cellulose by human's ancestors who fed on plants. Organs analogous to this one in some other animals still continue to perform that function.

Under some conditions, vestigial organs may develop new functions. Research suggests that the appendix may guard against the loss of symbiotic bacteria that aid in digestion. Another example of a vestigial structure is the remnant of a lost tail called coccyx or tailbone located at the end of the spine. Even today, all mammals have a tail during some stage of their development. We humans have a tail for a period of 4 weeks during some stages of human embryogenesis; the tail is most prominent in 31-35 day old human embryos. The original function of the tail in our ancestors was to assist with balance and mobility. The tail has been lost now, but it has not degraded further from its present form, the tailbone, because it still serves some secondary functions such as being an attachment point for muscles.

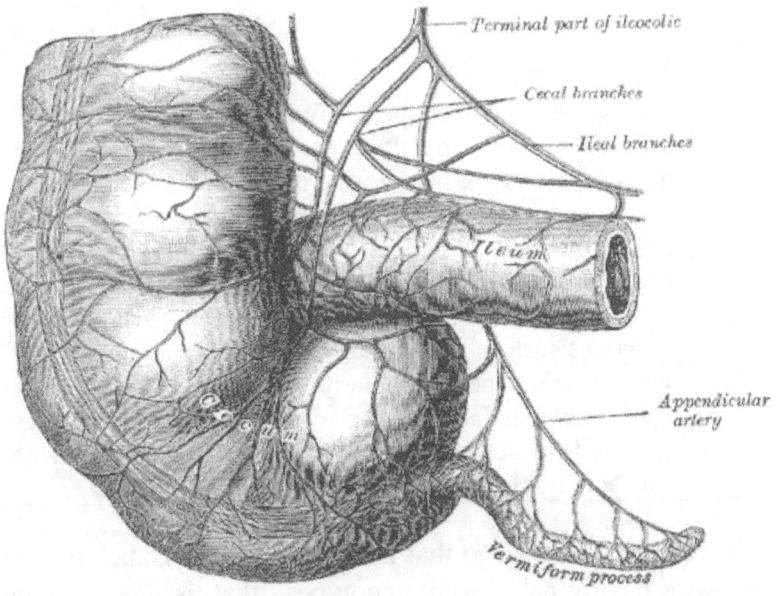

Figure 6.12 Human vermiform appendix, which is a vestigial structure, that is, it no longer retains its original function.

Note. In rare cases, human babies are born with a short tail-like structure; it's a known medical disorder called a congenital defect. Since 1884, at least 23 such cases have been reported in the medical literature.

And God Said: Let There Be Evolution

Think About It!

Many animal species such as ducks, ostrich, and penguins have wings but they do not fly.

A. In the light of the theory of evolution, what can you say about their common ancestors?

B. Does the theory of evolution say that these vestigial wings should have no use?

Answer

A. Ancestors of these species were birds who used the wings to fly.

B. No. It's possible that species find a new use for vestigial traits. They are vestigial not because they do not perform any function at all, but because they are unable to perform the function for which they originally evolved. So, wings are not necessarily useless, they have evolved to perform new functions. For example, the ancestral wings in penguins have evolved to flippers to perform new functions such as enable the penguins to swim underwater with amazing speed. Similarly, the wings of ostrich help the birds maintain balance while running, mate, they spread them out to protect their chicks from the burning sun, and use them to threaten the enemy.

Recall from Chapter 6 that traits (phenotypes), such as tails and wings, originate from genotypes, that is, genes. This implies that when a trait becomes outdated or degraded, the corresponding genes from which the trait originates do not instantly disappear. The genes stay in the genome, but are deactivated. So, corresponding to the vestigial traits, there must be useless or dead genes, say vestigial genes. So a prediction of the evolutionary theory in this case is that genomes of evolved species should contain dead or useless genes that once were active and useful in their ancestral species. Biologists have already found that there is no shortage of such genes. So, this prediction has also been positively tested in favor of the theory of evolution.

Think About It!

If all species were created from scratch, as creationism or the doctrine of perfect creation claims, would you expect to find useless genes?

Answer. No. There would be no need for such useless genes in perfect creation, and there would be no ancestor in which these genes were active and useful.

So, vestigial structures in organisms are an evidence of imperfection, which is a natural consequence of evolution. This imperfection also shows up in the structures that are currently fully functional. So prevalent is imperfection, which has no place in creationism or intelligent design.

6.7.4 Not so Intelligent Design

If a supernatural designer created all species of organisms from scratch, this intelligent designer would design them and their structure perfectly. However, in reality, there is no shortage of examples of imperfectness in the design of an organism and its components. Let's discuss one example here to illustrate this.

The larynx, commonly called the voice box and illustrated in Figure 6.13, is a muscular hollow organ in the neck of mammals, and is involved in sound production and passage of air to the lungs. This organ houses the vocal cords, which are an essential component of phonation. If the technical names of parts of the organ labeled in the figure do not make much sense to you, do not worry about it: The point here is to outline how the design of the left recurrent laryngeal nerve in us, humans, presents evidence that we evolved from fish-like ancestors.

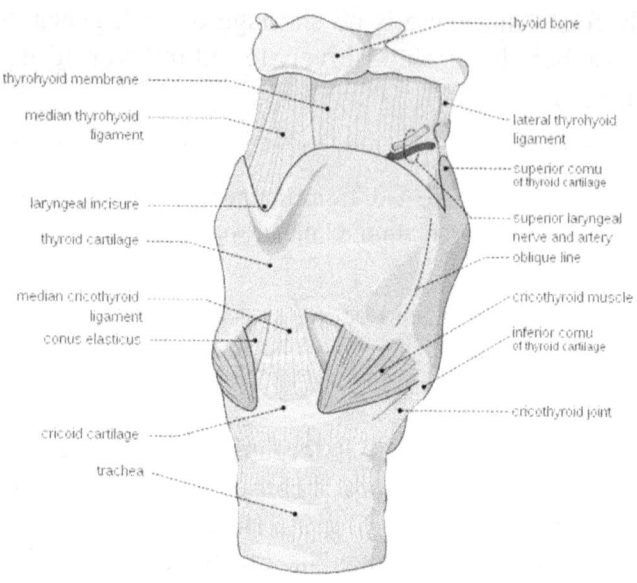

thyrohyoid membrane

median thyrohyoid ligament

laryngeal incisure

thyroid cartilage

median cricothyroid ligament

conus elasticus

cricoid cartilage

trachea

hyoid bone

lateral thyrohyoid ligament

superior cornu of thyroid cartilage

superior laryngeal nerve and artery

oblique line

cricothyroid muscle

inferior cornu of thyroid cartilage

cricothyroid joint

Figure 6.13 Anatomy of the larynx, anterolateral view

The left recurrent nerve (Figure 6.14), the imperfect feature under discussion, which runs from the brain to the larynx and enables us to speak and swallow, is much longer than it needs to be. To perform its functionality, it just needs to run directly from the brain to the larynx. Instead, as shown in Figure 6.14, starting from the brain it goes right past the larynx into the chest, loops around to travel and connect back to the larynx. This back and forth, rather circuitous, route earns it the title *recurrent* that appears in its name. It winds up being about a yard long, where a few inches could have done the job. The similar back and forth circuitous path for this nerve has been found in humans and giraffes. In giraffes (Figure 6.15), it runs about five yards longer than required. Such a design from scratch in a biology class will earn a C grade at best; a biologist with a physics or engineering background would be strongly inclined to hand out an F. However, here is how this F turns into an A: this design is the best we could do if our development as the human species went through evolution starting from a fish-like creature. Recall the six arches in the fish-like early embryos of all vertebrates discussed earlier. In those embryos, this nerve is a branch of the larger vagus nerve that runs from the brain along the back, alongside the sixth branchial arch. The fish embryo turns into an adult fish keeping this nerve in the same position connecting the brain to the gills. It helps the fish to pump water. In other vertebrates, it has evolved along with other components. So this bad design makes sense in the context of evolution.

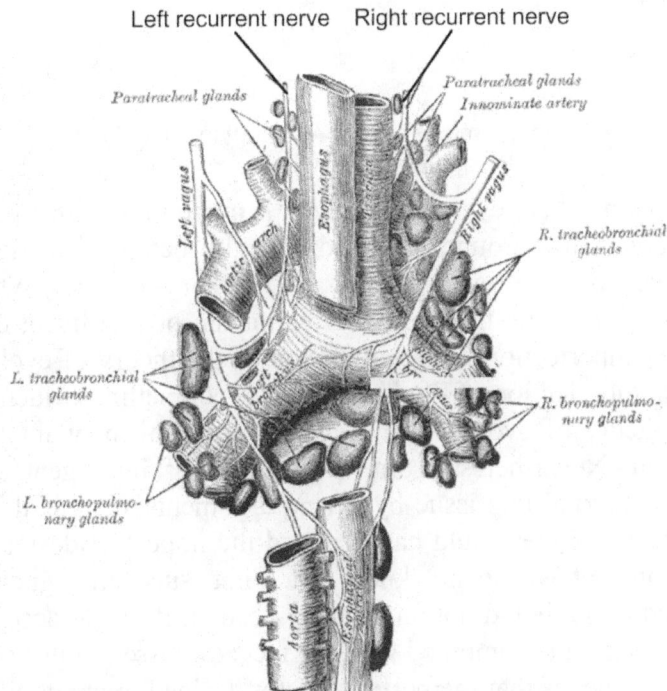

Figure 6.14 The posterior view of the tracheobronchial lymph glands. Recurrent nerves are labeled at the top.

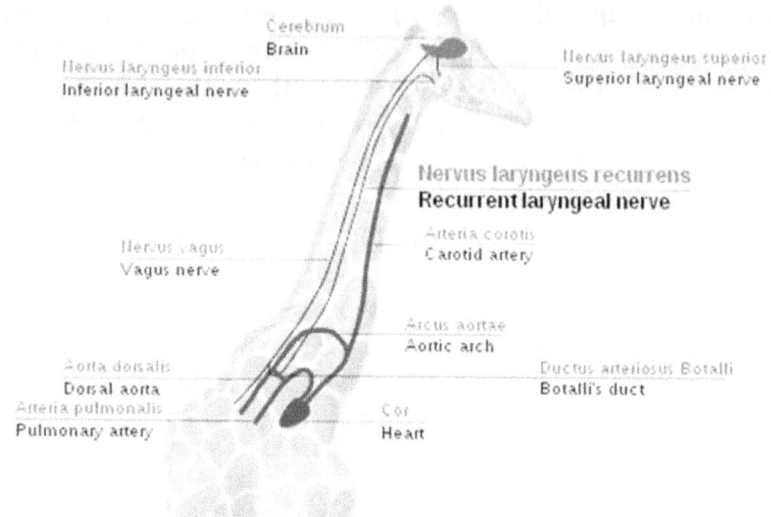

Figure 6.15 Scheme of path of recurrent laryngeal nerve in giraffe camelopardis. Image: courtesy of Dr. Vladimir V. Medeyko.

In a nutshell, even though there is much to admire about the beauty of life around us based on unity behind diversity, the perfect or intelligent design is nowhere to see. Whereas perfection is the hallmark of creationism or intelligent design (ID), imperfection is the prediction of the theory of evolution. This imperfection of life is yet another compelling evidence for the theory of evolution and against creationism or intelligent design. Nevertheless, some proponents of intelligent design dance around this issue by giving arguments such as this: the perfect designer could have created the imperfect design for a reason that we do not know yet. First, such an argument is based on faith and not on science. Second, if we do accept this argument, this combined with the facts discussed in this section simply means that the perfect creator decided to create life that would evolve. This is again an argument for evolution (in a twisted way) and not against it.

In this section (6.7) so far, we have largely discussed similarities of traits among evolutionarily-related species at organism or organ level, that is, macro-level for macro-organisms. Because macro-traits and trait variations originally arise from the microworld of molecules such as genes, theory of evolution predicts that there must be similarities among the evolutionarily-related species at molecular level.

6.7.5 Homologies: Cellular and Molecular Evidence

In spite of their remarkable diversity at macro (organism) level, all living organisms and species are fundamentally alike at the micro-level. These fundamental similarities point to the validity of an important element of the theory of evolution: all species originated from a common ancestor or ancestors. The fundamental similarities can be broken down to two complexity levels: cellular and molecular.

The cellular level. A cell is the smallest living entity. Here is the overall similarity: All organisms of all species are made of cells, which consist of common components such as membranes (filled with water, salt, carbohydrates, and lipids), a nucleus containing genetic material, proteins, and other substances. The cells of most living things use sugar (carbohydrates) for fuel while producing proteins based on the instructions chemically coded in genes. Proteins serve as building blocks for organs and are responsible for all dynamic functions of organisms.

Not to say that the cells of all animals or all plants are essentially identical, but the cells of animals are very similar to the cells of plants. Figure 6.16 shows the similarity between a typical animal and plant cell. Note that only three structures are unique to one or the other.

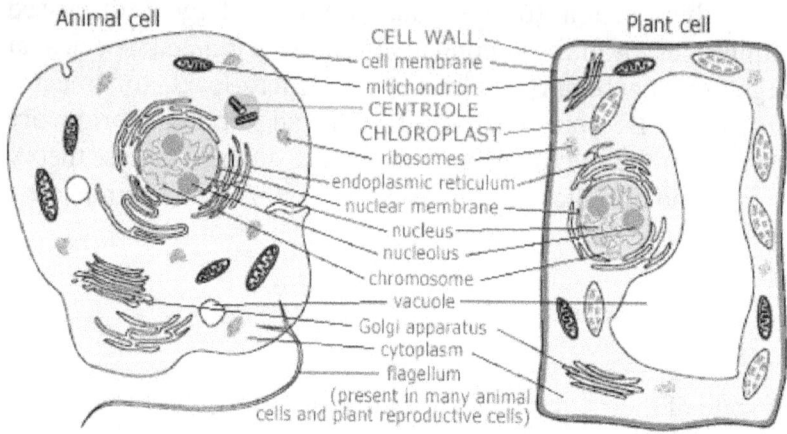

Figure 6.16 Compelling similarities between the typical cells of animals and plants.

Think About It!
From Figure 6.16, identify the three structures unique to animal cells and plant cells.
Solution
Animal cell: centriole
Plant cell: cell wall and chloroplast.

The cellular machinery deals with molecules.

The molecular level. Inside organisms, all inherited similarities and variations originally arise from genes. So, how do the genes of different species compare? In the case of all organisms, DNA possesses a simple four-base code that provides the recipe for developing the organism. The rules of this code are identical for all organisms. This way, the DNA code in itself is a homology (similarity) that links all life on Earth to a common ancestor (or ancestors). This similarity flows to the level of species and shows up at even the most unlikely places. For example, species as dissimilar as humans and roundworms share 25 percent of their genes. Similarly, humans and bacteria share genes inherited from a common ancestor in the very distant past. Even though these genes are slightly different in each species and, like the four limbs of tetrapods (such as human and whale), have often acquired different functions over time, their striking similarities nevertheless point to the common ancestry of species.

Note. The genetic code of life is essentially universal as all forms of life use the same genetic language of DNA. This strongly suggests that all species descended from a common ancestor (or ancestors) that used this code.

These genetic characteristics of life demonstrate the fundamental sameness of all living things on Earth. This sameness at molecular level is so fundamental that if we transfer genetic material from the cell of one living organism to the cell of another, the recipient cell will follow the instructions of the foreign genetic material as if it was its own. This sameness serves as the basis of all new fields such as biotechnology, genetic engineering, and regenerative medicine. This is very compelling evidence for evolution.

Some essential genes are still (at present) shared by all species. An example is the gene that encodes the protein called *cytochrome*, a crucial component of the electron transfer chains in species ranging from aerobic bacteria to humans. It's also found in plants. The electron transport chain is a part of a process called cellular respiration, which transforms consumed food to cellular energy molecules called ATP.

The cytochrome c (or cyt c) protein molecule has been studied and the results of these studies provide the incredible glimpse into evolution. Here are some of the results: It has been found that in both chickens and turkeys, this molecule has an identical sequence of amino acids (identical in the identity of amino acids and the order in which they are sequenced), which differs from that in ducks by just one amino acid. Similarly, both humans and chimpanzees have identical molecules, which differ from rhesus monkeys only by one amino acid. Also cows, pigs, and sheep share identical cytochrome c molecules.

Think About This!

Humans and chimpanzees have an identical sequence of 104 amino acids in their cytochrome protein. This sequence in humans and chimpanzees differs from that in rhesus monkeys, chickens, turtles, and yeasts by 1, 18, 19, and 56 amino acids respectively. Based on this biochemical information, can you arrange all these species in the increasing order of their relatedness to humans?

Answer

Yeast, turtle, chicken, rhesus monkey, chimpanzee.

Consider the following pattern of factual or experimentally-confirmed similarities:

1. Organisms and their traits (such as organs) originate from genes.

2. Some micro (or molecular) characteristics such as genetic code are shared by all species. This fact combined with fact 1 above implies that shared molecular characteristics date to very distant ancestral past.

3. In contrast to molecular characteristics, some organ-level or organism-level characteristics are shared only within smaller groups of organisms. This combined with facts 1 and 2 implies that these shared macro-characteristics developed more recently.

These three facts combined means that observed homologous characteristics form a nested pattern, that is, all forms of life share the deepest layer of characteristics, and each successive smaller group of organisms (species) add its own homologies to those it shares with the larger group. This pattern is called tree of life, the heart of the theory of evolution.

It is important to realize that there is another kind of similarity called an analogy, which is not the same as a homology.

6.8 Evolution of Analogies

Analogous traits are also similar traits but they are often contrasted to homologous traits due to their origin. Analogous similarities are the similarities between organisms that were not in their last common ancestor but rather evolved along separate lineages.

Convergent evolution. The evolution of similar characteristics among species in different lineages, called analogies instead of homologies, is called convergent evolution.

Similar environments can exist in different parts of the world. In the face of this fact, theory of evolution predicts that we must be able to find some similar characteristics among organisms from different species that do not share a common ancestor in the recent past, and therefore belong to different lineages of evolution. These similarities would have arisen from adaptations to similar environments. Such similarities are called analogies, and characteristics (traits) with these similarities are called *analogous characteristics*. This kind of evolution is called *convergent evolution*. Many examples of analogous characteristics have been discovered. For instance, the wing surfaces of bats and insects are analogous structures.

> **Comparative analogy.** Sometimes the evidence from homological, analogical, and vestigial structures is presented under the title of *comparative anatomy*, which is the study of comparing the anatomy of different organisms and species.

Homologous traits share common ancestry but not necessarily similar functions, whereas analogous traits share similar functions but not common ancestry.

In Section 6.8, we discussed how the nested pattern of homologous characteristics among groups of organisms points to the evolutionary tree of life. Homologies are one of many criteria that are used to construct the tree of life.

6.9 Evolutionary Tree of Life: History of Life

Fossil studies supported by techniques such as radioactive dating has helped scientists to reconstruct the history of life on Earth. Data leads to the evolutionary tree of life. Evolution emerges as a unified thread of life from the evolutionary tree of life that represents the history of life on Earth.

An evolutionary tree is a hierarchical diagram which summarizes our data-supported understanding of the pattern of evolution of a group of species. Each line in the diagram represents one lineage, and each branch point (also called node) represents a common ancestor. A cladogram is an evolutionary tree that shows whom is most closely related to whom. The collective tree that depicts the evolution of all life forms on Earth is called the evolutionary tree of life.

As an example, Figure 6.17 presents an evolutionary tree for tetrapods and their closest evolutionary relatives, lungfishes. This tree is based on anatomical and DNA sequence data. The tree shows the emergence of three significant homologies on the time scale: tetrapod limbs, amnion (a protective embryonic membrane), and feathers. Each of these homologies only evolved once. The tree depicts that tetrapod limbs appeared in common ancestor 3, and therefore are shared among the group of all the descendant species of that ancestor called tetrapods. Feathers appeared in common ancestor 6 and therefore are found only in birds, the descendants of common ancestor 6.

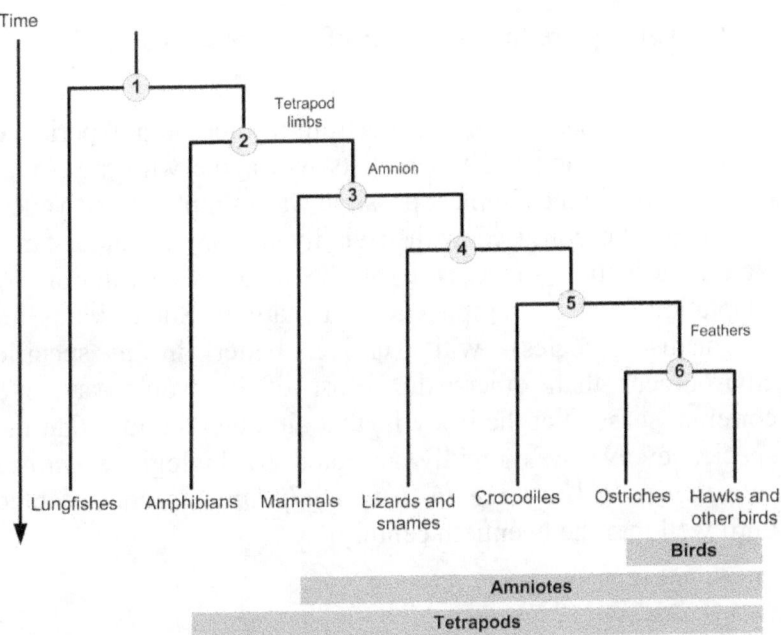

Figure 6.17 An evolutionary tree for tetrapods and their closest evolutionary relatives, lungfishes.

And God Said: Let There Be Evolution

> **Think About It!**
> Assume that based on fossil evidence a case can be made that some species such as gingko trees, garfish, and horseshoe crabs have barely changed over the last few millions of years. Would this observation be inconsistent with the theory of evolution?
>
> **Answer**
> No, not if we understand what the theory of evolution is not. The theory of evolution neither says all species evolve at the same rate nor does it say that a given species will constantly evolve. The rate of evolution for a species depends on the evolutionary pressure the species experiences, which is determined by factors such as genetic mutations and environmental changes.

We will explore the history of life in more detail in Chapter 7.

One thing that I have learned from my research experience in science is – and many scientists will agree with me – that nature works in the simplest possible way: simplicity, however, should not be confused with triviality or easy to understand. Natural selection – the core of the theory of evolution by adaptation – is a simple law of nature: Some kinds of organisms (species) will survive better in a specific environment than others do. Most of us would say: it's common sense. Yet the irony is, that although the idea that the species evolve was rapidly accepted by biologists, *natural selection as a key force to drive evolution* was not accepted until well into the twentieth century.

6.10 Summary

Here is the basic idea at the heart of the theory of evolution: Life on our planet has existed for billions of years and has changed over time. The history of living things is documented through multiple lines of evidence including fossil records, biogeography, comparative embryology, homologies and analogies, experiments in indoor and outdoor labs, vestigial structures, and field studies such as the evolution of antibiotic-resistant pathogens. These lines of evidence cover life from micro (molecular and cellular) level to macro (organism and species) level. All these lines of evidence converge to reveal a common thread running through the 3.5 billion-year-long story of life that unfolded on our planet. The story or the history of life is written by dead organisms on rocks in the language of fossils. This history is also written in all living organisms, which even today develop from the same four bases (A, C, T, and G) of DNA and from a single cell. This history unfolds itself during the development of embryos of different species, as they go through a series of changes that resemble an evolutionary sequence of the species.

All lines of evidence converge to reveal this: Starting from the single-celled ancestors, lifeforms although made from the same basic stuff, have gradually diversified in a branching pattern throughout billions of years. As a result, living forms of today are in some ways very different from the lifeforms of the past, and yet very similar in other ways. Biological evolution accounts for the pattern of similarity and differences among the grand diversity of life through its core concept: descent with modification from common ancestors.

Evolution connects the phenomena involved in the growth of an organism from alleles to gametes, to cells, and to organisms; the phenomena on the size scale from unicellular organisms to multi-cellular organisms of different sizes; and phenomena across different species. In short, the tremendous diversity among living organisms superimposed on the underlying unity is explained by the theory of evolution. Overwhelming evidence supports this fact.

In addition to the simultaneous existence of different local environments, it is not that difficult to understand that environment on Earth has changed severely and multiple times on a global scale. Life has survived through these changes due to trait variants, with adaptable traits and corresponding genes passing through the filter of natural selection. If there were no trait variants rooted in genes, and therefore no evolutionary process, life could have come to an end on this planet a long time ago.

In a nutshell

✓ In spite of their remarkable diversity at macro-level, all living organisms and species are fundamentally alike at the micro-level. For example, all organisms of all species are made of cells and the blueprints for the development of their bodies are written in their DNA, the same genetic code.

✓ Descent with modification from common ancestors, that is the theory of evolution, explains the unity behind diversity of life.

✓ The fossil record shows that species have evolved over a long period of time and that some species have become extinct.

✓ Homology shows us that organisms share some characteristics due to their common ancestry, an important component of evolution.

✓ Organisms can also share some characteristics even if they don't have a common ancestor in the recent past but because they evolved independently in similar environments. Such an evolution is called convergent evolution.

✓ As the early embryos of different vertebrate species develop further, they go through a series of changes that resemble an evolutionary sequence of species. This is strong evidence for evolution and against perfect creation.

✓ Geographic distribution of species of organisms is consistent with the predictions of the theory of evolution.

✓ Evolution is a 3.5 billion-year-long common thread running through all forms of life on our planet.

6.11 Behold

Multiple lines of evidence converge to make a joint statement: Evolution is a 3.5 billion-year-long common thread that runs through all levels of life from cells to organisms and species. If this is true, it would be impossible to describe the history of life on our planet without using the concept of evolution. In other words, the theory of evolution should be able to explain the history of life on Earth. We explore this topic in the next chapter.

And God Said: Let There Be Evolution

SEVEN

A Brief History of Life on Earth

And God Said: Let There Be Evolution

History may not repeat itself, but it does rhyme a lot.

Mark Twain

7.1 Once Upon a Time

Once upon a time, a four-legged creature, probably a homo-erectus ape, on Earth stood up on its two hind legs, looked around, looked up, and gazed at the sky wondering: *where did it all come from*? Of course it did not happen as suddenly and literally as the statement may suggest, but this is one way, rather philosophically, to describe our entry into the party of life on Earth. As curious and ambitious as we might be, we are the late comers to this party. A geologic time clock, presented in Figure 7.1, represents in brief the chronology of life on Earth.

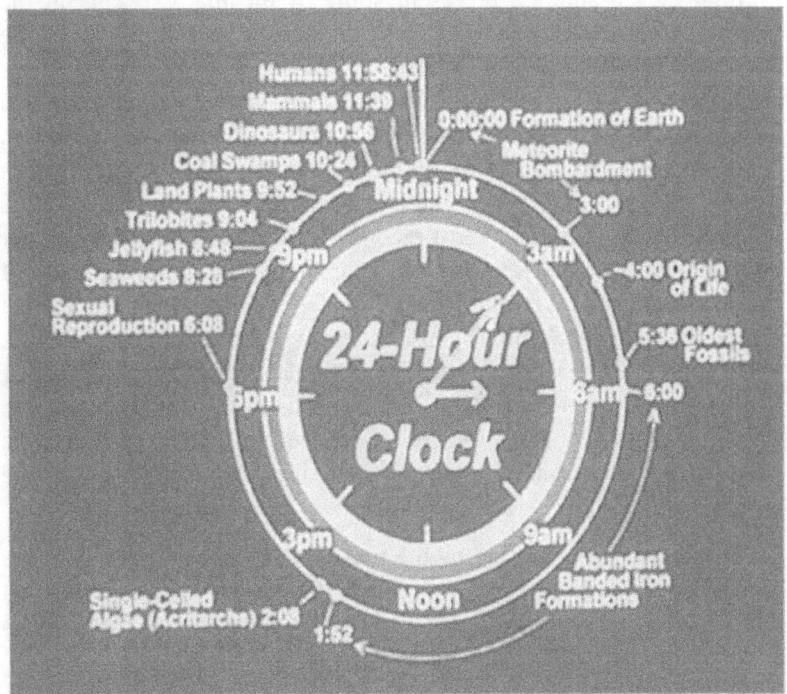

Figure 7.1 24-hour clock representing the relative timing of some of the events during the 3.5 billion-year-long evolution of life on Earth.

And God Said: Let There Be Evolution

As this clock shows, we humans, the macrobeings, as opposed to microbes (bacteria), the microbeings, evolved to a relatively advanced level, and are for worse or better late comers to the party of life on Earth. On the timescale of this 24-hour clock, if the party started at 0 hour – that is midnight – with the solidification of the Earth's crust, and if it is about midnight again, we have just arrived at 11:58:43 pm, less than two seconds ago. However, now that we have arrived, we are the most curious and ambitious creatures at the party. We want to know everything about everyone and everything at the party: Who (or what) are you, what are you made of, how did you get here, where are you from, no I mean where are you originally from, and where are you headed, that is, what is your future?

Figures 7.1 and 7.2 represent some of the major events in the history of Earth and the evolution of life on it. The chronology of life presented by these clocks is based on evidence from fossil records combined with other techniques such as radioactive dating discussed in Chapter 6. The largest defined unit of time is the supereon, composed of eons, which are divided into eras. The eras are in turn divided into periods, epochs and ages. The *Hadean eon* in Figure 7.2 refers to the time before fossil record of life on Earth with an upper boundary of 4.0 Ga [Giga (billion) years ago]. Based on the study of fossils, scientists have established a geologic record of the history of Earth divided into three eons reflecting the evolution of life: Phanerozoic, Proterozoic, and Archaean. The Palaeozoic, Mesozoic, and Cenozoic (shown in Figure 7.2) are eras of the Phanerozoic eon. The 2 million-year-long Quaternary period (part of the Cenozoic era), when recognizable humans appeared, is too small to be visible at this scale.

These clocks, supported by the fossil record and other studies, are consistent with the underlying idea of the theory of evolution that different species share common ancestors, which implies that life has a history, a history of change.

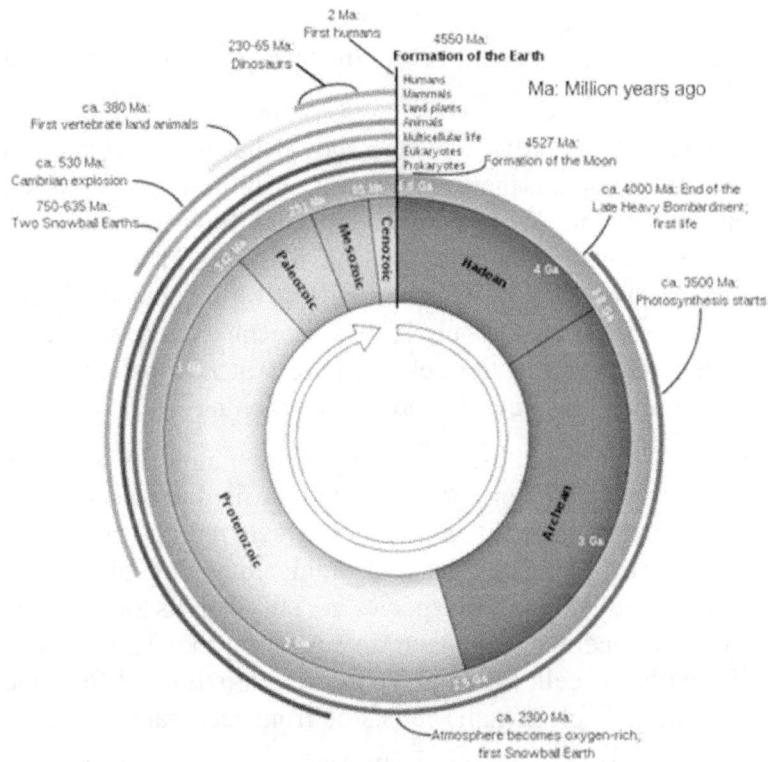

Figure 7.2 The clock representing the relative timing of some of the events during the 3.5 billion-year-long evolution of life on Earth.

7.2 In the Beginning

In Chapter 1, when I said 'our story begins in 1831,' I was referring to the story of discovering and understanding evolution and the scientific theory of evolution. Evolution, that is, biological evolution itself, has been operating since the beginning of life on our planet. In this chapter, we present a very brief history of life on Earth and we explore how this history has been determined by physical laws of nature and by evolution. We accomplish this task by venturing through three main avenues: origin of the Universe and life, origin of species of organisms, and evolution of life through given ecological conditions.

A body of scientific evidence shows that life originated in water and evolved there for about 3 billion years before extending to land. We can see even today how life is closely tied to water. For example, we can live without food much longer than we can live without water. All life is made of cells and most cells are about 70-95 percent water, as many chemical reactions in the cells require water. Most cells are also surrounded by water for proper functioning.

In the beginning, life was simple, as simple as a single living cell. Starting from primitive cells (called *protocells*) and unicellular organisms, life subsequently evolved into about 100 billion species of unicellular and multicellular organisms, out of which about 100 million species are still with us today. Even today, a multicellular organism, in sexual reproduction, begins its life with one cell, called a zygote, which is formed from the fusion of two haploid (half) cells, one from each parent.

A cell itself is matter, and all matter is composed of about 100 types of elements or atoms. An element is a pure substance, such as iron, hydrogen, and oxygen, that cannot be decomposed into other pure substances by chemical methods. An atom is the smallest particle of an element that can possibly combine with other atoms of the same or a different element to make substances. For example, a water molecule is made of two hydrogen atoms and one oxygen atom. At a very fundamental level, all living organisms, including humans, are made of atoms selected from the same set of about 100 elements (or atoms) from which any other material thing in the Universe is made of. As an example, Table 7.1 presents a list of major elements which form the human body. Note that hydrogen, oxygen, carbon, and nitrogen make up about 96 percent of your body by weight. The rest of the elements listed in the table make about 4 percent of your body by weight. Less than 0.01 percent of your body is made of elements called trace elements, such as copper, iodine, iron, silicon, tin, and zinc. This atomic composition is more or less true for all living organisms. It has been discovered that little more than 30 elements (including those in Table 7.1) perform key functions in helping plants and animals to live a healthy life.

Table 7.1 Major Elements of the Human Body

Element	**Percentage of Body by Weight**	**Some Functions**
Oxygen	65.0	Major contributor to molecules in the body, and as a gas is required by cellular processes such as respiration that turns consumed food into cellular energy.
Carbon	18.5	Essential element of all organic molecules in the body including the molecules of life: proteins, lipids, carbohydrates, and nucleic acids (DNA and RNA).
Hydrogen	9.5	A major component of all molecules of life, and also plays crucial role in its ionic form for biochemical reactions to occur.
Nitrogen	3.3	Essential component in nucleic acids, amino acids, and proteins.
Calcium	1.5	Critical role in bone and teeth formation, blood clotting to stop bleeding, various muscle and nerve functions.
Phosphorus	1.0	Essential component of nucleic acids, bones, and teeth.
Potassium	0.4	Critical for the functioning of cells, conduction of impulses in the nerves, acid-base balance, and muscle contraction.

Sulfur	0.3	Important component of certain proteins.
Sodium	0.2	Critical for the functioning of cells, conduction of impulses in the nerves, muscle contraction, acid-base balance, and water balance.
Chlorine	0.2	Supports the cell environment (most abundant ion outside the cell), acid-base balance, digestion in the stomach, and nerve function.
Magnesium	0.1	Found in bones, plays an assisting role in many biochemical (metabolic) reactions in the body.

Note

Trace elements get their name from the fact that they are needed by the organism only in very small quantities.

Organic substances are the substances whose molecules consist of carbon and possibly some other atoms such as hydrogen, oxygen, and nitrogen. Carbon is the essential element of an organic substance.

Not only are the molecules of life (carbohydrates, proteins, lipids, and nucleic acids) fundamentally composed of atoms, atoms also directly play significant role in living organisms. So, although this is not a book on physics, but to cover the context at hand, let us ask a courageous question: where did atoms come from? They certainly came before life.

7.3 First There was Matter: Back to Big Bang

Living organisms are material entities and they need matter in terms of food and environment to exist, maintain themselves, and reproduce. Earth and its environment provide that matter.

308

Earth is a tiny part of the universe we live in. Matter that makes the Universe is fundamentally made of atoms. So the question of where atoms came from translates into the question: Where did the Universe come from?

From the beginning of modern science in the sixteenth century, scientists have been working on figuring out the origin of the Universe. Through observations and experiments, physicists have developed a theory called the *big bang theory*, also called the *standard model*, in order to understand the origin and evolution of the Universe. According to this theory, in the beginning, matter was much more concentrated than today, extremely concentrated, and it was blown apart in an immense explosion referred to as the big bang. Explosion, as we know from our experience on Earth, includes the rapid expansion of matter. However, the rapid expansion in the big bang was not like familiar explosions on Earth starting from a single center of matter and spreading out. In the big bang, the explosion (expansion) occurred everywhere, every particle running away from every other particle. If this is how it started, it obviously does not make much sense to think that Earth (or even our galaxy) is at the center of the Universe. In this model, regardless of where in the Universe we were (center or not), all other galaxies would seem to be moving away from us and the Universe in general would look the same. In other words, the Universe looks about the same from all points in space and in all directions. This is called the *cosmological principle*, which is well supported by experimental evidence.

> **Caution!**
> The dictionary meaning of *evolution* has a broader scope than biological evolution. In this book, however, we use the term *evolution* to refer to the evolution of life, biological evolution, unless stated otherwise.

And God Said: Let There Be Evolution

Oh, did I say every particle was running away from every other particle? Yes, there were no chunks of matter, only fundamental particles in the beginning because it was so hot that even an atom would not be able to hold itself together. For example, at very early times, at about one hundredth of a second, the temperature of the Universe was about a hundred thousand million degrees Celsius (10^{11} °C), so hot that even the nuclei of atoms could not be held together. So, the Universe was a mixture of only a few species of particles, called fundamental particles: six quarks, six leptons such as electrons, and a few particles to facilitate interactions (forces) between these particles, such as photons to facilitate electromagnetic interaction between charged particles and gluons to facilitate strong interaction between quarks. At that high temperature, all the four natural forces (strong nuclear, weak nuclear, electromagnetic, and gravitational) were unified together into a single force. It means that all these four forces would behave the same way. From its initial extremely hot and dense state, the Universe continued to expand and cool down, the four natural forces progressively got uncoupled at certain temperatures, and each of them acquired its distinguished behavior. During this whole process, a few types of fundamental particles formed countless types of material and objects. To start with, quarks bonded through the strong nuclear force to make nucleons, that is, protons and neutrons; electrons bonded to nuclei through electromagnetic force to make atoms such as hydrogen and helium; atoms bonded together through different forms of electromagnetic force to make molecules such as water molecules and crystals such as sodium chloride (table salt) and diamonds. Simple molecules combined through forms of electromagnetic force to make macromolecules such as DNA molecules, protein molecules, and so on. It took about a few hundred thousand years for the

Universe to cool enough to allow the formation of atoms such as hydrogen and helium. The nucleus of an atom has a positive charge due to the positive charge of protons, and each electron around the nucleus has a negative charge.

The charges of the nucleus and electrons around it cancel out, and therefore an atom overall is neutral. However, if atoms get too close to each other, they can experience each other's charge and interact electromagnetically to make molecules. Gravity (gravitational force) pulled the neutral atoms together to form clouds of gas, which eventually turned into stars and galaxies. Scientific evidence supports that one of these stars, the Sun, formed about 5 billion years ago, and that about 400 million years after its formation, the planets of the solar system including Earth formed. A little bit more detailed timeline is shown in Table 7.2.

Table 7.2 A brief history of the Universe. Size is given as the diameter of the spherical universe.

Time Window from the Beginning	Age	Events
$0 \rightarrow 10^{-43}$ s	Age of Cosmological singularity. Many unknowns in this time window; scientific work still in progress.	Size of the Universe: 10^{-52} m. Temperature: $> 10^{30}$ K. All four natural forces (strong, weak, electromagnetic, and gravity) unified into one single force.
10^{-43} s $\rightarrow 10^{-35}$s	Age of particles called quarks and gluons.	Rapid expansion (big bang). Gravity decouples as a separate force. Size of the Universe: 10^{-30} m. Temperature: 10^{28} K

And God Said: Let There Be Evolution

10^{-35} s $\rightarrow 10^{-13}$s	Age of fundamental particles: Quarks and leptons.	Strong nuclear force decouples as a separate force. Fundamental particles (quarks and leptons) have formed. Size of the Universe: 0.1 m. Temperature: 10^{16} K
10^{-13} s $\rightarrow 10^{-3}$ s	Age of nucleons: neutrons and protons. By the end: soup of protons, neutrons, electrons, neutrinos, and photons.	Quarks bind together to form nucleons (neutrons and protons). Electromagnetic and weak nuclear forces appear as separate forces. Size of the Universe: 10^3 m. Temperature: 10^{11} K
10^{-3} s $\rightarrow$ 3 min	Age of deuterons (nuclei made of a proton and a neutron)	Beginning of the formation of nuclei. Size of the Universe: 10^{10}m. Temperature: 10^9 K
3 min $\rightarrow$ 300,000 y	Age of nucleosynthesis and ions. By the end: photons, electrons, protons, helium nuclei.	Light atomic nuclei such as helium form. Atoms did not form because the intense electromagnetic radiation ionized them as soon as they formed. Size of the Universe: 10^{21}m. Temperature: 10^4 K.
300,000 y $\rightarrow$ 30 million y	Age of atoms	Neutral atoms began to form. Release of cosmic microwave background radiation (CMBR) and beginning of the dominance of matter over radiation in the Universe. Size of the Universe: 10^{23} m Temperature: 1000 K
30 million y – 13.7 billion y (present)	Age of galaxies, stars, planets	Material particulates such as atoms began to clump together to form molecules, gas, gas clouds, and galaxies. Size of the Universe: 10^{27} m Temperature: 3 K

> **Note.**
> Remember that Kelvin (K) is a unit of temperature which is about 273 degrees higher than Celsius (C). For example, 25° C is equivalent to 298° K, obtained by adding 273 to 25.

According to Einstein's famous equation, $E=mc^2$, energy denoted by E and matter with mass m are equivalent and inter-convertible: You can create equivalent amount of energy from matter and vice versa. Throughout the history of the Universe, there has been a great display of interplay between electromagnetic radiation (light or photons) and matter. Today the Universe is dominated by matter, and only a very small fraction of it is in form of radiation. To the contrary, the early Universe, until it was about half a million years old, was dominated by the electromagnetic radiation (light). This means that most of the energy of the Universe existed in form of photons (light is composed of massless particles, the packets of energy, called photons) which were continually being absorbed and emitted by ions. When the temperature of the Universe lowered down to about 3000 K, protons were able to combine with electrons to form neutral hydrogen atoms. Photons (electromagnetic radiation) facilitate electromagnetic interaction between charged particles. The formation of neutral atoms created the neutral (charge free) environment that freed the electromagnetic radiation, which then spread throughout the Universe. This freedom (release) of the radiation led the Universe to become matter dominated, that is, more energy in form of matter than in form of radiation. Here is an important point to note: Because at 3000 K photons became free to spread throughout the Universe, that radiation should persist for ever. This is what is called cosmic microwave background radiation (CMBR), and we can test the big bang theory by attempting to search for this radiation. The CMBR has actually been detected and it provides astronomers with a time window to look into the early Universe.

Caution!

Remember that only a very small part of the electromagnetic spectrum is visible, that is, most of the light or electromagnetic radiation is not visible; it either has lower wavelength (ultraviolet light) or higher wavelength (infrared light) than the wavelength of the visible light.

Like any other scientific theory, although the big bang theory is still under scientific scrutiny, its main ideas have been well tested. Here are some examples:

History written on the light. If you go out and gaze at the sky, you are actually looking at how the sky looked in the past, and not how it looks right now. This is because of how we see things. We see things through the light that comes from those things and reaches our eyes. Light takes time to travel. For example, the star closest to us is our Sun, and it takes light about eight minutes to reach us from the Sun. We say that Sun is eight light minutes away from us. This means that if we see the Sun exploding, it actually happened eight minutes ago. Similarly if you are looking at a star that is one million light years away from us, the star you are seeing is really how the star looked a million years ago, not how it looks right now. With powerful telescopes such as the Hubble space telescope, you can see at star and galaxies as they were millions or even billions years ago. This way, light acts as a telescope into the past. We learn about the history of the universe by studying the light that was emitted by objects in the Universe in the past. Data collected so far supports the big bang theory. Cosmic microwave background radiation (CMBR) is an excellent example.

Cosmic microwave background radiation (CMBR). One of the greatest powers of the big bang theory is its prediction of an almost perfect blackbody radiation spectrum in the cosmic microwave background radiation as a result of freed electromagnetic radiation at about 3000 K, as mentioned earlier. In other words, the big bang theory predicts that there must be leftover radiation from the early stage in the evolution of the Universe, called cosmic microwave background radiation (CMBR). By now, according to the big bang theory, the CMBR has cooled to about 2.725 K. One of the greatest successes of the big bang theory is that this prediction is supported by experimental results. Also according to the big bang theory combined with the quantum theory, this radiation, if detected, should have a specific intensity-wavelength curve calculated for the temperature of 2.725 K because the intensity and frequency (and hence wavelength) of blackbody radiation should depend on temperature. Figure 7.3 presents the comparison of the CMBR data for the intensity of radiation (along y-axis) as a function of its wavelength (along x-axis) collected by the *Far-Infrared Absolute Spectrophotometer* (*FIRAS*) instrument on the NASA Cosmic Background Explorer (COBE) satellite, which has carefully measured the spectrum of the cosmic microwave background. The solid line represents the theoretical prediction of the big bang theory calculated according to the quantum theory of radiation called Planck's Law. The 34 data points equally spaced along the curve, match the curve so exactly with error uncertainties less than the width of the theoretical curve, that it is impossible to distinguish the data from the theoretical curve.

And God Said: Let There Be Evolution

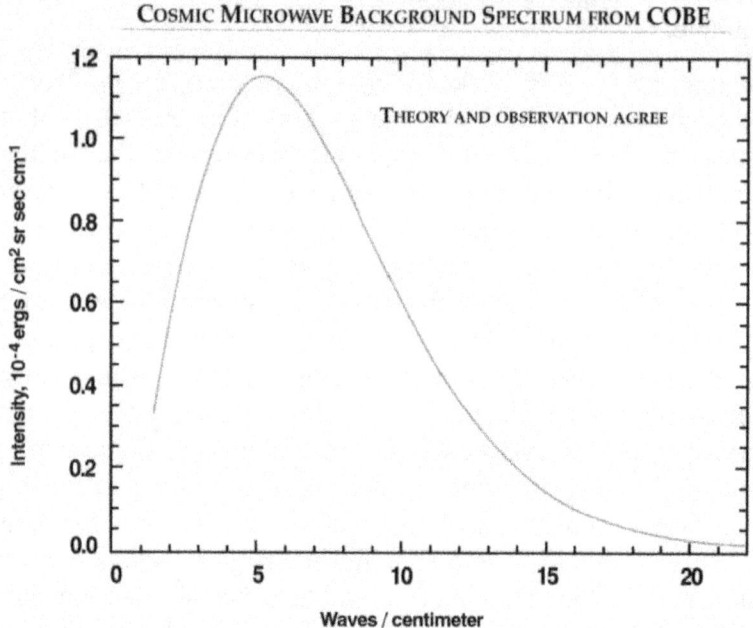

COSMIC MICROWAVE BACKGROUND SPECTRUM FROM COBE

THEORY AND OBSERVATION AGREE

Intensity, 10^{-4} ergs / cm² sr sec cm⁻¹

Waves / centimeter

Figure 7.3 The cosmic microwave background spectrum measured by the FIRAS instrument on the COBE satellite. Image: courtesy of NASA.

> **Blackbody radiation.** This is the technical term that refers to the light (or radiation) emitted by hot objects. It turns out that the intensity and frequency/wavelength of blackbody radiation emitted by an object depends on the temperature of the object. This means that the intensity-wavelength curve of radiation in the Universe at a specific time will have a specific shape depending on how hot the Universe was at that time.

The observed rate of expansion of the Universe helps in determining the age of our Universe since the big bang (the beginning), which turns out to be about 13.75 billion years.

Some predictions of the big bang theory can be tested in the lab by using particle accelerators and colliders.

Accelerators and colliders. Some of the fundamental particles such as quarks that existed in the early Universe, according to the big bang theory, have already been created at high energy particle accelerators under conditions similar to those that existed in the early Universe, and their behavior has been tested by using colossal particle detectors. One such accelerator is the hadron collider at CERN (Figure 7.4). The lab includes a circular tunnel 17 miles in circumference beneath the Franco–Swiss border near Geneva, Switzerland; used to provide protons an energy of 7 TeV before they are smashed together head-on. To study the behavior of quarks, I myself had the opportunity to work on an experiment set on the previous version of this accelerator called the Large Electron Positron (LEP) collider.

Figure 7.4 Large hadron collider: The world's most powerful particle accelerator used to recreate the conditions that existed just after the Big Bang, by colliding two beams of protons accelerated to energy 7 TeV/ proton. Image: Courtesy of CERN, The European Organization for Nuclear Research.

And God Said: Let There Be Evolution

Galaxies running away. By using powerful telescopes, Edwin Hubble (1889-1953) observed that the galaxies of our Universe are actually moving away from each other with high speed. This supported the expanding universe component of the big bang theory.

Nucleosynthesis. The known abundance of chemical elements in the Universe is in very good agreement with the predictions of the big bang theory about the primordial nucleosynthesis of the elements during the early Universe.

Both the evolution of the Universe and the evolution of life on Earth, biological evolution, have intimate connections and parallels: one (or few) evolving into many; or unity evolving into diversity. During the evolution of the Universe, one type of force evolved into four types of forces. Only a few species of fundamental particles formed about 100 atoms with different properties, which in turn formed a countless types of molecules and substances. For example, tens of millions of different types of molecules called organic molecules are largely composed of four types of atoms: carbon, hydrogen, oxygen, and nitrogen. Bottom line: The seemingly infinite diversity of physical materials and their properties that we see around us stem from four fundamental interactions and about 100 different kinds of atoms, which in turn are formed from a few fundamental types of particles, so few you can count on your fingertips. Physical laws of nature are driving the evolution of the Universe, whereas evolutionary mechanisms such as natural selection acting on variations are largely driving biological evolution. These variations are originally generated by genes and mutations in genes, which in turn are influenced by or driven by physical laws of nature. Changes in environment which play crucial role in evolution are also driven by physical laws.

And it is out of these laws that life emerged on Earth from the existing matter.

7.4 Then There was Life: The Origin of Life

As already discussed, matter and energy are interrelated through the physical laws of nature. These are scientifically known facts: the Earth spins around its imaginary axis and revolves around the Sun; the Sun and everything else in the solar system moves along as parts of our galaxy; all galaxies are moving apart from each other as part of the expanding Universe. The energy for all this motion comes from the big bang and the fundamental interactions of nature such as gravity, and the motion is governed by the physical laws of nature. These are the same physical laws that governed the formation of clouds of atoms and molecules, and then stars and planets from those clouds. One of those planets is our Earth, which according to scientific evidence formed about 4.6 billion years ago when a vast cloud of dust and rocks surrounding our Sun condensed under the influence of the physical laws.

So, how then did life originate on Earth? Well, you already know that cells are the anatomical and physiological fundamental blocks of life. Cells are largely composed of water and four types of molecules called the molecules of life: carbohydrates, lipids, proteins, and nucleic acids (DNA and RNA), which in turn are made of smaller and simpler molecules; simple sugars, fatty acids, amino acids, and nucleotides respectively. As depicted in Figure 7.5, these molecules formed due to some forms of electromagnetic interaction between appropriate atoms, just like all other molecules in our Universe formed. There is enough scientific evidence to support the idea that the molecules of life spontaneously formed from these simpler molecules, again under the influence of natural interactions and physical laws of nature, mostly some forms of electromagnetic interactions and bonding based on these interactions.

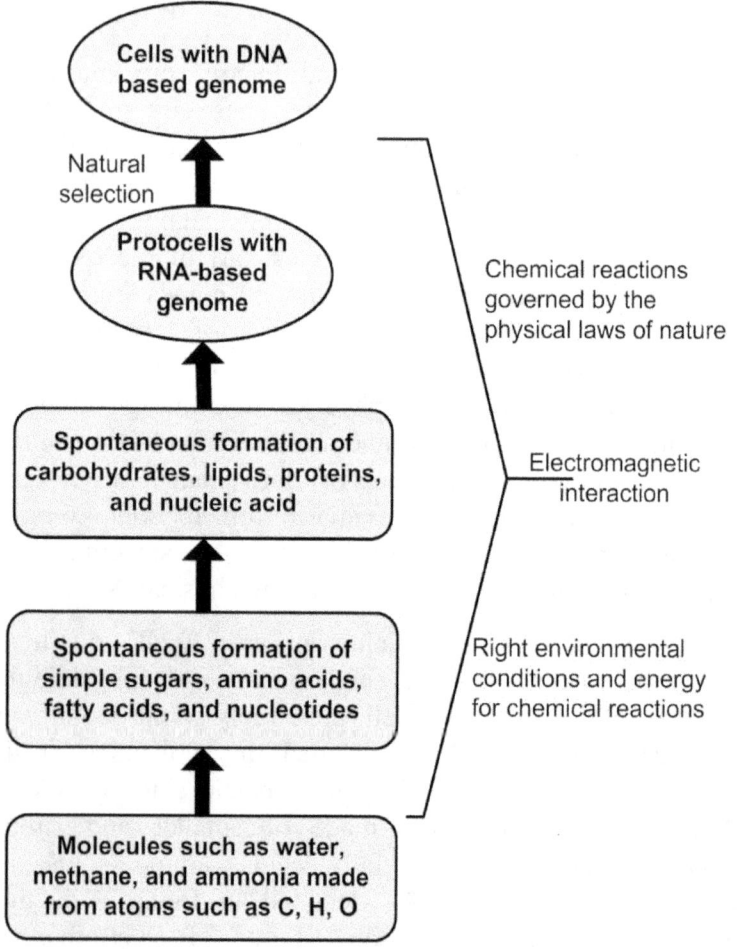

Figure 7.5 Emergence of living cells.

Based on careful observations and experimental studies in the fields of physics, chemistry, and geology, scientists generally agree that the process that gave birth to first cell (or cells) was driven by physical and chemical laws and went through the following sequential steps or stages: synthesis of small organic molecules, synthesis of the molecules of life, synthesis of pre-cells, and evolution of metabolism and genetic self-replication. Let's explore these four stages:

1. **Synthesis of small molecules.** Atoms such as carbon, hydrogen, nitrogen, and phosphorous bonded together under the influence of electromagnetic interaction to form molecules such as methane, water, and ammonia, which in turn interacted to make more complex molecules (but yet simpler than the molecules of life) such as simple sugars, amino acids, and nucleotides. The energy required for this synthesis was provided by the right environment such as lightening and sunlight. Scientists have proven this hypothesis by creating these organic molecules in the lab by re-creating the early Earth atmosphere in reaction chambers. For example, in the 1950s, Stanley Miller put hydrogen gas, methane gas, ammonia, and water vapors (known to exist in the early atmosphere) into a chamber. He then zapped the chamber with electric sparks to simulate lightening. The formation of expected organic compounds such as amino acids was observed. Since then, scientists have repeated this experiment under varied atmospheric composition (gases) and environment (energy source), and have concluded that spontaneous synthesis of these organic molecules is possible.

2. **Synthesis of macromolecules.** After their formation, the small organic molecules such as amino acids bonded together under the influence of natural forces, largely some

321

kinds of electromagnetic interaction, to form macromolecules such as proteins. Scientists have tested this hypothesis, for example, by dripping solutions of different amino acids onto sand and observing the formation of amino acid polymers, which are the primary structure of proteins.

3. Synthesis of protocells. After their formation, the molecules of life clustered together into a package that formed a membrane-like structure around itself, forming a sac-like structure. Because the membrane was only selectively permeable, the molecules inside the membrane were able to maintain their own internal chemical environment different from that of their surroundings. Such an environment would be necessary to develop two capabilities which are necessary conditions for life: metabolism and accurate genetic self-replication. These sacs were primitive cells called *protobionts* or *protocells*. The spontaneous formation of protocells has been proven in the laboratory. The structure and functioning of cell membranes, by now, are very well understood, and it is not hard to see how in the presence of water, lipids could have spontaneously organized themselves into these membranes, again under the influence of electromagnetic interaction. It is a well-known fact that when you mix phospholipids (type of lipids in cell membranes) with water, a lipid bilayer forms due to the electromagnetic properties of lipids and water. This is the same bilayer that forms the structural basis of cell membranes.

4. Origin of metabolism and genetic self-replication. The protocells, once formed, developed into cells with two essential capabilities: metabolism and genetic self-replication. Metabolism refers to the sum total of all

biochemical reactions that cells use to manage their energy resources. It includes creating energy by breaking down large molecules, such as carbohydrates, and then using that energy to synthesize needed molecules such as proteins, support genetic replication, and enable the organism to perform activities. To occur, these reactions require the suitable internal environment of the cell protected and maintained by the selectively permeable membrane.

> **Note.** In science, a *spontaneous* event means an occurrence within a system under the influence of the physical laws of nature and without the interference of any external agency.

Cells make copies of themselves in a process called cell division. During cell division, the genetic information (genome) is replicated. In this sense, a cell or its genome is called a DNA-based self-replicating system because the genetic information resides in the DNA molecules. In sexual reproduction, a multicellular organism, like a human, begins its life with a single cell, the zygote. To develop into a full organism, self-replication of the genome (set of genes or DNA) is essential because each cell of a multicellular organism needs this genome in it. So, DNA molecules must be able to replicate themselves. This would be necessary for the development of an organism, and for reproducing and passing the genes down to the offspring.

As you know from Chapter 4, there are two types of nucleic acids: DNA and RNA. There is enough scientific evidence to support that first genetic material was RNA and not DNA. Thus protocells with RNA-based genomes were able to reproduce and pass their genome to the next generation. Why then in today's world are the self-replicating systems DNA-based instead of RNA-based? There are three main reasons:

And God Said: Let There Be Evolution

First, there is a well-understood law of nature that states that physical systems tend to change toward greater stability. A double-stranded DNA molecule is more stable than a single-stranded RNA molecule, which breaks apart more easily and mutates more often. Second, DNA can replicate more accurately than RNA can. As genomes grew larger, greater accuracy in replication became necessary for the survival and reproduction of protocells in order to code the genetic characteristics and pass them down to the offspring. Third, RNA viruses could have attacked RNA-based genomes, and in that event, DNA-based genome could have experienced a strong selective advantage. This explains how even if life started with RNA-based self-replicating systems (genomes), it ended up with DNA-based self-replicating systems in which RNA acquired the present-day secondary but important role as an intermediate in the translation of genetic information into proteins. Due to the reasons just discussed, the RNA world evolved into the DNA world. This evolution from the simple to the complex is also observed at organism level, as you will see later in the chapter.

Evolution comes in when we try to link the origin of life, that is, the appearance of the protocell, to the cells that exist today. As discussed in Chapter 6, today, the cells of all organisms that we know of are anatomically and physiologically more similar than different from one another. For example, all cells are composed of the same four types of molecules (molecules of life); all cells have a biological membrane largely composed of lipid bilayer and proteins; all cells translate the genetic information in their DNA into proteins; and all cells support and run metabolism and genetic self-replication. These remarkable similarities among cells of all organisms not only point to the common origin of cells, but also common ancestors of organisms because organisms are made of cells. This diversity of cells and organisms from a common origin or ancestor can be successfully explained by the theory of evolution.

7.5 And Life Evolved

Life evolved, which made the history of life. Cells evolved before organisms did. Even today, the physiological and functional building blocks of life are cells, which are of two categories: *prokaryotic cells*, which do not contain internal membrane-enclosed structures called *organelles* such as a cellular nucleus; and *eukaryotic cells*, which contain organelles including a cellular nucleus. Organisms made of prokaryotic cells such as *bacteria* and *archaea* are called *prokaryotes*, and those made of eukaryotic cells such as plants, fungi, animals, and protists are called *eukaryotes*. Let's first make sense of these terms by putting them into bigger context.

An approach common to all scientific fields is to group things together based on their properties because it would be impossible, redundant, and unwise to study every single thing individually. The old system of classification, put into motion by Carolus Linnaeus (1707-1778), grouped all organisms into five higher-level groups called kingdoms: *Monera* consisting of prokaryotes, *Protista* consisting of a diverse group of mostly unicellular organisms, *Plantae* consisting of plants, *Animalia* consisting of animals, and *Fungi*. Recently, based on further similarities and differences revealed by studying enormous amount of genetic data, biologists have classified all living organisms into three broad highest-level groups called domains: *Bacteria, Archaea,* and *Eukarya*. These domains are taxonomically at a level higher than that of the five kingdoms. The domain Bacteria consists of the most diverse and widespread unicellular organisms, whereas the domain

And God Said: Let There Be Evolution

Archaea consists of unicellular organisms that live in extreme environments such as hot springs and salty lakes. The domain Eukarya consists of all organisms (unicellular and multicellular) whose cells contain a cellular nucleus and possibly other organelles. Note that the two prokaryotic domains, Bacteria and Archaea, consist entirely of unicellular (single-celled) organisms, whereas Eukarya consists of both unicellular and multicellular organisms. This underlies the fact that unicellular organisms have played a dominant role in the history of life on Earth.

In Chapter 4, we presented a tree of life based on rRNA gene sequencing (Figure 4.9), which displays the evolutionary relationships among different groups of organisms. Another version of that tree of life in terms of the three domains summarizing the history of life on Earth is presented here in Figure 7.6. This tree, supported by data, indicates that the first major divergence or branching in the history of life occurred when the ancestors of Archaea and Eukarya separated from prokaryotes (labeled as 1 in the figure). Subsequently, prokaryotes evolved into bacteria, whereas the organisms in the lineage that split from prokaryotes evolved into

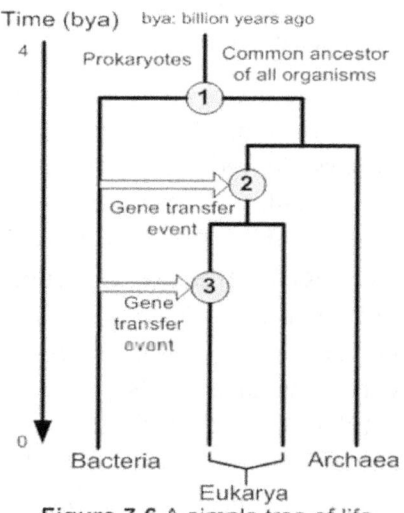

Figure 7.6 A simple tree of life based on the three domains

Eukarya and Archaea. The gene transfer events (labeled 2 and 3 in the figure) from prokaryotes to this lineage contributed to this evolution. Such a transfer of genes between genomes of

326

different lineages is called *horizontal gene transfer*. For example prokaryotes can directly inject their genes into another microbe through a sexual act. The injected genes become part of the recipient microbe, which passes them down to its offspring.

A brief history of life, that is, the timeline of the evolution of life, on Earth is presented in Table 7.3. By 3.8 billion years ago, the first living cells evolved as prokaryotic cells, the cells that had outer biological membranes but no internal membraneous structures called organelles.

Caution!
If you are not familiar with some of the terms, for example, the names of some groups of organisms used in Table 7.3, check with a biology book if you are curious or don't let it bother you, because memorizing them is not crucial for understanding the material and for the logical flow in this book.

Most of the functions of an organism, such as moving and synthesizing proteins, are facilitated at the cellular level, and cells need energy to perform these tasks. The energy to support cellular work is produced in the cell in form of molecules called ATP in a process called cellular respiration. For example, oxygen from the air we breathe in is used to support this process. The breathed-in oxygen reacts with the organic compounds from the food we eat such as glucose to produce carbon dioxide (that we breathe out), water, and energy in form of heat and ATP. The ATP molecules produced in the cellular respiration process are used by the cell as a source of energy to perform its work. Cellular respiration that occurs in our cells is called *aerobic respiration* because it consumes oxygen. Some organisms use a form of cellular respiration that does not consume oxygen. This kind of respiration is called *anaerobic respiration.* Today, aerobic respiration is by far the more common form of cellular respiration. However, this was not the case in the beginning of life on Earth.

When prokaryotic cells emerged, both the atmosphere and the sea had very low levels of oxygen gas. There is enough scientific evidence to assume that the first prokaryotes were anaerobic. During the first major evolutionary divergence, the ancestors of archaeans and eukaryotes separated from the lineage of prokaryotes, which evolved into modern bacteria. Just before 2.5 billion years ago, the process of photosynthesis evolved in a lineage of bacteria called cyanobacteria. Photosynthesis is a cellular process in which carbon dioxide, water, and energy from light such as sunlight are consumed to produce oxygen and carbohydrates such as glucose.

Photosynthesis evolved in cyanobacteria as oxygen levels on Earth were very low at that time. Cyanobacteria, by performing photosynthesis, began producing oxygen as a byproduct of creating food for themselves, carbohydrates; this is what plants do today. Free oxygen produced by photosynthesis accumulated in the atmosphere and paved the way for the evolution of eukaryotic cells and aerobic respiration. Over time, increased oxygen levels favored aerobic respiration over anaerobic respiration.

Think About It!

What is the relationship between aerobic respiration and photosynthesis?

Answer

The two processes are the reverse of each other. In cellular respiration oxygen and organic molecules such as glucose are consumed and carbon dioxide and energy are produced; whereas in photosynthesis, carbon dioxide and energy are consumed and oxygen and organic molecules such as glucose are produced.

However, be mindful of the fact that the energies consumed in photosynthesis and produced in cellular respiration are of different types.

Table 7.3 Timeline for history (evolution) of life on Earth based on hypotheses supported by a body of scientific evidence.

Time	Events
4.6 bya ↓	**Formation of Earth**. Material body revolving around the Sun due to the gravitational force. Earth surface too hot for life to begin.
4.5 bya ↓	**Formation of moon.** The collision of Earth with another planet (giant impact hypothesis) shot multiple material bodies into the space which due to gravitational force began revolving around the Earth. These revolving bodies eventually combined into one body that we now know as the moon.
4.1 bya ↓ 3.8 bya ↓	**Emergence of conditions for life**. Surface of the Earth cooled enough for the crust to solidify. **Molecular evolution.** By 3.8 bya, the simple organic compounds present on Earth had formed the molecules of life: complex carbohydrates, lipids, proteins, and nucleic acids (RNA and DNA). **Emergence of earliest life** in the form of self-reproducing RNA molecules and RNA-based protocells which eventually evolved to DNA-based protocells, followed by prokaryotic cells.
3.2 bya ↓ 3 bya ↓	Evolution of the three domains: **Bacteria.** Divergence in the prokaryotic lineage gave rise to bacteria and to the common ancestors of archaeans and eukaryotic cells. **Achaeans and eukaryotic cells**. The ancestors of archaeans and eukaryotic cells diverged into two separate lineages (domains): Archaea and Eukarya. **Photosynthesis and aerobic respiration** evolved, changing the Earth's atmosphere by continually adding oxygen to it. Around 2.4 bya, cyanobacteria supporting photosynthesis appeared.

And God Said: Let There Be Evolution

2 bya ↓ 1 bya ↓	**Emergence of cell nucleus.** Number of genes and the overall size in the lineage of cells that would eventually become eukaryotic cells kept increasing. Between 3 bya and 2 bya, the nuclear membrane emerged enclosing the genome and giving rise to the cell nucleus. Eukaryotic cells, the cells containing nuclei and organelles, appeared around 1.9 bya. **Emergence of mitochondria.** By 1.2 bya, a would-be eukaryotic cell engulfed a small prokaryotic cell, a process called endosymbiosis. The smaller internal cell eventually became a mitochondrion, an organelle that carries on cellular respiration in eukaryotic cells. **Multicellular organisms.** Sexual reproduction and simple multicellular organisms appeared around 1.2 bya. **Emergence of chloroplasts.** Between 1.5 bya and 1.0 bya an evolving eukaryotic cell engulfed a cyanobaterium, a bacterium capable of performing photosynthesis, which evolved into a chloroplast, an organelle that serves as the site for photosynthesis inside the cell.
900 mya ↓ 200 kya ↓ 50 kya ↓	**Evolution of plants, fungi, and animals.** By 900 mya, all the major lineages of the tree of life such as animals, fungi, and algae evolved into plants and diverged. The organisms representing these lineages appeared along the shores of the first known supercontinent called Rodinia. **First human.** Around 200,000 years ago, anatomically modern humans appeared in Africa. Around 50,000 years ago, they began colonizing other continents including Asia and Europe.
10 kya ↓ Today	**Human civilization** develops; began affecting world climates.

One obvious observation from this timeline is the change from unicellular to multicellular and from simple to complex organisms. But, how do we know that the timeline presented in Table 7.3 is true, even approximately? We know it because it is supported by pieces of scientific evidence from different sources such as experiments, genetic data, and fossil records.

7.6 Life History on the Rocks: Revelations from Geologic Records Based on Fossils

Nature has several ways to give away its secrets to those who can ask the right question and can recognize patterns. In Chapter 6, we explored fossils and radiometric dating and how they help to reconstruct the major events and changes in living organisms over a time span of billions of years, and this way help reconstruct the history of life on Earth. This history revealed by the fossil record has the signature of evolution everywhere. The fossil record clearly shows that over time there have been great changes in the types (such as species) of organisms that lived on Earth. In general, the fossil record suggests that the evolution of life on Earth has been from microscopic organisms to macroscopic organisms. The evidence of the oldest life forms, as old as 3.5 to 3.7 billion years ago (bya), which have been found in Greenland and Australia, is that of microscopic life. Consider the timing (dates) of the following pieces of fossil record:

1. **A strand of prokaryotes.** Dates back to 3.5 bya; found in Australia.

2. *Grypania spiralis.* Eukaryotic species; dates back to about 2.1 bya.

3. *Tappania,* **an alga.** A unicellular eukaryote from northern Australia; dates back to about 1.5 bya.

4. *Bangiomorpha pubescens,* **a red alga.** Multicellular species; dates back to 1.2 bya.

5. **Dickinsonia costata.** A soft bodied organism, a few cm thick animal; dates back to 570 million years ago (mya).

6. Hallucigenia. An animal found in the Canadian Rockies; dates back to 525 mya.

7. Coccosteus cuspidatus. A fish that had a bony shield covering its front end including head; dates back to 400 mya.

8. Group of malluscus. Size from few cm to 2 m; dates back to 375 mya.

9. Dimetrodon. Largest known carnivore of its time, more related to mammals than to reptiles; dates back to 270 mya.

10. Rhomaleosaurus victor. Large marine reptile; dates back to 175 mya.

In this record, have you noticed the evolution from unicellular to multicellular and smaller size to larger size with the passage of time?

It is important to note that fossils not only show when a certain species of organisms lived on Earth, but also their pattern tells us how new, different organisms arose from the previously existing ones. To a geologist, a layer in a sedimentary rock containing fossils represents a slice of geologic time. By using fossil data, geologists carefully reconstructed the history of life on Earth by counting backward through these layers. They found that the pattern of transitions represented by the fossil layers were consistent among the rocks found in different parts of the globe. These transitions represent the boundaries for time slices such as eons, eras, and periods.

Table 7.4, based on the fossil record, summarizes important geologic and biological events during these eras and periods. As shown in the table, life originated in the Archaean eon/era with the appearance of prokaryotes. Groups of organisms

332

called protists, fungi, plants, and animals appeared with the evolution of eukaryote cells in the next eon, Proterozoic. This record presents an unambiguous picture of how life evolved starting with simple forms and gradually evolving into complex or sophisticated forms. However, the evolution from small to large and simple to complex is not that simple and linear; for example, many animals that evolved before humans are larger than humans. But we can argue that humans are more sophisticated (complex in that sense) than them in many ways. Nevertheless, remember, evolution is not goal-oriented and not purpose driven.

Table 7.4 Geologic timescale derived from the fossil record by using techniques such as radiometric dating. Mya: million years ago.

Era	Period	Events
Cenozoic 65 mya to today	1. Quaternary (1.8 mya to today) 2.Tertiary (65 to 1.8 mya)	1. Modern humans evolve. 2. Climate cools; grasslands and woodlands emerge. Apelike ancestors of human appear. Major radiations of mammals, birds, and pollinating insects appear.
Mesozoic (248 to 65 mya)	1. Cretaceous (144 to 65 mya) 2. Jurasic (206 to 144 mya) 3. Triassic (248 to 206 mya)	**1. Early:** Very warm climate. Flowering plants and modern inset groups such as ants, bees, butterflies, and termites appear. Dinosaurs continue to dominate. **Late:** Major extinction event possibly caused by asteroid impact. Mass extinction of many organisms including dinosaurs. 2. Lush vegetation. Gymnosperms (cone-bearing plants) continue to be the dominant plants. Age of diverse dinosaurs. 3. Gymnosperms dominate landscape. Origin of many new groups including dinosaurs, turtles, and mammals. Dinosaurs radiate.

And God Said: Let There Be Evolution

Paleozoic (543 to 248 mya)	1.Permian (290 to 248 mya) 2.Carboniferous (354 to 290 mya) 3.Devonian (417 to 354 mya) 4.Silurian (443 to 417 mya) 5.Ordovician (490 to 443 mya) 6.Cambrian (543 to 490 mya)	1. Supercontinent Pangaea and world oceans form; adaptive radiation of conifers, cycads, and ginkgos appear. Dry climate; adaptive gymnosperms and insects such as beetles and flies appear. Origin of most present-day insects and radiation of reptiles. 2.High atmospheric oxygen level supports giant arthropods. Extensive forests of vascular plants form. First appearance of seed plants and reptiles. Ears evolve in amphibians, which dominate the period. Penises evolve in reptiles; vaginas evolve later (in mammals only). 3.First tetrapods and insects appear. Explosion of plant diversity leads to the formation of forests and trees. Many new plant groups appear including ferns with complex leaves, lycophytes, and seed plants. Bony fishes diversify. 4.Diversification of vascular plants and marine invertebrates. Appearance of vascular plants, bony fish, and land fungi. 5. Fish and reef-forming corrals appear. Land plants appear. Gondwana moves toward the South Pole. Colonization of land by plants and animals. 6. Earth thaws. Cambrian explosion: Sudden increase in animal diversity; many animal groups appear (in the oceans).
Proterozoic (2500 to 543 mya)	1.Neoproterozoic (900 to 543 mya) 2.Mesoproterozoic (1600 to 900 mya) 3.Paleoproterozoic (2500 to 1600 mya)	Oxygen accumulates in atmosphere. Origin of eukaryotic cells followed by protists, fungi, plants, and animals.
Archaean (3800 to 2500 mya)		Origin of prokaryotes.
Hadean (4500 to 3800 mya)		Origin of Earth's crust. Seas and atmosphere form. Molecules of life evolve leading to the origin of life.

As explained in Chapter 4, a population is the smallest unit of evolution. All organisms in a population, by definition, belong to one species, even though a species may be living in several populations at a given time. This is how evolution is connected to species, through populations.

7.7 Origin of Species

As already mentioned in this chapter, unity behind diversity does not limit itself to the non-living entities or living entities; it equally applies to both. About 100 billion species of organisms have lived on Earth so far, and about 100 million of them are still with us today. All of these organisms are made of cells, which are more similar to one another than dissimilar, and all the genes in these cells are made of the same DNA molecules which use the same language to code genetic information. The cell and the genome (or genetic code) represent unity, whereas the multitude of species represent diversity. As shown in this chapter, life on this planet began with a cell and unicellular organisms and complex organisms emerged from the simple organisms. The history of life on this planet is the journey from one or more cells to a multitude of species of organisms and evolution explains each step of this journey. In other words, cells evolved to a multitude of species. The question arises: How does evolution give rise to a new species?

First, what is a species, anyway? Biologists use more than one definition of species. The definition that helps understand the origin of species is called *biological species*, which is defined as a group of populations whose members:

1. Have the potential to interbreed and produce viable, fertile offspring, and

2. Do not produce viable and fertile offspring with members of any other such group.

For example, even though humans living in Seattle, Washington, United States and in Shimla, India, make two different populations, they belong to the same species (*Homo sapiens*) because a man in Shimla and a woman in Seattle (and vice versa) have the ability (or potential) to mate and produce a viable, fertile child. However, the man in Shimla has no potential to mate with one of the female monkeys of Shimla and produce a viable, fertile organism. Within the framework of the theory of evolution, a new species forms from an existing one when an ancestor species splits into two or more species in a process called speciation. This explains the differences between different species as well as similarities among them, the unity of life. They are similar because they share a common set of characteristics that they inherited from a common ancestral species. In Chapter 4, we introduced the concepts of microevolution, changes in allele frequencies in a population, and macroevolution, the evolution visible at a macro-scale over longer periods. The evolutionary concept of speciation (a species splitting into two or more species) also connects microevolution with evolution, as you will see further on in this section.

The gene pool of a population, discussed in Chapter 4, is the genetic identification of the population. It is also the genetic identification of the species to which the population belongs, and which may have spatially spread out over multiple populations. As we know from Chapters 4 and 6, evolution begins at the microscopic level of genes. It is the gene pool of a species that binds the organisms of the species together as one species. However, here is a question: what keeps the gene pool together? The answer is: gene flow. The frequent gene flow (random mating) within the same population causes its members to resemble to each other lot more than they do with any other species, and stay sexually (or reproductively) compatible. Gene flow can also occur within geographically different populations of the same species if they continue to interact with each other in the process of sexual reproduction and hence gene exchange.

Think About It!
What is the process that facilitates gene flow within the members of the same species?

Answer
Sexual reproduction

Here is how species connect microevolution with macroevolution. Because the ability to produce viable and fertile offspring binds the members together as one biological species, and therefore the gene pool of the species defines the species, the split of the gene pool (microevolution) will give rise to new species (macroevolution). A split of the gene pool can occur as a result of reproductive isolation, that is, members of one section of a species become unable to produce viable and fertile offspring with the members of another section of the same species due to some barriers such as geographic or habitat isolation. In this case, as depicted in an example illustrated in Figure 7.7, speciation may occur in the following steps:

1. Three populations (P_1, P_2, and P_3) are bound together as a species due to the gene flow (represented by arrows) among them.

2. Due to a barrier, gene flow between P_1 and P_2, and between P_1 and P_3, stops.

3. Population P_1 begins to genetically diverge from the populations P_2 and P_3.

4. Population P_1 evolves a gene pool which is different from the gene pool of populations P_2 and P_3.

5. Population P_1 over generations has given rise to a new species.

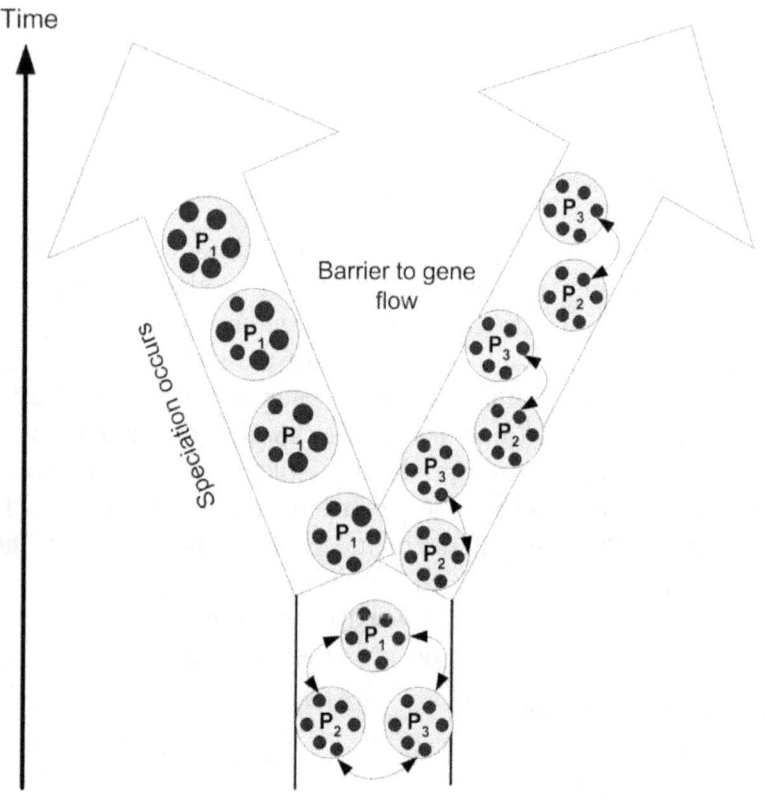

Figure 7.7 Origin of a new species due to reproductive isolation.

The gene pool of a species represents the reproductive compatibility of the members of the species. The diversity of looks (a combination of traits) among the members of the same species represents the genetic fluctuations within the same gene pool. The differences in looks, traits, and body forms between two species reflect the differences between the gene pools of those two species. Experimental studies in genetic biochemistry confirm that differences in morphology (body form) between different species do represent the differences in their gene pools. Therefore, the morphological definition of species, based on similarities in body shape and structural features, is also a valid definition. Historically, this definition has been more practical for biologists to use.

Think About It!

If all species evolved from a single common ancestor, and if evolution is just a gradual genetic change within a species, then shouldn't we have just one evolved species today?

Answer

Yes, we would if the organisms of the ancestor species randomly interbreed within the species and the species never got split, that is, it lived as one population. However, when a species gets split geographically or otherwise into two groups that cannot interbreed, the groups start following two different paths of evolution and eventually we have two species different from each other.

The keyword in speciation is *reproductive barrier*, which may or may not be due to geographic separation. For example, some accidental genetic events occurring during the reproduction process can give rise to a new species within the same population, that is, without any geographic separation. You know by now that an offspring gets one set of chromosomes from each of its parents during the reproduction process. Due to an accident, it is possible that the offspring gets both sets of chromosomes from both parents, which would double the number of chromosomes in the offspring. This offspring, if viable and fertile, is definitively a new species. The process in which a new species forms within the same population without the geographic split is called *sympatric speciation*. These kinds of speciation events do happen and have been observed.

Think About It!
Which one of the following conditions is necessary for speciation to occur?
  A. Reproductive isolation
  B. Sympatric speciation
  C. Adaptive radiation
  D. Mass extinction
  E. Interbreeding among neighboring populations

Answer
A

In a nutshell, microevolution is connected to macroevolution through speciation. Due to microevolution, the changes in the gene pool keep cumulating, eventually giving rise to a new species. Even though evolution begins at micro-scale with genes and mutations, environment plays a crucial role at macro-scale to set in motion the mechanisms or processes of evolution such as natural selection. It's important to understand the role of environmental stress in evolution at different complexity levels of life: individual, population, community, and ecosystem.

7.8 Ecological Dance of Life: Big Picture

In Chapter 1, we discussed different organizational levels of life: cell, organ, organ system, organism, population, community, and ecosystem. We also noted how at each level, life is more than sum of its parts due to the new properties that emerge at each level and do not belong to a specific part. Organisms interact with their environment to exchange energy (I just say energy instead of saying matter and energy because matter and energy are interconvertible). For instance, look at this ecological dance of life from our vantage point: We humans need other organisms of our species to survive and reproduce. Also to survive, we consume meat from a variety of animal species and non-meat food from a variety of plant species. We also interact with our physical environment, sometimes badly; for example, polluting it, which has its consequences such as extinction of a species. After all, environment does play an important role in the evolution of life on this planet.

Recall from Chapter 1 that an ecosystem in a given area is the system that consists of all organisms in the area and the physical (abiotic) environment with which they interact. The study of the flow of energy in an ecosystem is called ecosystem ecology, or just ecology. For example, if organism A eats organism B,

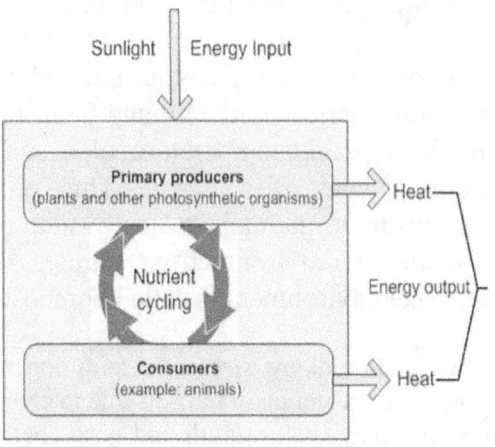

Figure 7.8 Energy flow in an ecosystem

energy flows from B to A. Figure 7.8 presents the big picture of this energy flow. Energy input from the Sun flows through primary producers (plants and other photosynthetic organisms), which convert it to carbohydrates through the process of photosynthesis. Producers also produce other molecules of life such as proteins, lipids, and nucleic acid for their own survival. Consumers such as animals consume this energy produced by plants in the form of carbohydrates and other molecules of life. Some nutrients released by the consumers such as carbon dioxide and decomposed mass are recycled back to the producers. Energy enters an ecosystem as sunlight and leaves it as heat both from producers and consumers.

In this ecological dance of life, changes in the environment occur which lead to environmental stresses on populations living in the ecosystem. In the following, we explore how organisms deal with environmental stress.

7.8.1 Homeostasis and Environmental Stress

The key to your (or any other organism's) survival is that you must maintain a constant internal environment defined by a constant or narrow-range of values of certain factors or variables. For example, your internal body temperature must be in a very narrow range around 98.6 °F, regardless of if you are in the freezing temperature of a snowy mountain or in the burning heat of a desert. This property or capability of organisms to maintain a steady-state (constant) internal set of conditions (environment) regardless of the changes in the external environment is called homeostasis.

Stress. In biology, *stress*, the term borrowed from physics, means the pressure put on organisms due to changes in their environment. Stress may cause stimuli and pose challenges for the survival of the organisms in the environment.

To stay alive an organism must maintain its homeostasis, for which they need to live in an external environment in which the environmental factors, such as temperature, vary from the optimal value only within a certain range called the range of tolerance. As depicted in Figure 7.9, the range of tolerance for a given factor belongs to the population, that is, a population survives an environmental factor as long as it is within the range of tolerance. This is because different individuals within the population like to be at different point within the range of tolerance. For example, not all individuals like to be at exactly the same temperature; some like cold weather, others like hot weather, yet others a moderate temperature. However, none of us can survive extreme cold or extreme

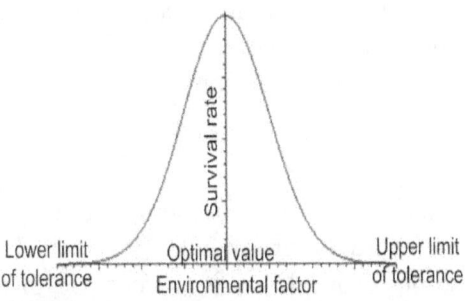

Figure 7.9 Survival rate of individuals in a population within a tolerance range

hot outside the range of tolerance. Some of us die even within the range of tolerance. The maximum number of people in a population survive at an optimal value of the given factor. As we move from the optimal value toward the extremes of the range of tolerance, the environmental stress increases and the rate of survival drops. Nevertheless, the population survives within the range of tolerance, and the population vanishes outside the range of tolerance. So, the range of tolerance defines the conditions needed for the survival of the population. However, note that even within the range of tolerance, there is environmental stress. This is because all individuals of the population have to maintain the same homeostasis, the range for which is much narrower (almost a fixed value for each factor) than the range of tolerance. For example, think of the hottest and the coldest areas you have ever visited or lived in, and you will realize that humans can survive in quite a wide range of external temperature; even though all of us have to maintain the internal body temperature very close to 98.6 °F, as mentioned earlier.

Within the environmental stress, life survives at different levels of complexity by using different mechanisms. In the following, we explore how an individual organism survives through acclimation, populations survive through adaptation and evolution, and communities and ecosystem survive through ecological succession.

7.8.2 Dealing with Environmental Stress

There is no environment that is stable forever. Change and therefore environmental stress is a law of nature. Here are some examples of sources of environmental stress: flood, drought, fire, hurricanes, disease, competition, erosion, overhunting, pollution, pesticides, and predation. Environmental stresses may occur gradually or quickly, and it can affect life at multiple levels such as organism, population, community, and ecosystem.

Organism level. Environmental stress can cause changes in the behavior, physiology, and reproductive ability of an organism; and can even kill it. An individual organism survives environmental stress through *acclimation*, which is the process of physiologically adjusting to a change in an environmental factor, such as temperature. A mammal growing fur in the winter and shedding it off in the summer is an example of acclimation. Acclimation usually works well when it is performed as a *quasistatic* process, that is, a gradual process in which change occurs in small incremental steps. For example, you can tolerate very hot water if you climb into the bathtub with water at moderate temperature and then add hotter water gradually. In each step, you make the water a little hotter so that the temperature before and after adding the hot water is approximately equal, that is, the change is infinitesimal. You get used to this new temperature and repeat the step of adding a little more hot water; repeat it over and over again. This way, you will be able to tolerate hot water at much higher temperature than you will by just jumping into a hot bathtub. The same holds true for many other external environmental changes that cause stress.

But acclimation can only go so far. There reaches a point in the value of the environmental factor, called threshold value, beyond which the organism cannot survive. So, under an environmental stress, an individual organism survives through acclimation, or it dies, depending on the level of stress.

Think About It!
Is acclimation equivalent to adaptation?

Answer
No. Acclimation causes a temporary change brought about in the lifetime of an organism, whereas adaptation is a process that brings change in a population over multiple generations.

If an organism dies at a given level of environmental stress, it does not necessarily mean that the whole population dies at that level of stress.

Population level. Environmental stress can cause a decrease in population resulting in a decrease in its genetic diversity, or can even cause extinction of the population altogether. A population, if not gone extinct, adapts to the environmental

345

stress and therefore evolves. In other words, a population under environmental stress survives through evolution. Consider an example of a farmer who had a population of insect pests in his farm, and he sprayed a pesticide periodically several times to keep his crop safe of the pests. What he noticed is that after a substantial decrease the number of pests in the population it rose back to right where it was before the first spray and it did not decrease with further sprays. A biologist at such a situation would say that the population has become resistant to the pesticide. What does it mean? It means that right from the beginning there were a very few insects which had the genetic capability to survive the pesticide. Those individuals, not killed by the pesticide, proliferated over generations. This way, the population adapted to the pesticide and evolved. So, a population under environmental stress survived through adaptation and evolution, even though majority members of the population in the earlier generations died.

A population is generally a part of a community and an ecosystem.

Community level and ecosystem level. Environmental stress on a community or on the entire ecosystem can cause a disruption in the flow of energy in a food chain and in the cycling of essential nutrients; and can also cause the entire ecosystem to collapse. As said earlier, populations, if they do not go extinct, adapt to the environmental stress and therefore evolve. Communities and ecosystems under environmental stress can survive through *ecological succession*, which refers to the transition in the species composition of a community or an ecosystem. The transition occurs as a gradual change in the composition and function of various populations within the community or the ecosystem. Ecological succession after a wildfire is an example (Figure 7.10). Consider a fire that spreads through a primary (or an old-growth) forest and burns down all of its trees. What would follow are the growth of grasses and weeds followed by small shrubs. After a while, a few species of trees will begin appearing, and eventually, after a long time, the forest may acquire the state in which it was before the fire. So, communities and ecosystems under environmental stress survive through ecological succession.

7.9 Summary

The two expressions *history of life on Earth* and *evolution of life on Earth* are synonymous because life evolved. In other words, evolution defined the history of life on Earth. Earth is part of the Universe and is subject to the same physical laws of nature as any other part of the Universe.

a)   b)

Figure 7.10 Ecological succession after a wildfire. Two photos of the same place in Lahemma National Park, Estonia: a) one year after the fire, and b) two years after the fire.

The universe we live in began with a big bang about 13.75 billion years ago and has been expanding since then. According to the well-tested *big bang theory*, the Universe was extremely hot and dense in the beginning and therefore it only consisted of a few fundamental particles such as quarks, leptons, and gluons. As the Universe progressively cooled down, these particles combined under the influence of laws and forces of nature to form atoms, molecules, planets, stars, and galaxies. For example, our Sun is a star that formed about 5 billion years ago, and our Earth is a planet that formed about 4.6 billion years ago. Just like the Universe, life on our planet has also evolved in space and time.

Under the influence of forces (interactions) and physical laws of nature, atoms spontaneously bonded to form simple organic molecules such as simple sugars, fatty acids, amino acids, and nucleotides, which in turn spontaneously formed the molecules of life: carbohydrates, lipids, proteins, and nucleic acids (DNA and RNA). The spontaneous development of selectively permeable membranes from lipids and water gave rise to primitive cells called protocells, which could replicate or reproduce themselves.

The pattern in the timeline, derived from the fossil record of the appearance of different groups or species of organisms, clearly indicates that new species originated (evolved) from pre-existing (ancestor) species via gradual modifications over generations. The origin of these modifications can be traced to genes. The reproductive isolation of a population from the rest of the species separates its gene pool and puts it on the path to evolve into a new species. As a big picture, all the three domains of life (Bacteria, Eukarya, and Archaea) evolved from prokaryotes.

This picture of the history of the Universe and the history of life on planet Earth is well supported by scientific evidence.

In a nutshell

✓ Earth, formed about 4.6 billion years ago, is not the center of the Universe.

✓ Self-replication and reproduction are essential properties of life.

✓ Life originated spontaneously under the influence of physical laws and forces of nature in the right environment on early Earth.

✓ The fossil record and genetic and biochemical studies reveal the history of life on Earth.

✓ Cells, the basic units of life, originated as RNA-based primitive cells, called protocells, which evolved into DNA-based protocells and subsequently into the DNA-based cells because DNA molecules had a selective advantage over the RNA molecules.

✓ Life evolved from prokaryotic cells to eukaryotic cells, from unicellular organisms to multicellular organisms, and in general from simplicity to sophistication and complexity.

✓ The early prokaryotes evolved as anaerobic because the early atmosphere had very low level of oxygen gas.

✓ The evolution of photosynthesis in some bacteria increased the oxygen level in the atmosphere, which then favored aerobic respiration in organisms, which is by far dominant today.

✓ The cause of speciation, the split of one species into two or more species, can be traced down to the change and split in the gene pool of the ancestor species.

✓ Species, a group of organisms that can interbreed to produce viable and fertile offspring, connects microevolution with macroevolution.

7.10 Behold

Life on Earth began with cells, which have their own self-replicating genome, set of genes. The genes are not only the origin of life, they are also the origin of evolution in many ways. These are genes that are passed down from generation to generation, and these are different versions of genes (alleles) that originate variations on which natural selection operates to facilitate evolution. Change in the allele frequency in the gene pool of a population is called microevolution; and the change and split in the gene pool gives rise to new species, the beginning point of macroevolution. This is how speciation connects microevolution to macroevolution.

Here is an issue: All organisms of a species have to perform some functions in order to survive and reproduce. Some of these functions are shared with other species (unity) and others are unique to the species (diversity). In both cases, the morphology (external body form) of the members of a species must support these functions. In other words, the form must fit the function. What makes sure that this happens?

In the next chapter, we explore this issue.

EIGHT

Form Fits Function: What a Perfect World- Or Is It?

And God Said: Let There Be Evolution

It is my belief that it is of the very essence of every problem that it contains and suggests its own solution... the life is recognizable in its expression, that form ever follows function. This is the law.

Louis H. Sullivan (1856-1924), Architect

Form and function are a unity, two sides of one coin. In order to enhance function, appropriate form must exist or be created.

Ida Pauline Rolf (1896-1979), Biochemist

8.1 Once Upon a Time

Once upon a time, there emerged the need for tall office buildings, which gave rise to the idea that *form follows function,* which is now a well-established principle in modern industrial design and architecture. The underlying idea of this principle is that the form of an object, such as a building, should be designed based on its planned function, so that it will then serve the function. The principle was first presented by an American architect, Louis Henri Sullivan (1856-1924), in 1896 in an article titled *The Tall Office Building Artistically Considered*, in which he wrote:

Figure 8.1 The Empire State Building in New York City, built in 1931, is one of the oldest and tallest skyscrapers.

"...It is the pervading law of all things organic and inorganic, of all things physical and metaphysical, of all things human and all things superhuman, of all true manifestations of the head, of the heart, of the soul, that the life is recognizable in its expression, **that form ever follows function. This is the law...**"

And God Said: Let There Be Evolution

Sullivan saw this law in action all around him in nature when he wrote:

> Whether it be the sweeping eagle in his flight or the open apple blossom, the toiling work horse, the blithe swan, the branching oak, the winding stream at its base, the drifting clouds, over all the coursing sun, **form ever follows function,** and this is the law. Where function does not change, form does not change. The granite rocks, the ever brooding hills, remain for ages; the lightning lives, comes into shape, and dies in a twinkling.

By that time (1896), the convergence of social, economical, and technological conditions such as industrialization had happened or had been happening. Other factors that were in progress included population rise in the cities, development of steel manufacturing, and maturation of high-speed elevators. Within this kind of environment, the construction of tall office buildings not only became feasible, but also became necessary. Realizing the newness of the idea of tall, lofty office buildings, Sullivan advocated that the form of these buildings should emerge from their intended purpose and function, and should not come from the established book of patterns and styles of the past. In other words, he proposed that *form follows function*, as opposed to *form follows precedent*. Based on his principle, Sullivan designed the shape of the tall steel skyscraper in late nineteenth century Chicago.

In modern manufacturing and technology industries, the *form follows function* principle has developed into the *F3 principle*: form, fit, and function. *Form* of an entity consists of the visual parameters such as dimensions, shape, size, and mass that define the look or morphology of the entity. *Fit* refers to the ability of the entity to perform a specific task or to physically interconnect with or become an integral part of another entity. *Function* is defined as the set of actions that the entity is designed to perform. Both the *form follows function* principle and the *F3 principle* mean that function precedes form in the sense that you should know the function so that you could design the form based on it. Once the form is designed, it should fit (facilitate or support) the function.

The principle of *form fits function* is also in action in biology even though there are no predetermined functions in the natural course of life: evolution is not goal oriented. In other words, form, including structure (anatomy or morphology), plays a crucial role in determining the function (physiology) of life at all levels. For example, it's a well-known fact that the shape of a protein and DNA molecules are crucial in determining their functions. What determines form?

8.2 In the Beginning

In the beginning of life, there were physical entities (matter and energy), physical laws of nature, and environmental conditions; which determined form, and the function followed the form in such a way that the form fit the function. It is not a secret that the form (including structure) of atoms and molecules is determined by well-understood physical laws. These laws apply equally to living and non-living material entities. For example, the well-celebrated double-helix form of the DNA molecule is formed under the influence of different forms of electromagnetic interactions, which is also applicable to non-living material. Millions of biochemical reactions in our body that keep us alive are governed by the same physical laws of nature that govern the non-living world. Even at organism level, for example, during hiking or looking down from a high peak, we realize that we are as vulnerable to the laws of gravity as a rock. The only difference is that we, the living, can respond to our environment in ways that the non-living cannot. But even that we do within the constraints of physical laws. To start with, our form (or body plan), like any other organism, arises from the pattern of development programmed in our genome, which itself has resulted from billions of years of evolution influenced by physical laws, as described in

And God Said: Let There Be Evolution

Chapters 5, 6, and 7. The diversity built into the genome expresses itself as the diversity in the form.

As discussed in Chapter 5 (Section 5.4), the diversity or variation in the form (for example: shape and size) of living organisms is further limited or constrained by the physical laws that govern factors such as diffusion rates, movement, strength, resistance, gravity, and energy exchange. For example, in order to stay alive, animals need to exchange matter and energy with their environment. Here they need to balance two conflicting factors:

1. The rate of exchange of material such as nutrients and waste products is directly proportional to the surface area of the cell membrane.

2. The amount of material that must be exchanged is directly proportional to the volume of the organism.

These two factors are conflicting because surface area decreases with increase in volume. As the volume of an organism increases, the number of cells in the body increases, and the surface area per cell exposed to the external environment decreases. In other words, as the size of the organism increases, the ratio of the outer surface area of the organism to its volume decreases. A larger body means less surface area exposed externally for exchange; the amount of material that must be exchanged increases whereas the outer surface area available to make the exchange decreases. This influences the shape or body plan of the organism because a certain amount of material must be exchanged with the environment in order to stay alive. This condition or requirement has played an important role in the evolution of internal surfaces in the form of circulatory, digestive, and respiratory systems in large multicellular organisms such as humans. The cells of these systems are exposed internally for exchange purposes through extensively folded surfaces such as small intestine, lungs, and kidneys.

In a nutshell, if there is any designer of form, it is nature itself equipped with the laws of nature including the laws of physics and chemistry, and the theory of evolution. In this chapter, we explore how theory of evolution predicts the relationship between form and function, and how that prediction compares with the facts. To accomplish that, we venture through three avenues: relationship between form and function in the bio and non-bio world, evidence for *form fits function* in life, relationship between the theory of evolution and the principle of *form fits function*.

You know from Chapter 4 that a consequence of adaptive evolution is the improved fit of organisms with their environment. You may look at *form fits function* as a part of this natural consequence of evolution.

8.3 Form Fits Function: A Natural Consequence of Evolution of Life

Although Louis Sullivan coined the phrase *form follows function* toward the end of the nineteenth century in context of architecture, the principle of *form follows function* or *form fits function* has been active in our applications since humans began developing tools and other objects. For example, in olden times when crops were cut manually with hands, the form of a sickle used for this purpose was such that it could be conveniently held in the hand to cut the crop. The same is true for scissors. Think of anything around you such as paper clip, shoes, shirts, pants, bike, lawn mower, and so on. You will immediately realize that *form fits function* is common sense.

And God Said: Let There Be Evolution

The function (or working) of a tool, device, or some other object is strongly correlated to its form (including structure).

In the world of life, the principle of *form fits function* has been in action since the dawn of life on this planet. This can be seen in terms of the often elegant match between the anatomy and physiology of organisms at every organizational level. For example, thin and flat plant leaves help to maximize the absorption of sunlight, which is converted to carbohydrates in the process of photosynthesis. As another example, a streamlined body contour helps animals from different groups such as fish and penguins to swim. The wings of a bird have a form (such as shape and structure) that is efficient for flying according to the laws of aerodynamics. In general, in biology, it is no secret that by studying anatomy, we learn about physiology, and knowing the function of a biological entity gives us a clue about its structure.

Note the difference here in how the principle of *form fits function* applies to living and non-living entities. For human-created non-living entities, we design the form based on the function that the entity is supposed to perform. We know the function before we design the form. In this sense, function precedes form. But once the form is there, it fits the function. In living organisms, form originates from the molecules of life in accordance with the laws of nature including the theory of evolution. The dynamics of the principle of *form fits function* in the context of physical laws and natural selection is illustrated in Figure 8.2. The forms of organisms develop from the genomes of organisms, which are subsets of the gene pool of a population. The gene pool and hence the genomes are the results of billions of years of evolution. The variations in the genome reflect in the variations in form within the same species or population. The functions follow these forms, that is, certain forms enable or fit certain functions. Recall that natural selection favors those organisms that are best suited to their environment. This means that the organisms with those functions that favor survival and reproduction are selected by natural selection. This means that the genes that give rise to the forms that support these functions persist in the gene pool, whereas the genes that gave rise to unfavorable (or not favorable) forms, or forms that do not support the required functions well, disappear from the gene pool because their carrier organisms do not survive and reproduce well. This brings about changes in the gene pool of a population, which is evolution, as discussed in Chapter 4. At the end of each of these cycles (Figure 8.2) *form fits function* more and more. Therefore, you can see that *form fits function* is a natural consequence of evolution.

Think About It!

What would be the relationship between form and function, for example, which would precede the other, according to:

A. Lamarck's already scientifically-discredited hypothesis of evolution

B. Darwin-Wallace theory of evolution?

Answer

A. According to Lamarck's hypothesis, function precedes form (and structure) based on use and need. For example, a giraffe's neck is long (or grew long) because it reaches (or needed to reach) the leaves of tall trees. In other words, form is modified by the required function.

B. By contrast, in Darwin-Wallace theory of evolution, form (selected by natural selection) *precedes* function. In light of this theory, in case of giraffes, there already were a very few long-neck individuals among the normal-neck-length majority, the new environment favored the long-neck individuals, and they multiplied over generations.

As illustrated in Figure 8.2, physical laws involving forces of nature such as gravity and electromagnetic interaction are in action at each step of this dynamics. For example, the form of your hands fits the function of holding objects, but it is the electromagnetic interaction between the particles of your hand and those of the object that make this function possible. Similarly, the wings of a bird fit the function of flying, but when the bird is flying, the physical laws of aerodynamics are at work. Physical laws are also behind environmental changes and the resulting operation of natural selection. The physical laws act as

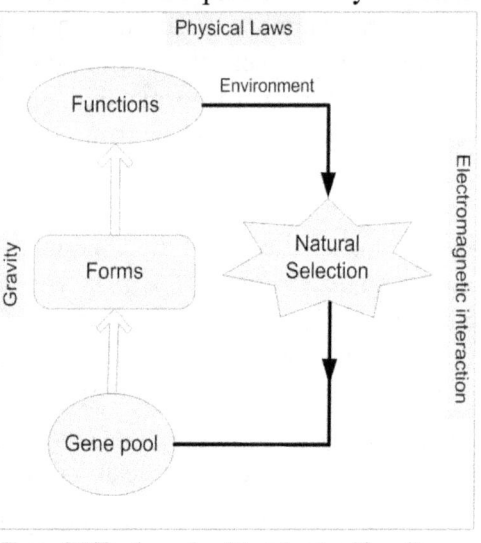

Figure 8.2 The dynamics of the principle of *form fits function* in context of physical laws and natural selection

constraints and limit the variety of available of forms, as discussed in Chapter 5.

The principle of *form fits function* runs across species. For example, streamlined bodies allow fast movement through water whether the animal is bird, fish, or mammal. This contributes to convergent evolution, the evolution of similar traits among organisms of different evolutionary lineages. This is an example of how physical laws demonstrate their power by imposing physical limits on evolution of different aspects of life such as form.

The principle of *form fits function* reveals that form and function are related. Evolution adds another element into this relationship: adaptation.

8.4 Triple Constraint of Life: Form, Function, and Adaptation

Form, function, and adaptation are interrelated concepts. An organism is defined by its form and function; form in a broader sense represents anatomy and functions represent physiology. The traits of an organism can be linked to anatomy (form) and physiology (functions). If a form trait supports a function trait in a given environment, selective advantage follows, and the traits turn into adaptations. Let us explain.

As illustrated in Figure 8.3; form, function, and adaptation influence each other in a triangular relationship. Physical laws and natural selection bind them into this relationship. Recall that form (or anatomy) is the shape and structure of an entity as distinguished from its content or material. A given form

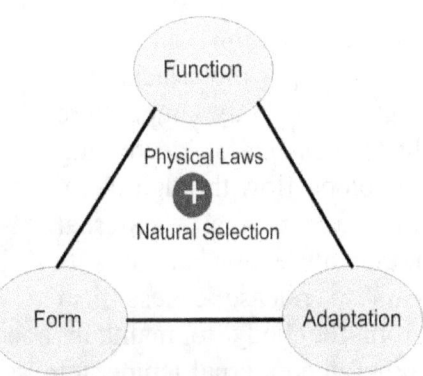

Figure 8.3 The triangular relationship among form, function, and adaptation.

consists of a set of traits or characteristics. An adaptation is an inherited characteristic of an organism selected by natural selection because it enhances its survival and reproduction in the given environment. So, an adaptation is a characteristic selected by natural selection for its current needed function or for supporting such a function. As already mentioned, all this happens within the constraints of the physical laws of nature. If a trait related to form supports (fits) a function (another trait) needed for survival and reproduction in a given environment, both traits provide selective advantage for the organism, and are adapted.

In the following, we cite a few examples out of countless cases of *form fits function*.

Think About It!

We said in the body of this chapter that a form consists of a set of traits. Does it mean all traits come from form?

Answer

No, traits can also come from function, including behavior.

Jackrabbit. As shown in Figure 8.4, thin and large ears help the jackrabbit regulate its body temperature by increasing or decreasing the blood flow through them. In a normal external temperature, this hare uses its ears to release excess heat

Figure 8.4 The large thin and flat ears of the jackrabbit are there to serve some functions.

from the body to maintain a constant internal temperature. When the external temperature is much hotter than normal for the hare, the blood vessels in the ears become narrow and thereby reduce the blood supply through the ears. This means that the ears now absorb very little heat and keep the body from overheating. When the external temperature cools down to normal, the blood vessels dilate, and the ears begin releasing any extra heat from the body again. Note how the form (anatomy) of the ears here supports the function of thermoregulation.

Leafy seadragon. Leafy seadragon (Genus: Phycodurus), shown in Figure 8.5, is a marine fish. As the name suggests, the shape of its body looks like a mythical dragon with leaf-like projections stemming out of the main body. It also has a long pipe-like snout, a forward projection of the head containing the mouth part. Even when it is swimming, it uses its leafy projections to disguise itself as floating seaweed. It is also able to change its color to blend into the surroundings. This fish uses these form

Figure 8.5 Leafy seadragon: Is this a plant or an animal?

features to prey on small fish. It uses the long pipe-like snout to suck up its prey along with water.

Stone plants. Stone plants (Genus: Lithops), also called pebble plants or living stones, are native to the Kalahari Desert of South Africa. As the name suggests, they have stone-like appearance, which helps them to blend in with the surrounding rocks and avoid being eaten by thirsty animals. Furthermore, as part of their form, their leaves are fused together, a form feature that supports the function of minimizing evaporation and therefore conserving water, which is crucial in desert-living. The

Figure 8.6 Stone plants: Are these objects stones or plants?

leaves are mostly buried under the soil surface with a narrow opening for sunlight to enter for the purpose of photosynthesis.

Cacti with spines. As mentioned earlier in this chapter, the thin and flat surface of plant leaves maximizes the absorption of sunlight, which is used in photosynthesis. As shown in Figure 8.7, the leaves of most cacti species are reduced to spines. It is the wrong form for the function of maximizing the absorption of sunlight. However it is the right form for another function: minimizing the loss of water from the plant through the leaves via evaporation. This function is critical for the survival of cacti in the desert. Then, what happens to photosynthesis? Well, the

Figure 8.7 The spines of this cactus are its leaves with wrong form; what happens to the function?

green stem of a cactus performs photosynthesis. So, this adaptation in the form of leaves has enhanced the survival and reproduction of the cacti by giving it a selective advantage.

Examples of *form fits function* in nature are countless and everywhere from bottom up and top down.

8.5 Form Fits Function from Bottom Up and Top Down

Anybody who has taken a course in biology on anatomy and physiology realizes that the *form fits function* principle is as pervasive in life as the air in our atmosphere. It is in action across species and across all organizational levels of life from molecules to organisms to ecosystems. So pervasive and almost perfectly implemented, it is no wonder that on the surface it looks like that some designer has designed life from molecules to cells, to organs, to organisms, to species, and to ecosystems. Because this is not a course in anatomy and physiology, without going into much detail, we can explore the principle of *form fits function* at all levels of life by citing just a few examples as in the following.

Cellular/molecular level. Let's begin from the forms of entities with which life begins: the haploid cells called gametes, that is, sperms and eggs. They are designed so that they fit together to fuse into one cell called a zygote. If these complementary forms and structures of male and female gametes (sperm and egg) did not fit the function of fusing together, sexual reproduction would not be possible. As another example, the form (double-helix) of a DNA molecule helps to fold it into a structure called a chromosome, which resides in the nucleus of a cell where it functions as a blueprint for the organism. Without this form, it would be impossible to keep the long DNA, about 6 feet long, in one human cell, which is only about 10 micrometers in size. No DNA in a cell means no life as we know it.

Organ and organ system level. We described earlier how crucial it is for sexual reproduction that the forms of sperm and egg fit the function of fusing into a zygote. Even before that, the male and female reproductive organs (penis and vagina) are formed to fit together appropriately to begin the reproduction process, that is, to bring the sperm and the egg together. All other organs in our body, or in the body of any other organism, such as the heart, lungs, kidneys, and liver, are designed in such a way so that their form fits their function. The same is true with organ systems such as the digestive system, circulatory system, respiratory system, and nervous system. For example, digestive systems across species vary according to their diets. If the form stops fitting the function well, it often results in disorder or disease.

And God Said: Let There Be Evolution

Organism level. *Form fits function* works between two organisms as well as between an organism and its external environment. For example, look at the long, thin beak of the humming bird, perfectly designed to get the nectar out of a flower; and a long, brushy-tipped tongue to lap up the nectar. And now look at the flower whose bright colors are designed to attract pollinators such as the hummingbird. The flower needs pollinators to facilitate reproduction. When a pollinator such as the hummingbird interacts with the flower, some amount of pollen sticks to it, and the pollinator pollinates flowers as it roams from flower to flower. As another example, the bald eagle (Figure 8.8) has forward-facing eyes with wide-field binocular vision that helps it to see as far as about one and

Figure 8.8 American bald eagle: From toes to head, formed to kill its prey.

half miles away to search for its prey, strong feet and sharp talons to capture and kill the prey, and a razor sharp dagger-like beak to tear apart the prey. The wide, long (with a span of up to about 8 feet) wings help them stay in the air in search of prey.

Think About It!

The hard coat of a plant seed fits (or serves) which function?

Answer

It protects the embryo inside, and it holds the food for the embryo.

As an example of organisms interacting with their external environment, the trunk of an elephant is well-suited to gather food, drink water, and dig. As another example, some animals grow thick warm fur in the winter, which help maintain their body temperature required for survival. We can go on forever citing examples. As mentioned earlier, the fit between form and function is generally so excellent that at the first sight it looks like a designer designed life. Well, who is then the designer? From scientific perspective, there is no need for assuming the existence of a supernatural being to explain this excellent fit of form and function. In nutshell, nature is the designer (nature includes natural forces, physical laws, environment, and mechanisms, such as natural selection). Variations in the genome and the gene pool give rise to various forms (designs), and natural selection picks the one that best suits the given environment. In other words, a form (or a feature of the form) that did not fit the function would eventually be eliminated by the filter of natural selection. And the selected form or design is the adaptation. Physical laws limit the choices available at different levels of life. The physical laws and evolution keep this universe and this world including life, running spontaneously.

Think About It!

How do animal forms (or structures) well-suited to specific functions come about?

A. Natural selection favors the most functional structure for a particular environment.

B. Mutations arise to provide required structures for survival in a particular environment.

C. An animal that needs a new function will develop it to meet the need.

D. Animals invent structural designs that enhance their fitness.

E. Animals continually improve their structures in order to improve their functions.

Answer
A

We have explained in the previous chapters, for example in Chapters 1 and 6, how the doctrine of perfect creation by a creator does not hold water in the presence of the known facts. One of those many facts is that despite the perfect looking match between form and function, life or its design is still not perfect.

8.6 Impressively Functional, Yet Not Perfect

As discussed in Chapter 4, there are three main mechanisms of evolution: genetic drift, gene flow, and natural selection. But only natural selection consistently improves the match between the population and its environment leading to what is called adaptive evolution. Even adaptive evolution was based on the best (or the better) traits in a given population, that would only mean relatively best and not on the absolute best. Also as described in Chapter 4, nature is full of examples of imperfections in life such as vestigial structures in organisms. Theory of evolution predicts the existence of these imperfections, as they are built into the process of evolution for the reasons discussed in the following:

1. **Natural selection operates on the existing set of variations.** Natural selection does not develop the best traits. This is because it only selects the best trait from the available set of traits in a given population in a given environment. This best may not be the best of the species in absolute sense or the best possible.

2. **Evolution is not purpose driven or need based.** In designing artificial objects, you can predetermine the purpose and base your design on the purpose to make the perfect (or almost perfect) product. However, evolution (or any other law of nature) is not purpose or goal driven. It is not like an organism or even a species needs something, and it consciously or intentionally develops or evolves it.

3. **Evolution remembers history.** Recall the core theme of evolution: descent with modification from common ancestry. This means it does not design everything from scratch but builds upon what is already there, for example, by modifying the existing traits. A new trait either develops from the old trait, such as the two of the already existing four limbs evolving into wings, or at least the new trait has to evolve in the framework of existing form and function. This way, each species lives with a legacy of its evolutionary history.

4. **Adaptations are superposition of many things.** As you already know, an adaptation is a heritable trait of an organism that improves its probability for survival and reproduction in a given environment. A trait is an aspect of organism's form or function, including behavior. We can build things that are designed to do one thing really well. For example, an airplane is designed to fly, a car is designed to run on the roads, a clock is designed to display accurate time, and so on. However, we, the living organisms, have to do many things to stay alive. We are not designed (by nature or evolution) to do just one thing really well. Not only do we have many traits to do different things, very often we use the same trait for multiple purposes as well. For example, we use our legs for standing, walking, running, swimming, and so on. So our

adaptations are the result of superposition of many things, that is, the compromise between many factors rather than perfection.

5. **Evolution is the result of interactions between multiple mechanisms.** As discussed in Chapter 4, evolution is the result of interactions among many mechanisms such as genetic drift, gene flow, natural selection, and environment. There are multiple possibilities and probabilities for the output of this interaction, and hence no perfection.

Due to these limitations built into the process of evolution, we do not have perfect organisms and species. Therefore, as opposed to the doctrine of perfect creation by a supernatural being, theory of evolution is able to successfully explain the abundant imperfection found at all organizational levels of life.

In nutshell, the thread of evolution connects form with function: form and functions are correlated. This is true for all organisms and at all levels of life, all the way from the molecular level to organism level.

8.7 Summary

Form fits function is a ubiquitous principle that spans across living and non-living entities. In life, the form or body plan of an organism originally emerges from its genome, which itself is a result of billions of years of evolution. The variation in form is further constrained by the physical laws of nature. Even within the same species, there is still always a variation in the form at a given point in time. Across generations, under changing external environment, natural selection selects the features of the form (traits or characteristics) that best fit the function required for the organisms to survive and reproduce. This way, natural selection, over generations, improves the fit between the form and function. *Form fits function* is a prediction of the theory of evolution and its abundant examples in life offers validity to the theory.

According to the doctrine of perfect creation, some supernatural power (such as God) designed form perfectly to fit the function. But such a conjecture offers no explanation for the variations in form. It also offers no explanation for the imperfections found in organisms despite the excellent fit between form and function. However, these imperfections arise naturally in evolution because within the framework of evolution, form features arise not from scratch but from the already existing form.

Therefore, in contrast to creationism, theory of evolution provides a very convincing explanation for the principle of *form fits function*.

In a nutshell

✓ *Form fits function* is a general principle of nature applied to both living and non-living entities.

✓ Theory of evolution predicts that form should fit the function, and the fact that it does, validates the theory.

✓ Form plays a crucial role in determining the functions of life at all organizational levels from cells to organisms.

✓ The form of atoms and molecules is determined by physical laws, whereas the form (or body plan) of organisms is determined by their genome, a result of billions of years of evolution; and physical laws also play their role in all of this.

✓ Variations in the genome appear in variations in form, and these variations are further constrained by physical laws.

- ✓ There is generally an excellent fit between form and function at all organizational levels of life, and theory of evolution explains that in terms of adaptations by natural selection.

- ✓ Despite the excellent fit between form and function, there are imperfections in living organisms, which are successfully explained within the framework of evolution.

8.8 Behold

According to the theory of evolution, the organisms that adapt well to a changed environment survive to regenerate, whereas the organisms that cannot adapt to a changed environment vanish. This natural selection improves the fit between form and function spontaneously because the fit helps the organisms to survive and reproduce. Theory of evolution explains the observed variations in form, the excellent fit between form and function, and also the observed imperfections of organisms despite the excellent fit between form and function.

How does this set with the opponents of the theory of evolution? Well, the problem is that in large part what the opponents of the theory of evolution oppose and what the theory of evolution actually is are two very different things.

In the next chapter, we dispel some myths about the theory of evolution. In other words, to understand the theory of evolution you need to know what theory of evolution is not: the things that have been erroneously attributed to the theory of evolution. We address this issue in the next chapter.

NINE

In the Name of Evolution: Eugenics and Other Deviations

And God Said: Let There Be Evolution

Gravitation is not responsible for people falling in love.

Albert Einstein

9.1 Once Upon a Time

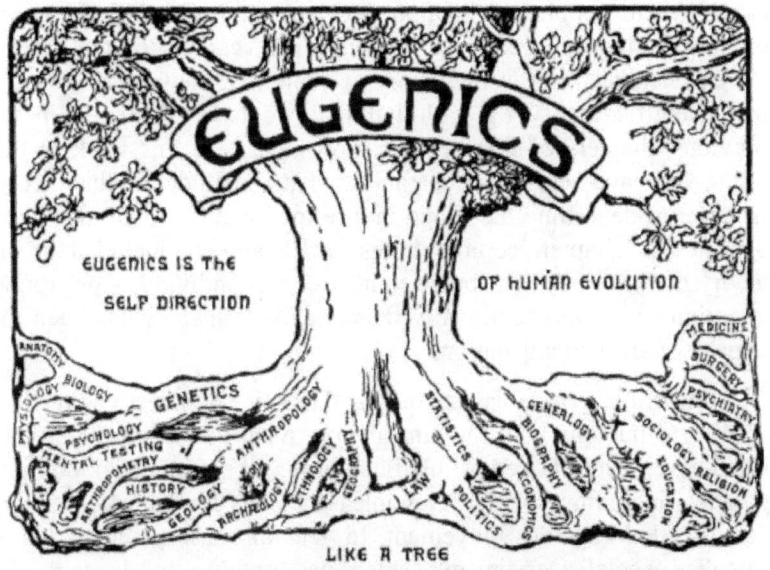

Figure 9.1 Logo of the Second International Conference of Eugenics, Museum of Natural History, New York, September 25-27, 1921.

When Einstein said, "Gravitation is not responsible for people falling in love," he was intentionally or unintentionally referring to the scope of gravity. Scope management is one of the major knowledge areas on which the whole field of project management is based. If you fail to properly determine and manage the scope of your project, the project is most likely doomed to fail. Properly trained or experienced project managers know that in order to determine the scope of a project, they need to determine both what is included in the project and what is not. This way, they draw boundaries around the scope.

And God Said: Let There Be Evolution

Because the theory of evolution has been at the center of social controversy in some areas of the world since its introduction by Darwin and Wallace about one and half centuries ago, many things have been said about it by many people with different levels of expertise and with different intentions and sociopolitical agendas. So often when people are supporting or opposing the theory of evolution, they are referring to their own versions of the theory. When scientists refer to the theory of evolution, they are referring to the scientific theory of evolution that we have discussed in the previous chapters of this book. If you have read this book to this point, you have very good idea about what is included in this theory. In order to determine the scope in a refined way, we will also point out, in this chapter, certain things which are not included in the theory but have been attributed to it by certain individuals or groups, intentionally or unintentionally. Historically, *eugenics* has been the biggest of all such attribution.

Eugenics in general is the study and practice of selective breeding applied to humans with the aim of improving the species. Darwin's ideas have been invoked as justification for all sorts of sociopolitical policies, including some very unpleasant ones implemented during the so-called eugenics movement. In general, these practices led to so-called social Darwinism, which has nothing to do with the scientific theory of evolution. Social Darwinism including eugenics flourished in the late nineteenth and early twentieth centuries. Rather than arguing against the theory of evolution directly, some opponents of the theory try to trash it by association. Before it went haywire, eugenics started off as a research effort to apply Mendel's laws to the inheritance of human traits. As different sections of the society joined eugenics with different intentions, it turned into a movement that remained widely popular in the early decades of the twentieth century, but committed suicide along with Hitler after being criminally used or having become associated with Nazi Germany. Eugenics is associated with evolution through Darwinism or social Darwinism by the opponents of the concept or theory of evolution. The fact is that neither Darwin's work nor the theory of evolution recommends the inhuman practices carried on under eugenics. Even at its pre-war height, the eugenics movement often pursued notions of racial supremacy and purity, presenting pseudoscientific arguments at best.

9.2 In the Beginning

In Chapter 5 (Section 5.13), we integrated all the well-tested components of the theory of evolution into what we called the standard theory of evolution (STE). From the previous chapters, you know the story of how the standard theory of evolution developed from 1858 − when it was originally introduced by Darwin and Wallace − to find its roots in Mendel's genetics, then in gene mutations and molecular biology, and then in quantum physics. While the roots of the theory of evolution were being discovered in molecular biology and quantum physics in the first half of the twentieth century, in parallel to all this there was hocus pocus called eugenics going on in the name of biology, evolution, and Darwinism.

In this chapter, we will consider some of the myths or hearsays, including eugenics, that are attributed to the theory of evolution, often to make a case against it. By comparing these myths to the theory of evolution, we will examine their real connection to the theory.

9.3 Busting the Top Ten Myths About the Theory of Evolution

In the following, we will discuss the top ten myths about evolution or the theory of evolution. We have selected them based on their commonality and significance. Some of them have already been discussed or pointed out in the previous chapters of this book; but here they are:

And God Said: Let There Be Evolution

1. Evolution is just a theory. First, as we explained in Chapter 2, there is no such thing in science as *just a theory*; a theory unsupported by evidence does not exist in science. Second, the theory of evolution is supported by a multitude of evidence even right from the beginning when it was introduced by Darwin and Wallace in 1858. To start with, Darwin himself, by using data he collected and analyzed, presented compelling evidence for evolution in *The Origin of Species,* and since his time, further evidence has been piling up starting from macro-organisms to micro-organisms to cells and to molecules. We have already discussed this body of evidence in this book.

Those who say evolution is just a theory obviously have made no serious effort to find out what a theory means in science, rest aside the effort to understand what the theory of evolution is. Because for some reason they believe it is *just a theory*, they do not care about its scope; they take some version of it, twist it around, create their own straw man (as they call it in philosophy or critical thinking) and start beating on it believing they are combating the theory of evolution, which they obviously are not. If we do this, we are just kidding ourselves and in the process we are doing serious damage to our next generation by confusing them on a very important science subject in this increasingly competitive world of science and technology.

2. Evolution is the theory about how life was created. Let me be very clear: evolution is not the theory of the origin of life; it is the theory about the origin of species, about how life evolved once it was here. The origin of life is a different, though related, topic (subject for another book). So a question like whether a creator created life or it evolved is the wrong question, whereas the right question is: Did a creator create (or design) species, such as human beings and fish, from scratch or did the species evolve from existing life?

3. **Individual organisms evolve.** This is a common misconception in understanding the theory of evolution. As we emphasized in Chapters 1 and 4, it is a population and not individual organisms that evolve. In a given environment, natural selection operates at organism level, which over generations results in evolution of the population. As an example, consider an island with all kinds of seeds for finches to feed on. There are finches with beaks of all sizes, each size suitable for a specific kind of seed. Now consider a change in the environment: a few years of drought. This change favors the finches with larger and sharper beaks that are able to crack hard and large seeds because these are the only seeds which are abundantly available during the draught and short and soft seeds are in short supply. Now, the finches with short beaks will not grow their beaks larger; they are at a disadvantage due to the change in the environment, and their chances for survival and reproduction are diminished. As a result, their numbers in the population decrease. In other words, the average beak size of the finch population increases in a few years over few generations. So, it is the population that has evolved and not the individuals. Individuals are selected; the population evolves as a result of this selection. As described in Chapter 6, these kinds of examples of evolution have been observed by researchers.

Another misconception that is an extension of this misconception: not only that organisms evolve but they evolve at their own will.

4. **According to the theory of evolution, organisms modify traits or develop new traits to adapt to the environment.** The opponents of evolution, while beating on the theory of

evolution, still assume that according to the theory, organisms modify traits and develop new traits in order to adapt to the changed environment. In other words, individual organisms evolve at will. You know from the previous chapters that this idea is more related to Jean-Baptiste Lamarck's discredited ideas of evolution which never became part of the theory of evolution. In the framework of the theory of evolution, there is no will or consciousness involved in the process of evolution. Adaptation only happens when the right kind of trait variation based on genetic mutation is already there, and in that case the adaptation happens in a natural way. According to the standard theory of evolution, traits have their origin in genes. In a given environment, the genes corresponding to the traits favorable to the environment dominate the gene pool of a population. There may be some rare genes which are not favored by the environment. When the environment changes, traits corresponding to rare genes may become favorable and the traits corresponding to some dominant genes may become unfavorable in the changed environment. Over generations, the individuals with previously dominant genes which are no longer favorable will disappear because they will not reproduce as much as the organisms with previously rare genes, which are now favored by the environment. Eventually, the traits corresponding to previously rare genes will become dominant in the population. This is the process of natural selection that occurs naturally, independent of desire or want.

5. Evolution occurs for the good or betterment of the species. It is neither the claim nor the prediction of the theory of evolution that natural selection operates for the good of the species. Yet this false notion exists. The fact is that evolution or natural selection has no mind or intention to act for the good or the bad of a species. It is a process that occurs inevitably in accordance with fundamental scientific laws. So, adaptation happening in the framework of natural selection does not always increase the quality or abilities of the species in the absolute or global sense, but it always increases the fitness of the individuals of the species with favorable traits and of the resulting population over generations within the existing environment. There is no such thing as absolute fitness; it's always related to the given environment. Natural selection favors those individuals who are able to adapt for fitness to the environment, and eliminates those who are not able to adapt. Over generations this results in changed (or new) traits of the population based on a changed genetic pool of the population; and hence we have a new species evolved from the old species.

As discussed in Chapter 8 and elsewhere in the book, evolution is not purpose or goal oriented.

6. Everything in evolution is based on mere chance or accident. In order to mock the theory of evolution, its opponents often oppose it by saying how we, humans, cannot be a product of chance or a mere accident. The premise of the argument is that according to the theory of evolution, everything in evolution occurs by chance, by a mere accident. It should be clear from the previous chapters of this book that this premise about the theory of evolution is plain false. As you learned in Chapters 4 and 5, the engine of evolution includes genetic mutations and different versions of each gene, which generate variations among traits or individuals; and the process of natural selection filters these variations to facilitate adaptations. There is no chance or accident in the process of natural selection. As explained Chapters 4 and 5, the randomness involved in mutations is also not chance or accident in the sense the creationists present it. It is scientific chance called probability, and its very nature is not any different from the uncertainty involved in quantum physics that explains the atomic nature of all things material. And there is nothing uncertain about the uncertainty principle of quantum physics. It plays an important role in keeping atoms intact inside molecules and inside non-molecular matter and keeps the Universe going. The same kind of role randomness (or uncertainty) in genetic mutations and in the reproduction process plays in keeping the living world going. So, the randomness involved in genetic mutations and reproduction, components of the evolutionary engine, is very much a part of scientific law. It also means that the probability (or randomness) in evolution-related life processes occurring spontaneously inside an organism are independent of whether they would be helpful or harmful to the organism. This indifference or spontaneity is also a part of the meaning of *randomness* here.

7. **Darwin is the ultimate authority on evolution.** Opponents of evolution such as *creationists* often mistake Darwin and evolution as synonymous, as if one person called Charles Darwin made up this theory called evolution. Nothing could be further from the truth. As explained in Chapter 1, various ideas about the evolution of life were already there long before Darwin came up with his theory. Darwin's contribution was doubtlessly great as he presented compelling evidence for evolution assembled in his book *On the Origin of Species*, and provided a testable mechanism for evolution: natural selection. The theory was originally introduced not just by Darwin, but by Darwin and Wallace. By analyzing data, both had independently reached the same conclusion: evolution from common ancestor (or ancestors) by natural selection. As explained in this book, neither Darwin nor anybody else at that time had the answers to the questions that were needed to advance the theory further. For example, nobody knew the cause of variations (on which natural selection operates), which

was later discovered in genes and gene mutations; well, Mendel did, but his work did not come to enough light to meet the theory. Long story short, what we call the theory of evolution today is an extension and refinement of the ideas of Darwin and Wallace.

Note. Darwin and Wallace mark the time before which almost all people were creationists who believed in the doctrine of perfect creation of species by divine power. The introduction of the theory of evolution by Darwin and Wallace and the publication of *The Origin of Species* by Darwin had such a revolutionary impact on us by dispelling this centuries-held notion of perfect creation that some of us have still not recovered from this shock, even after one and a half century of continuous testing and piling up of the evidence for the theory. This is one way of understanding the fierce and irrational opposition by many to the concept of evolution.

Some supporters of the theory, even some scientists, add to confusion by using the terms Darwinism and the theory of evolution synonymously. By doing so, they themselves open the door to confusion and misunderstanding and help a process like the following to occur:

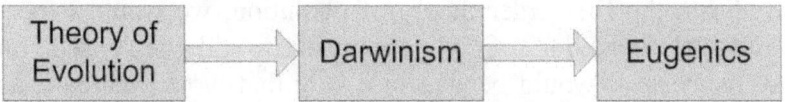

The fact is that the theory of evolution was originally built on some (but not all) ideas of Darwin and Wallace but has advanced far beyond them.

8. Humans evolved from apes or chimpanzees. This misconception leads to questions like this one: If humans evolved from apes, monkeys, or chimpanzees; why are these animals not evolving into humans anymore? Such a question also exposes another misunderstanding or lack of understanding of another aspect of evolution: a given species evolves from another species under certain environmental conditions. Also, the fact is that humans did not evolve from apes, monkeys, or chimpanzees; instead the claim of evolutionary theory supported by scientific findings is that these animals are evolutionary cousins of humans, and humans along with these animals evolved from a common ancestor that lived millions of years ago in the past and does not exist anymore.

9. Evolution cannot account for morality. A significant number of people who believe in religion usually assign their morality to their belief and are unable to understand how one can be moral without believing in God. An extension of this notion is that evolution cannot account for morality in humans. Even though the definition of morality can vary over individuals, religions, and cultures; you do not need to believe in religion or God in order to have morality or a sense of right and wrong. Moreover, morality, or sense of right or wrong, is very compatible with and to some extent part of evolution. We will illustrate this through an example.

From the first order theory of evolution, we would expect that the behavior of an organism would be based on selfishness: It would behave in a way that would increase its own fitness, the ability to survive and reproduce. This expectation works most of the time. However, in nature, organisms, not only humans but also other animals, behave in a way that can only be understood if we apply second order (perturbative) effects or corrections to the first order theory of evolution. For example, there is a body of evidence to support that sometime some organisms behave in a way that goes against selfishness, and increases the fitness of other individuals in the population at the expense of their own individual fitness. Such a behavior is called *altruism*. According to the theory of evolution, the evolutionary basis for altruism is *inclusive fitness*, which is measured as the sum of an organisms own offspring (direct fitness), and of the equivalents of its own offspring (indirect fitness), such as nephews and nieces, that it can add to the population by helping others.

The point here is that morality including altruism and the sense of right and wrong, even parenting, are traits that have evolved in certain animals including humans to increase their overall (inclusive) fitness. This is a well-studied area of evolution.

Think About It!
Most (if not all) of us have a moral sense inside of us that tells us that things like lying, stealing, cheating, and adultery are wrong. How is this sense related to evolution?

Answer
We evolved this sense as a part of our evolution into a social primate species. Immoral things such as lying, stealing, cheating, and adultery damage or destroy the trust of other people in us and it decreases our fitness. Some moral sense becomes necessary for survival and fitness of a social primate species like us. So, the cooperative social behavior that includes some moral sense built into all human societies is based on human nature that has evolved according to the principles of evolution.

Another deviation of the theory of evolution, a big one, is social Darwinism including eugenics.

10. Social Darwinism including eugenics is part or an application of evolution. Social Darwinism including eugenics, which flourished in the late nineteenth and early twentieth centuries, was at best a pseudoscientific attempt to

And God Said: Let There Be Evolution

apply Darwin's ideas to social planning. The extremes of distortion and misunderstanding of the theory of evolution are represented by Nazi ideologues and neoliberal economists, reflected from comments such as those made by American novelist Kurt Vonnegut that Darwin, "taught that those who die are meant to die, that corpse are improvements."

The opponents, especially the vocal opponents, of the theory of evolution often focus on the stuff that is being said in the name of theory of evolution, and in fact is not part of the scientific theory of evolution. Most of that stuff has become part of what is called Darwinism or social Darwinism. They (opponents) present arguments against social Darwinism or eugenics, beat it with their arguments, and then think or claim that they have beaten the theory of evolution. As mentioned earlier, in philosophy or critical thinking, there is a name for this kind of activity: beating the straw man.

Think About It!

Are Darwinism and the theory of evolution one in the same?

Answer

No.

Unfortunately, by not being precise with their language, some biologists are also responsible to promote this myth that Darwinism and theory of evolution are synonymous.

Some elements of social Darwinism are used to try to trash the theory of evolution by association rather than directly arguing against the theory itself. It is true that Darwin's ideas have been invoked as justification for all sorts of policies, including some very unpleasant ones implemented during the so called eugenics movement. But keep in mind, the theory of evolution, just like any other scientific theory, is a descriptive science. It cannot tell us what to do or what is right and what is wrong.

So, what really is eugenics?

9.4 Eugenics: The Ugliest Face of Social Darwinism

Social Darwinism in general is an ideology that endorses applying concepts of biological evolution to non-biological areas such as sociology and politics. The arguments of social Darwinism go, for example, like this: competition or conflicts among social groups lead to social progress by selecting the superior groups against the inferior groups. But remember, this is at best an analogy of a science principle used in society, and not an application or a scientific extension of the principle.

Eugenics, a part of social Darwinism, in general is the study and practice of applying selective breeding to humans with the aim of improving the species. The term eugenics was coined by an English scientist, Sir Francis Galton (1822-1911), in 1883. The word *eugenics* means well born and has a Greek origin: *eu* for *good* or *well* and the suffix *-genēs* for *born*. It is fair to say that before it went haywire, eugenics started off as a research effort to apply Mendel's laws to the inheritance of human traits. As different sections of society joined eugenics with different intentions and agendas, it turned into a movement that remained widely popular in the early decades of the twentieth century, but committed suicide along with Hitler after being criminally used or having become associated with Nazi Germany.

To set the record straight, eugenics was practiced in different forms at different levels around the world and was promoted by governments, well-reputed institutions, and socially influential and respected individuals. Even its advocates regarded it as not so much a science, but as a social philosophy for the improvement of human hereditary traits. To implement this philosophy, they promoted higher reproduction of certain people and traits, and the reduction of reproduction of certain people and traits. And when you indulge in that kind of practice, where do you draw the line?

Over time, eugenics in general has encompassed a range of views and practices including the following:

✓ Eliminate hereditary diseases such as hemophilia and Huntington's disease.

✓ Encourage the "fit" in breeding and breeding more. The definition of "fit" varied from the educated classes such as the eugenicists themselves to specific races determined by racial supremacists.

✓ Discourage marriages by those who carried a bad gene; again the definition of a bad gene went beyond the realm of science.

✓ Impose reproductive restrictions on those who are "unfit" such as the "feebleminded" in order to promote fitness in society.

✓ Eliminate an entire ethnic group that is deemed to be "unfit." This view was implemented in Nazi Europe, part of Europe occupied by Nazi Germany, to the Jews, Gypsies, and slaves.

Even though eugenics is associated with evolution or Darwinism by the opponents of the concept or the theory of evolution, the fact is that neither Darwin's work nor theory of evolution recommends the inhuman practices carried on under eugenics, as we will discuss further on in this chapter. Even at its pre-war height, the eugenics movement often pursued notions of racial supremacy and purity, presenting pseudoscientific arguments at best.

Then what kinds of people were involved in eugenics anyway?

9.5 Faces of the Eugenics Movement: Darwin and Wallace are Missing

The Three International Eugenics Conferences, which took place between 1912 and 1932, are a good representation of the high times and the global scope of this movement. These conferences served as the global venue for scientists, politicians, and social leaders to come together to discuss and plan the application of programs to "improve" human heredity.

First Eugenics Congress. The *First International Eugenics Congress* (Conference), which was organized by the British Eugenics Education Society, took place at the Hotel Cecil in London, England on July 24–29, 1912. This five-day-long conference, attended by about 400 delegates, was dedicated to Francis Galton, the scientist who coined the term eugenics, who had passed away the year before. It was presided over by none other than Major Leonard Darwin, son of Charles Darwin. Furthermore, the English Vice-presidents of the Congress included a future Prime Minister, First Lord of the Admiralty, Winston Churchill; the Lord Chief Justice, Lord Alverstone; the Bishop of Oxford; and the President of the College of Physicians, Sir Thomas Barlow. The German Vice-presidents included M. von Gruber, Professor of Hygiene at Munich; and Dr. Alfred Ploetz, President of the International Society for Racial Hygiene. The American Vice-presidents included Gifford Pinchot, a future governor of Pennsylvania;

And God Said: Let There Be Evolution

Charles W. Eliot, President Emeritus of Harvard University; Alexander Graham Bell, the inventor of the telephone; David Starr Jordan, President of Stanford University; and Charles B. Davenport, who was listed on the program as Secretary of the American Breeders Association. Ideas on better breeding procedures for humans were presented and discussed. The American exhibit at the conference, sponsored by the American Breeders Association, demonstrated the incidence of hereditary defects in human pedigrees (lines of descent or ancestry). A report by Bleeker van Wagenen presented information about American sterilization laws and promoted compulsory sterilization as the best method to cut off "defective germ-plasm."

Think About It!

A. Charles Darwin's son presided over the first Eugenics Congress. Does it mean that the theory of evolution supported eugenics?

B. Some scientists and university representatives were there at the Eugenics Congress; does it mean that activists in the eugenics movement were performing their scientific responsibilities?

Answers

A. No. Even if Darwin himself attended the conference, it does not necessarily mean that the theory of evolution supported eugenics.

B. No. An activity is deemed a scientific activity or not by its own virtue, and is determined by the definition of science, and not because the person who does it happens to be a scientist or not.

Second Eugenics Congress. The second Congress met at the American Museum of Natural History in New York during September 25–27, 1921 and was presided over by Henry Fairfield Osborn (1857-1935), a professor of biology and zoology at Columbia University and President of the Museum. Alexander Graham Bell was the honorary president of this Congress. The U.S. State Department mailed invitations around the world. With its lion's share of 41 out of 53 papers presented at the conference, the U.S. maintained its leadership and dominance throughout the conference. The meeting included delegates from Europe, North America, Latin America (Cuba, Mexico, San Salvador, Uruguay, and Venezuela), and Asia (India, Japan, Siam). Major Darwin participated as a prominent guest speaker. The ideas that were advocated and pushed forward in this conference included the encouragement of large families in the "well-endowed," the discouragement of large families in the "ill-endowed," and the "elimination of the unfit."

Third Eugenics Conference. The third Congress was also held at the American Museum of Natural History in New York City, on August 22–23, 1932. The conference was dedicated to Mary Williamson Averell who had provided significant financial support. The meeting was presided over by Charles Davenport, the founder of the U.S. Eugenics Cold Springs Research Facility. Osborn, in his address, emphasized birth selection over birth control as a method to improve the offspring. Furthermore, audience members were very comfortable with F. Ramos from Cuba proposing that immigrants should be carefully checked for harmful traits, and suggested deportation of their descendants if inadmissible traits would become later apparent. Major Darwin, 88 years old at the time, was unable to attend but sent a report, which was presented by Ronald Fisher, an English statistician and a biologist, predicting the doom of civilization unless eugenic measures were implemented. Ernst Rüdin (1874-1952), a Swiss psychiatrist who at the time was Director of the Genealogical-Demographic Department at the German Institute for Psychiatric Research, was unanimously elected President of the International Federation of Eugenics Societies. This specific event marked the beginning of the demise of the eugenics movement.

A few months later, when Adolf Hitler came into power in Germany in January 1933, Ernst Rüdin (same Rüdin who was elected president by the third congress a few months ago), Alfred Ploetz, and several other experts on racial hygiene were brought together to form the Expert Committee on Questions of Population and Racial Policy under Wilhelm Frick, the Minister of the Interior. The committee's ideas were used as a scientific basis to justify the racial policy of Nazi Germany. After 1933, the Nazi government and the Nazi party of Germany endorsed Rüdin's work by supplying financial and manpower support. The health policy of the Nazi government required a scientific basis (rather excuses) to justify its actions, and Rüdin proved to be a great asset. In 1934, Rüdin prepared (or wrote) the official commentary on the racial policy of Nazi Germany: *Law for the Prevention of Genetically Diseased Offspring*. He was awarded medals by the Nazis and Adolf Hitler personally. The Law was passed by the German government on January 1, 1934. It set off the march of Nazi Germany toward the Holocaust.

This is how Ernst Rüdin, who during the third conference was unanimously elected President of the International Federation of Eugenics Societies, played into the hands of the Nazi government. No further International Eugenics Conference was held.

In the U.S., eugenic supporters included Theodore Roosevelt, an American President, and even science organizations such as the National Academy of Sciences and the National Research Council. So-called eugenics research was funded by distinguished philanthropies and government funding agencies, and carried out at prestigious universities.

This brief description of the three eugenic conferences gives you some idea about the diversity of the people involved in the eugenics movement. Now, let's see the real impact of the movement on people's lives.

9.6 Implementation of Eugenics

To appreciate the wide scope of the eugenics movement, realize that eugenics was implemented in many areas of personal and social life including education, immigration, law, marriage, and birth control.

9.6.1 Eugenics Plaguing Education and Society

So prevalent was the influence of this movement, that by around 1914, America's leading universities including Harvard, Columbia, Cornell, and Brown had started offering courses in eugenics or courses that included eugenics. In the 1920s, the National Education Association's Committee on Racial Well-Being sponsored programs to help college teachers integrate eugenic content into the existing courses. By 1928, eugenics was a topic in hundreds of separate college courses being taken by thousands of students. Well, it is understandable that some universities and colleges catch up with current trends rather quickly and begin offering courses. What is surprising is that, according to content analysis of high school science texts published between 1914 and 1948, eugenics in these courses was being largely presented as a legitimate science. So, in the name of science, this education, by embracing Galton's concept of differential birthrates between the biological "fit" and "unfit," was teaching high school students in the U.S. that immigration restriction, segregation, and sterilization were worthy policies to maintain in American culture.

In many ways, eugenic ideology became a part of American popular culture during the 1920s and 1930s. As Dr. Steven Selden, professor at University of Maryland, put it:

> On Saturday night, high school students might go to the cinema to see "The Black Stork" – a film that supported eugenic sterilization. On Sunday, in church, they might listen to a sermon selected for an award by the American Eugenics Society – learning that human improvement required marriages of society's "best" with the "best." On a field trip to a state fair with their hygiene class, students might sign up for a eugenic evaluation at a Fitter Families Exhibit – hoping to win a medal claiming, "Yea I Have A Goodly Heritage." Back in school, these same students might open their biology textbooks to the chapter on eugenics – which recommended the eugenic policies of immigration restriction, sterilization, and race segregation.

This was happening in the 1920s. Those who try to tie eugenics with evolution should know that a subset of people who were supporting passing laws to implement eugenic policies were also in the same time period pushing to pass laws against teaching evolution in our schools. For example, as you will learn in the next chapter, on May 5, 1925, John Thomas Scopes, a biology teacher in Dayton, Tennessee, was arrested for teaching evolution, and thereby violating The Butler Act, an act of the state legislature which prohibited the teaching of human evolution in schools.

Eugenics was being used as a weapon to selectively and discriminately oppose immigration.

9.6.2 Eugenics Entering Law and Politics: Restricting Immigration

To be fair to eugenicists, race- or ethnicity-based selective immigration restriction did not begin with eugenics. However, eugenics offered an argument to justify and widen the scope of such restrictions already in place at that time. America's first naturalization law was passed by Congress in 1790, which limited the privilege of U.S. citizenship to "free white persons." Other people could become residents of the country but not citizens. About a century later, under the influence of eugenics, immigration laws began to restrict who could enter the country as well. The first American entity officially associated with eugenics was The Immigration Restriction League, founded in 1894. The League noticed the higher numbers of Southern and Eastern European immigrants entering the United States and Canada. The League made an organized effort to bar members of certain races from entering America because it viewed them as dysgenic and considered them diluting "the superior American racial stock" through procreation. Based on the belief that literacy rates were low among "inferior races," the League lobbied for a literacy test for immigrants, hoping that the test would work as a filter. The literacy test bills had been passed by the U.S. Congress but vetoed by the President in 1897, 1913, and 1915. Eventually, President Woodrow Wilson's second veto was overruled by the Congress in 1917. According to this act, Asians, including Indians, were altogether barred from immigrating to the U.S.

To give you some idea about how the Immigration Restriction League that championed this cause looked like, here is a list of some individuals included in the League: A.

And God Said: Let There Be Evolution

Lawrence Lowell, President of Harvard University; William DeWitt Hyde, President of Bowdoin College; David Starr Jordan, President of Stanford University; George F. Edmunds, U.S. Senator; and Madison Grant, famous author of *The Passing of the Great Race*. Demonstrating the clear connection with eugenics, the League made a coalition with the American Breeder's Association (ABA) to gain influence in order to further its goals. In 1911, the League President, Prescott Hall, asked Charles Davenport of the Eugenics Record Office (ERO) for assistance in influencing Congressional debate on immigration. Davenport designed a survey plan to determine the national origins of "hereditary defectives" in American prisons, mental hospitals and other charitable institutions. He appointed his colleague at ERO, Harry Laughlin, to manage this research program.

> Note. In this book, by Indians we mean people of India or people with India as their ethnic origin.

In 1920, Laughlin appeared before the U.S. House of Representatives Committee on Immigration and Naturalization. Based on the data from his "research," Laughlin argued that the "American" gene pool was being polluted by a rising tide of intellectually and morally defective immigrants – primarily from Eastern and Southern Europe (remember Asians by then were largely barred from entering America altogether). Influenced by Laughlin's argument and message, the committee chairman, Albert Johnson, a U.S. Representative from the State of Washington, appointed Laughlin as *expert eugenics agent*. In this capacity, Laughlin conducted research, which became the basis for his 1924 testimony to Congress in support of a eugenically-crafted immigration restriction bill. The whole idea was to pass an immigration law that would discriminate immigration to the U.S. based on a hierarchy of nationalities, rating them from the most desirable Anglo-Saxon and Nordic peoples to the least desirable Asians such as Chinese, Indians, and Japanese (who were already almost completely banned from entering the U.S. anyway).

> *Nordic*: of, relating to, or belonging to a subdivision of the Caucasoid race typified by the tall, blond, blue-eyed, long-headed inhabitants of N. Britain, Scandinavia, N. Germany, and the Netherlands

Some of the strongest supporters of The Immigration Act of 1924 were influenced by Madison Grant, a member of the Immigration Restriction League, and his 1916 book, *The Passing of the Great Race*. The Act was aimed at further restricting the Southern and Eastern Europeans who were immigrating in large numbers starting in the 1890s, as well as prohibiting (or continue prohibiting) the immigration of East Asians and Asian Indians. Congressman Albert Johnson and Senator David Reed, both Republicans, were the two main architects of the Act. The Act established the "national origins quota system" to deal with the threat of "inferior stock" from Eastern and Southern Europe. To be specific, the law was designed consciously to halt the immigration of supposedly "dysgenic" Italians, Eastern European Jews, Catholics, and some other non-Anglo-Saxon and non-Nordic groups whose numbers had been collectively "mushrooming" during the period from 1900 to 1920. Furthermore, the Act barred altogether people from specific origins from entering the country: origins from the Asia-Pacific Triangle, which included Burma (Myanmar), Ceylon (Sri Lanka), China, Dutch East Indies (Indonesia), France, India, Indochina (Cambodia, Laos, Vietnam), Japan, Korea, Malaysia, the Philippines, Siam (Thailand), and Singapore. Based on the Naturalization Act of

1790, these immigrants, being non-white, were not eligible for naturalization. The 1924 Act forbade further immigration of any person who was not eligible for naturalization. The new act, in a way, endorsed the eugenic belief in the racial superiority of "old stock" white Americans as members of the "Nordic race" (a form of white supremacy), and strengthened the position of existing laws prohibiting race-mixing.

Upon signing the 1924 Immigration Act into law, President Calvin Coolidge, a Republican, commented, "America must remain American." This presidential comment became the rallying cry of anti-immigration sentiments in the U.S. until after World War II. The eugenic implementation of the 1924 law including the quota system it established remained in place until they were repealed by the Immigration and Nationality Act of 1965. The 1965 act, which abolished the National Origins Formula, was proposed by Representative Emanuel Celler, a democrat from New York; co-sponsored by Senator Philip Hart, a democrat from Michigan; and championed by Senator Ted Kennedy, a democrat from Massachusetts.

While race-based laws were being passed and enforced to determine who could enter the U.S., people inside the U.S. were being told who could have children and who could not.

9.6.3 Eugenics Entering Law and Politics: Forced Sterilization

Sterilization mandated by law was perhaps the most radical practice that originated from the eugenics movement in the U.S. The first law allowing sterilization using the arguments of eugenics was enacted by the State of Indiana in 1907. The state of Connecticut followed suit soon after. Harry Laughlin, the director of the Eugenics Record Office, from its inception in 1910 to its closing in 1939, pushed for the law to force sterilization. As a matter of fact, this effort was Laughlin's first major project at the Eugenics Record Office. In 1914, he published a Model Eugenical Sterilization Law proposing to authorize forced sterilization of the "socially inadequate" who were the people supported in institutions or "maintained wholly or in part by public expense." The Model Law included, in the list of *inadequate people*, the "feebleminded, insane, criminalists, epileptic, inebriate, diseased, blind, deaf, deformed, dependents, orphans, ne'er-do-wells, tramps, homeless, and paupers." To give you an idea of how quickly things were moving: by the time the Model Law was published in 1914, twelve states had already implemented it into their sterilization laws.

Sterilization really gained popular social approval during the 1920s. Approximately 3,000 people had been involuntarily sterilized in the U.S. by 1924, in which California had its lion's share: 2,500. Also in 1924, Virginia passed a sterilization act based on Laughlin's Model Law, and was even named the *Eugenical Sterilization Act*. It was presented as part of a cost-saving strategy to relieve the tax burden in a state where public facilities for the "insane" and "feebleminded" had experienced rapid growth. The law asserted that "heredity plays an important part in the transmission of insanity, idiocy, imbecility, epilepsy and crime." It focused on "defective persons" whose reproduction represented "a menace to society." About 8,300 Virginians were sterilized.

Eventually thirty-three states adopted legislations allowing forced (or compulsory) sterilization of certain individuals or groups. In a unanimous decision, the U.S. Supreme Court, in the case of Skinner v. State of Oklahoma, found on June 1, 1942, that the Oklahoma law providing for involuntary sterilization violated the Equal Protection Clause of the

And God Said: Let There Be Evolution

Fourteenth Amendment of the U.S. Constitution. Here is an interesting fact: Despite the decision of the Supreme Court that involuntary sterilization was unconstitutional, sterilization of people in institutions for the mentally ill and mentally retarded continued through the 1970's. More than 60,000 Americans endured involuntary sterilization. The last known forced sterilization in the U.S. was performed in 1978 in the state of Oregon, even though the state did not repeal its forced sterilization law until 1983.

So, you can see that the Germans were neither the original creators nor the first to enforce state-sanctioned sterilization. By the 1920s, half of the states in the U.S. had already enacted sterilization laws which included forced sterilization of the criminally insane along with other groups. It was not until July 14, 1933, about six months after Hitler came into power (became Chancellor of Germany), that the first German sterilization law was enacted, which allowed the forced sterilization for anyone suffering from any of a wide spectrum of conditions or diseases including: alcoholism, congenital feeblemindedness, deafness, epilepsy, genetic blindness, hereditary manic depression, Huntington's chorea (a brain disorder), and schizophrenia (mental disorder). The Nazi government sterilized over 450,000 people in less than a decade. The process to rid society of "the unfit" was initiated in Nazi Germany by pointing to the related steps taken in other European countries and by referring to the leadership provided by the United States. During their trial for war crimes in Nuremberg after World War II, Nazi administrators tried to justify the mass sterilizations by citing the lead by the United States as their inspiration.

Ethnic prejudices and the triplet of economical, political, and racial interests played their roles in eugenics. For example, sterilization gained support as a means of reducing costs for institutional care and savings from relief funds for the poor and disabled. It's not a simple coincidence that sterilization rates climbed in some American states during the Depression. During the same period, forced sterilization laws were also passed in Finland, Norway, and Sweden. Nowhere, however, did the numbers of individuals sterilized come even close to the mass sterilization in Nazi Germany.

Not only was the freedom to reproduce under attack, but also the freedom to marry.

9.6.4 Law and Politics: Barring Interracial Marriages

Laws barring interracial marriages (marriages between people from different races) and interracial sex were common in America from the Colonial period through the middle of the twentieth century. For example, by 1915, twenty-eight states had laws prohibiting marriages between "Negroes and white persons" and six states included this prohibition in their constitutions. These laws practically banned the marriage between whites and non-white groups, which included primarily blacks, but often also Asians and Native Americans, which were also referred to as Indians. The scope (or implementation) of these laws often included criminalizing cohabitation and sex between whites and non-whites.

The eugenics movement supplied a new set of arguments not only to justify existing racism and restrictions on interracial marriage but to take them to the next level. Racism flourished in the name of science as alleged biological dangers of mixing the races were emphasized. Influential writers such as Madison Grant, who was also a leading eugenicist, branded racial mixing as "a social and racial crime." Grant warned that racial intermarriage would lead America toward "racial suicide" and the subsequent disappearance of white civilization. In his widely acclaimed best seller book at that time, *The Passing of the Great Race* (1916), Grant cautioned: "The cross between a white man and an Indian is an Indian; the cross between a white man and a Negro is a Negro." Note here how Mendelian genetics is being incorrectly used here to make racism look scientific.

Think About It!

Show how Grant's following claim is incorrect according to Mendelian genetics:

"The cross between a white man and an Indian is an Indian; the cross between a white man and a Negro is a Negro."

Answer

According to Mendelian genetics (and also modern genetics), the offspring inherits half of its genes from each the two parents regardless of their race.

Note that Madison Grant was not an isolated case. His views reflected the views of the eugenics movement and his racist warning mirrored the warning issued by other eugenics leaders such as Charles Davenport and Harry Laughlin, leaders of the American eugenic bureaucracy at the Eugenics Record Office. Furthermore, American political leaders expressed similar sentiments. For example, Vice-President Calvin Coolidge presented this sentiment as a scientific fact, when he said: "Biological laws tell us that certain divergent people will not mix or blend."

Like it or not, here is a clear link between what was happening in the U.S. in those days and what was happening in Nazi Germany: Adolf Hitler called Madison Grant's book "my bible." Beginning with Connecticut in 1896, many U.S. states enacted marriage laws with eugenic criteria, which included prohibiting anyone who was "epileptic, imbecile or feeble-minded" from marrying altogether. Eugenics was also used to formulate more restrictive definitions of white racial purity in existing state laws banning interracial marriage; these laws were called anti-miscegenation laws. The most famous example of using eugenics to promote strict racial segregation through such anti-miscegenation laws was Virginia's Racial Integrity Act of 1924. The legislative strategy was designed by three local Virginia eugenicists – John Powell, Earnest Cox and Walter Plecker – after consulting with Madison Grant and Harry Laughlin, the director of the Eugenics Record Office. Powell was the founder of the Anglo-Saxon Clubs of America, an elitist version of the Ku Klux Klan dedicated to maintaining "Anglo-Saxon ideals and civilization in America." Earnest Cox was the author of the book *White America*, which like *The Passing of the Great Race*, emphasized white supremacy and warned against the dangers of racial mixing. Walter Plecker was the registrar at the Bureau of Vital Statistics of the Virginia Board of Health. His ideas on racial interbreeding as the source of "public health problems" appeared in pamphlets which were published by the state of Virginia and distributed to all who were planning to marry. So, you see the clear collaboration here of eugenicists, governments, and the racists elements of society.

As a law, The Racial Integrity Act included provisions which required racial registration certificates, and strict definitions of who would qualify as members of the white race. The law claimed the "scientific" basis of race assessment, and the "dysgenic" dangers of race mixing. The major provision of the law declared: "It shall hereafter be unlawful for any white person in this State to marry any save a white person, or a person with no other admixture of blood than white and

And God Said: Let There Be Evolution

American Indian. ...the term "white person" shall apply only to such person as has no trace whatever of any blood other than Caucasian; but persons who have one-sixteenth or less of the blood of the American Indian and have no other non-Caucasic blood shall be deemed to be white persons...."

Adding to this ridiculous racial drama, at least sixteen members of the Virginia General Assembly claimed to be descendants of Pocahontas, the daughter of a Virginia Indian chief. Pocahontas (1595 – 1617) is notable for having assisted colonial settlers at Jamestown in present-day Virginia. She converted to Christianity and married the English settler John Rolfe. She was the daughter of Powhatan, the tribal chief of the Powhatan Confederacy, tributary tribes that numbered 14,000 to 21,000 people. In May 1607, the English arrived and began building settlements in Virginia. When a colonist, John Smith, captured by a group of Powhatan hunters, was about to be executed, Pocahontas saved him. This earned her respect from people of the English Settlements.

Figure 9.2 The 19th century illustration of Pocahontas saving the life of Captain John Smith, 1870. Source: Library of Congress, Prints and Photograph Division, LC- USZC4-3368.

So, the "Pocahontas legislators" objected to the first draft of the law that was proposed because it defined "non-white" as anyone with at least 1/64 of American Indian ancestry. So the limit on the maximum amount of American Indian ancestry was relaxed from 1/64 to 1/16 to accommodate the complaining legislators. Soon Alabama and Georgia followed suit by literally copying the Virginia law.

Within a decade, similar laws prohibiting inter-racial (or inter-ethnic) marriages were adopted by the government of Nazi Germany, which attempted to sort citizens by percentage of Jewish "blood." The Nazis gratefully borrowed the eugenics concept and ran with it. Generally speaking, the eugenics crusade started in the U.S., proliferated into a worldwide campaign, and came to the attention of Adolf Hitler in the 1920s. When Hitler came into power in the 1930s, eugenic principles already applied in the U.S. were applied under the Nazis, only to an extreme and without restraint, resulting in the infamous genocide.

Note. The U.S. Supreme Court overturned eugenics-based racially discriminating marriage laws in 1967 in *Loving v. Virginia*, as it declared anti-miscegenation laws unconstitutional.

After reading this brief account of eugenics, some of you may already be wondering what it has to do with the subject of this book, evolution.

9.7 Eugenics and Evolution: Where is the Connection?

Because eugenics is linked with evolution to criticize or invalidate evolution, the question arises: where is the connection? To establish a connection between evolution and eugenics, the best you can do is argue on shaky pseudoscientific grounds such as eugenics was based on Mendelian genetics. If you do connect it to evolution or theory of evolution, say based on the terminology used by eugenicists, it can just as easily be connected to creationism. So, such a connection is vague and meaningless. Here is a connection of eugenics with creationism: As a matter of fact at least one Young Earth creationist, William J. Tinkle, advocated eugenics and selective human breeding. Even if Darwin's son called eugenics an application of evolutionary principles, it does not become so: again critical thinking 101. Neither Mendelian genetics, nor theory of evolution tells you that Jews, Chinese, Indians, or people from Eastern Europe have bad genes. Nor does Einstein's theory of relativity tell you that it is fine to make an atomic bomb and drop it on human population. The point is that scientific theories and principles such as the theory of evolution or quantum mechanics tell us how certain aspects of nature work, and that is it. The results of how we apply or use that knowledge is up to us and it cannot be used to invalidate the theory itself. Scientific theories can only be invalidated by using the scientific method described in Chapter 2.

It is true though that some of those who supported eugenics cited the theory of evolution as inspiration or they used the theory as justification for their actions. However, this is also true, as mentioned above, that the theory of evolution, like any other scientific theory, tells us what is and how it is, and not what ought to be. It is descriptive and explanatory, but not prescriptive or normative. It can predict the likely outcome of certain actions and behaviors, but not which of these actions or behaviors are ethical or desirable.

The truth is, as mentioned earlier, that many policies implemented in the name of eugenic principles were motivated by certain economic and political conditions, expected advantages, and racial and social prejudices. Genetics and evolution were used as a vehicle, an excuse, with grossly erroneous interpretations. An excuse is not necessarily a cause or a justification. On the contrary, many of the most enthusiastic promoters of the eugenics movement in the U.S., which led to policies such as compulsory sterilization and barring interracial marriages, were evangelical Christians, who were also opposing evolution. For example, clearly under the influence of eugenics, Mary Teats wrote in her book *The Way of God in Marriage*, "The great and rapidly increasing army of idiots, insane, imbeciles, blind, deaf-mutes, epileptics, paralytics, the murderers, thieves, drunkards and moral perverts are very poor material with which to 'subdue the world', and usher in the glad day when 'all shall know the Lord'."

Darwin himself is often accused of being a racist by some in an attempt to associate him or his theory with eugenics. The fact is that Darwin went very much against the ideas of his time by dismissing some of the perceived differences between races. For example, in *Descent of Man*, he argues: "...this fact can only be accounted for by the various races having similar inventive or mental powers." Regardless, I don't need to state the obvious, that a scientific theory is not validated or invalidated by the social or political views of the scientist.

Many eugenics arguments, such as the expected effect of selective sterilization and the results of interracial mating, are based on faulty and erroneous research wrongfully attributed to biology. Scientifically correct and sound biology education based on proven fundamental principles such as molecular biology (including genetics) and theory of evolution can help

407

avoid misdeeds and crimes such as those committed by the eugenics movement from happening in the name of biology in the future.

Here is a warning: To a larger extent, biologists, unlike chemists, have successfully managed to avoid the widespread application of physics and math in their research. As a result, most of the biology education and research techniques are recipe based rather than on sound physical and mathematical principles. With the emergence of the field of molecular biology and the slow but sure recognition of the connection of physics (especially quantum physics) to biology, this is going to change. Until that happens, the recipe-based empirical approach to biology education and research always leaves a room for *hocus pocus* such as eugenics. This is not to say that empirical or phenomenological research does not have its role to play, but its limitations should be recognized.

In a nutshell, not me but the facts cry out loud that those who try to tie eugenics with evolution cannot make this attempt without being hypocrites.

9.8 The Hypocrisy Triplet

In America, we have an idiom for a person who says different or opposing things about the same subject: you are speaking from the both sides of your mouth. Being literal (and perhaps vocal as well) here, some creationists have historically spoken from the three sides of their mouths:

1. They (or their ideological ancestors) have actively participated in the eugenics movement to further their non-scientific agenda in the name of science.

2. At the same time when they were active in the eugenics movement, they were vigorously opposing the scientific

theory of evolution and fiercely advocating a ban on teaching evolution in public schools.

3. When eugenics hit the target where it was clearly headed and as a result got a bad name, they started associating it with evolution to condemn the theory of evolution.

After this, nothing much is really left to say on this topic. Nevertheless, we should always learn from the past, even scientists.

9.9 Eugenics and Modern Genetics

Genetics provides a basis for many current biological fields such as biotechnology, bioengineering (including genetic engineering), and molecular and cellular biology, with great implications and applications for human health such as a whole field of regenerative medicine with the promise of treating fatal diseases such as Parkinson's and Alzheimer's. Because the misdeeds of the eugenics movement were committed in the name of genetics, it is important for everyone, including scientists, especially scientists, to learn from the past and keep things in perspective. Here are some points to note:

Eugenics in the Rear View Mirror. There is some truth in the words of Spanish-American philosopher, poet, and novelist George Santayana (1863-1952), "Those who cannot remember the past are condemned to repeat it." This adage is appropriate in the current context of the molecular age or the gene age that we as a civilization are entering. If we do not learn from the eugenics chapter of our history and keep things in perspective, the chapter in one or another form may pop up to haunt us. On one extreme, some people may oppose progress in genetics and hence in science by associating it with eugenics; whereas on the other extreme, some people may fall into the same kind of fallacy or trap that the original eugenicists fell into before different interest groups joined them to hijack the whole thing.

In hindsight, it is rather puzzling to note how some highly educated scientists and well-reputed universities fell for the trap of the eugenics movement. One can try to understand this highly bizarre phenomenon by putting it into context. Every movement, good or bad, has some context in which it plays out. As pointed out earlier in this chapter, the eugenics movement also had a sociopolitical and economic context in which it flourished. This context consisted of turbulent economic and social conditions or problems in the U.S. after the Civil War; the first major migration away from rural areas (farms) into the cities due to rapid industrialization and as a result, expansion of cities faster than the availability of adequate housing; the rise of rather militant labor organizations as a result of increased exploitation of labor; bankruptcies of many businesses as a result of price fluctuations causing a series of depressions; and an ever increasing tide of immigrants mostly from Eastern and Southern Europe. However, be mindful that we can use the context to partly understand the eugenics movement but not to justify its actions. Speaking in terms of a chemical reaction, a context at best can act as a catalyst but cannot take the place of reactants.

So, despite these reasons and contexts, the fact remains that at the end of the day we are judged by our actions and their consequences and not as much by the circumstances that led to those actions. This is because we are living beings and have the capability of making decisions about how to react to the circumstances or the environment around us. So, it is neither eugenics nor any other circumstance but the U.S. politicians and legislators who were responsible for enacting discriminatory laws about barring interracial marriages, sterilization, racial segregation, and selective immigration restrictions; much the same way as it was not U.S. leadership but the Nazi government that was responsible for implementing eugenic principles to the extreme of committing genocide.

Now that we are reviewing the eugenics movement, let's ask a question: When all this was happening in the name of science, where were the scientists? Did all of them fall?

Where Was the Sound Science? It is not that there was no opposition at all to the implementation of eugenics, as a matter of fact there was scientific opposition from the very beginning when the eugenics movement was trying to organize itself into an allegedly scientific discipline. For example, by 1910, Godfrey N. Hardy and Wilhelm Weinberg had developed their scientifically sound genetic equilibrium model, which disproved the eugenic claim that degenerate families were increasing the societal load of dysgenic genes. Their work such as the Hardy-Weinberg equation also showed that sterilization of some individuals would not appreciably decrease the percentage of mental defectives in society. Their model and equation, discussed in Chapter 4, was firmly rooted in and validated by experimental data.

Furthermore, American geneticist George Harrison Shull (1874-1954), working at the Carnegie Station for Experimental Evolution, demonstrated that hybrid corn plants are more vigorous and better than pure-bred ones. He highly influenced agriculture on a global scale by playing an important role in the development of hybrid corns. It was found that hybrid plants will withstand harsh weather, and deflect diseases and pests better than non-hybrid brands. This discovery had a great impact upon global agriculture as farmers were able to produce more corn per acre with these vigorous hybrids. However, it had no impact on the thinking of those eugenicists who had

411

their racist agenda, even though this discovery refuted the notion that racial purity offers any biological advantage or that race mixing destroys "good" racial types.

Also, it was clear at that time that experimental research was invalidating the claim of the eugenics movement that a complex behavioral trait could be determined by a single gene. For example, American geneticist Herman Muller (who won the 1945 Nobel Prize in Physiology or Medicine) conducted a study of mutations in Drosophila and other organisms from 1914 to 1923. The study showed that a single gene might affect several characteristics (traits) and not just one at a time, and conversely mutations in several different genes can affect the same trait. It had also been scientifically demonstrated that genes alone do not determine all inherited traits, let alone all traits, and that environment also plays an important role along with the genes in determining even some inheritable traits. Here are two examples from those times:

1. In 1930s, Horatio Newman, Frank Freeman, and Karl Holzinger conducted a study on identical twins who were raised apart. It was demonstrated that the twins developed different IQs despite their identical genes.

2. Lionel Penrose found that most mental cases among the patients at a state-run institution in Colchester, England resulted not from genetic causes alone, but from a combination of environmental, genetic, and pathological causes.

Think About It!
Does the theory of evolution claim that all traits are generated by genes?

> **Answer**
> According to the theory of evolution only heritable traits are generated by genes, and even those can be influenced by the environment, as explained in Chapter 4.

Here is an example (or evidence) of hocus pocus research in the eugenics movement in the name of science. Recall that the Carnegie Institution provided Charles Davenport, a eugenicist, with funds to found the Station for Experimental Evolution in 1904. Charles Davenport also founded Eugenics Record Office (ERO) in 1910 as a center for eugenics and human heredity research. The center was financially supported by the Carnegie Institution among others. The same Carnegie Institution convened a review panel in 1935, which concluded that the vast majority of the work sponsored by the Eugenics Record Office was without scientific merit. Consequently, the review panel recommended a halt to the ERO's propaganda for eugenic social programs, such as sterilization and immigration restriction. In the 1930s, leading American and British geneticists started increasingly criticizing eugenic organizations for freely mingling prejudices with a simplistic and problematic understanding of human heredity.

However, this opposition did not prove to be enough or timely to stop the eugenics movement. The Catholic Church opposed eugenics pretty much from the beginning, and succeeded to block eugenic social legislations in much of Europe. However, the Catholic viewpoint held little water in dominantly Protestant America. With Buck vs. Bell providing the full approval of the U.S. Supreme Court, state legislatures continued to enact new eugenic sterilization laws up until World War II when they ran into a dead end: the Holocaust and the association of eugenics principles with Nazi Germany.

And God Said: Let There Be Evolution

> **Buck v. Bell** 274 U.S. 200 (1927). A decision of the United States Supreme Court ruling that a state statute permitting compulsory sterilization of the unfit, including the mentally retarded, "for the protection and health of the state" did not violate the Due Process Clause of the Fourteenth Amendment of the United States Constitution. The ruling was largely seen as an endorsement of a eugenics policy: improve the human race by eliminating "defectives" from the human gene pool.

With most scientific fields developing to the level of molecules and as a result giving rise to nanoscience, the twenty-first century is often referred to as the molecular age.

The Road Ahead: Genetics in the Molecular Age. We should notice that our understanding of genes is much better and at a different (deeper) level now than it was during the time when the eugenics movement flourished. But still learning from the eugenics movement, the biological research community needs to remain on guard and counter any effort of hocus-pocus research such as *intelligent design* discussed in the next chapter.

Remember, eugenics started off as an effort to breed better human beings by encouraging people with "good" genes to reproduce more and discouraging those with "bad" genes. To meet this end, eugenicists used means such as legislations to keep different racial and ethnic groups genetically segregated, to bar immigration from certain regions such as Asia, to restrict immigration from Southern and Eastern Europe, and to sterilize people considered "genetically unfit." This social engineering attracted individuals and organizations with racist agendas so much so that elements of the American eugenics movement initially became models for the Nazis, whose radical adaptation of eugenic principles resulted in the Holocaust.

However, in this age, it will not be as easy to misinterpret and hijack evolution or genetics to implement discriminatory and inhuman policies like those during the eugenics movement. The reason is that now evolution and genetics are based on much firmer scientific grounds and are secured by the bigger context of molecular age that we are entering. With the tools of nanotechnology such as electronic microscopes, we can look into the structure of matter (living and non-living) to the details of molecules and atoms. We are also increasingly gaining the capability of controlling and manipulating individual molecules and atoms. Theory of evolution and genetics are already in the process of finding their roots in the well-established theories of physics such as quantum mechanics. There is no room in science for pseudoscientific theories such as eugenics or intelligent design, nor is there any room for any *ism,* even if it is Darwinism.

So far in this chapter, we have been referring to social Darwinism. Eugenics was the cancerous form of social Darwinism. There is another kind of Darwinism that is still well and alive in the hustling and bustling environment of our college campuses, and we would be ignoring the elephant in the room if we close this topic without talking about it.

9.10 Scientific Darwinism: A Disease that Must Be Cured

With a background in physics and computer science, when I entered the field of biology, it did not take me long to notice that a significant number of biologists are infected with a disease called Darwinism. I call it scientific Darwinism to distinguish it from social Darwinism to which we have been referring so far in this book. Here are some of the symptoms of this disease of scientific Darwinism. If you are infected with this:

♦ You use the terms evolution and Darwinism equivalently, that is, you equate evolution with Charles Darwin.

♦ You usually replace the term *adaptive evolution* or sometimes even *evolution* with Darwinian evolution.

♦ You often use the term Darwinian selection instead of natural selection, that is, you ignore the historical fact that the principle of natural selection was discovered independently by both Charles Darwin and Alfred Wallace.

♦ Considering evolution from molecular to species level, you only see one mechanism at play: natural selection. You cannot even imagine the hands of some physical or chemical laws playing any significant role if at all.

♦ Along with science, the theory of evolution has marched on a 150 year-long road of progress by developing through continuous scientific scrutiny, evidence, and discovery; but you seem to stuck on its founder, I mean one of its founders, that is, Darwin.

This disease must be cured, if for nothing else, just because it leads to inaccurate presentation of the theory of evolution even to some extent in the classroom. For the public, this approach certainly helps shrink the theory of evolution to one man and one book, which the opponents of evolution would love to do anyway. It scientifically misrepresents evolution by ignoring the 150 years of biology research that contributed to understanding evolution, which includes:

♦ Mendel's work on heredity that explained the cause of trait variations, which are the raw material for natural selection to operate on and hence crucial to understand evolution. It provided a genetic base for the theory of evolution as discussed in Chapter 3.

♦ Discovery and understanding of the DNA molecule, which elucidated evolutionary lineages that we use to draw the evolutionary trees of life, as discussed in Chapter 4.

♦ Discoveries in almost all fields of biology, so much so that it is not possible to teach almost any topic in biology without calling in the concept of evolution.

♦ A continuous stream of fieldwork studies and documenting evolution in nature, which added to the understanding of evidence for evolution, as discussed in Chapter 6.

♦ Applications of evolution in multiple fields including health and medicine, which we explore in Chapter 11.

For last about one and a half century, scientists have been testing and developing the theory of evolution. Where we stand now, most if not almost everything we understand about evolution came from scientific studies performed by Mendel and other countless scientists after Darwin. To start with, as mentioned in Chapter 1, evolution was not even Darwin's idea, and he was not the only one to come up with the idea of natural selection. I am not asking to belittle the great contribution of Darwin to the theory of evolution, I'm asking to put things in perspective and get rid of Darwinism. The cult of Darwinism in science is as harmful to the advancement of science, if not more, as the irrational opposition from creationists or intelligent designers.

Any *ism* including Darwinism implies an ideology such as Marxism or Capitalism and not a scientific theory validated through continuous scientific scrutiny. It carries with it an aura of a belief system rather than a scientific theory. So, by using

417

terms such as Darwinism, Darwinian selection, or Darwinian evolution, we provide an opening for this line of argument: well Darwinism or Darwinian evolution is just a theory; here is an alternate theory and it's called intelligent design or creationism. This is not to say that Darwinism gave rise to intelligent design or creationism.

Darwin's genius can only be fully understood and appreciated in the light of evolutionary research performed after Darwin, which we cannot see clearly if we are suffering from this disease of scientific Darwinism.

9.11 Summary

Darwinism (or social Darwinism) is neither part nor a scientific implication of the theory of evolution. Theory of evolution has been at the center of social controversy in some areas of the world for about one and half centuries now, because some people believed it challenged their religious beliefs. Some of these people threw whatever they could at the concept of evolution and believed whatever anybody said against the theory. This gave rise to myths about and against the theory and resulted in associating elements of social Darwinism such as eugenics to the theory in order to discredit it.

The field of eugenics started toward the end of the nineteenth century with Francis Galton, who coined the term eugenics, and practically ended with Adolf Hitler who implemented his racist agenda in the name of eugenics. Along the way, a host of different things happened to eugenics as different sections of society ranging from prestigious universities, to governments across the world, to racist organizations joined it with different intentions and for different purposes. At least initially, it picked up the support of luminaries such as Winston Churchill and Theodore Roosevelt.

Governments participated in the movement by implementing the movement's agenda based on pseudoscience at best, for example by making and enforcing laws barring interracial marriages, allowing forced sterilization of certain groups, and racial- and ethnicity-based immigration restrictions.

Yes, the field of eugenics started with Francis Galton, a cousin of Charles Darwin. Yes, it is a fact that Major Darwin, son of Charles Darwin, presided over the First International Eugenics Conference, and actively participated in the field of eugenics. However, it does not mean that evolution or theory of evolution is responsible for the misdeeds of eugenics: critical thinking 101. A scientific theory describes what is and how it is, it does not tell what ought to be, or what is right and what is wrong. Yes, the field of eugenics may originally be based on Mendel's genetics and may mean to help the human species; however, it went off the track pretty quickly, went haywire, and turned into a sociopolitical movement, which was hijacked largely by racist individuals and organizations. The actions of ethnic and racial discrimination such as immigration restrictions and inhuman actions such as forced sterilization of certain social groups in the name of science or biology started off in the U.S. and proliferated to other countries around the world including Germany, where the Nazi Government implemented its genocidal agenda in the name of eugenic principles and by referring to American leadership in implementing these principles.

In a Nutshell

✓ Evolution is not just a theory. Instead, evolution is a scientific theory supported by a wide range of compelling evidence from macro-organisms to micro-organisms, to cells, and to genes.

419

✓ Darwinism or social Darwinism is not part or a scientific implication of the theory of evolution.

✓ Scientific Darwinism is misrepresentation of the theory of evolution and is not healthy for the progress of science.

✓ The theory of evolution was originally built on some (but not all) of Darwin's ideas but has gone far beyond them.

✓ Theory of evolution or genetics cannot be used to justify the misdeeds of the eugenics movement.

✓ Substantial bodies of scientific evidence presented against the arguments of the eugenics movement were ignored.

✓ Eugenics movement was effectively driven by sociopolitical and economic forces and not by science.

✓ The school of creationism first supported the eugenics movement and its misdeeds, and opposed evolution at the same time, and later began associating evolution with eugenics to attack and discredit the concept or the theory of evolution.

9.12 Behold

In this chapter, we discussed eugenics in rather detail for several reasons: It played a significant historical role in the history of opposition to evolution and in this process it exposed the contradiction and hypocrisy of this opposition. Furthermore, it's important to remember it while we are entering the molecular age so that we stay alert or on guard against the possibility of such tragedies happening again in the name of science.

It is a fact that Christian evangelists and some churches in the U.S. and elsewhere promoted the implementation of the inhumane and racist eugenic policies and deeds. Some of the same people, who promoted racially discriminative eugenic policies, were also opposing evolution at the same time. How can you simultaneously support eugenics and oppose evolution and then later link eugenics to evolution in order to discredit evolution? In science such a contradiction can kill a theory, but in religion such contradictions survive and thrive because religion is based on beliefs and not logic. This is one of many differences between religion and science.

So, there were religious people who on one hand supported eugenics and on the other opposed evolution. Their opposition to evolution was not rooted in any scientific evidence or lack thereof, but rather in their belief in creationism.

In the next chapter, we explore how creationists have been opposing evolution. Interestingly, one thing you will notice is that their support for eugenic policies and opposition to evolution were at their peaks at about the same time.

And God Said: Let There Be Evolution

TEN

Evolution of Creationism

And God Said: Let There Be Evolution

Great is the power of steady misrepresentation; but the history of science shows that fortunately this power does not long endure.

Charles Darwin

Science has proof without any certainty. Creationists have certainty without any proof.

Ashley Montague

10.1 Once Upon a Time

Figure 10.1 a) John Scopes in 1925   **b)** Clarence Darrow and William Jennings Bryan chat in court during the Scopes Trial.

A high school science teacher in Dayton, Tennessee, named John Thomas Scopes was arrested on May 5, 1925. His alleged crime was that he taught evolution to students and therefore violated the Butler Act, an act passed by the state legislature which had prohibited the teaching of human evolution in state-funded schools. According to the Butler Act, it was unlawful "to teach any theory that denies the story of the Divine Creation of man as taught in the Bible, and to teach instead that man has descended from a lower order of animals" in any Tennessee state-funded school or university. The importance of the trial was clear due to its high profile from the outset reflected

And God Said: Let There Be Evolution

by the facts that the American Civil Liberties Union (ACLU) decided to take up the case and America's top-tier and most famous criminal defense lawyer of that time, Clarence Darrow, offered to defend Scopes without a fee. At the end of the trial, even Scope's bail was paid by Paul Patterson, the owner of the *Baltimore Sun* newspaper.

William Jennings Bryan, a three-time presidential candidate, led the prosecution team along with A. T. Stewart, the District Attorney. Bryan, who believed in the literal interpretation of the Bible, had been asked by the World Christian Fundamental Association to participate in the trial. He had already been leading the Fundamentalists' crusade to expel evolution from the classrooms of America. After all, it was William Bell Riley, head of the association, who had lobbied legislatures in many states to pass anti-evolution laws. He succeeded in Tennessee when the Butler Act was passed there.

The Scopes trial, also known as the Monkey Trial, began in Dayton on July 11, 1925, and claimed incredible publicity across the nation, as modernists with science on their side were pitted against traditionalists armed with the literal interpretation of the Bible. The town of Dayton witnessed over 100 journalists arriving in the town to report on the trial, which also made history by becoming the first trial in America that was broadcast to the nation.

Perhaps good material for a comedy show, the court audience watched three high school students testify on their presence in the class where Scopes had taught evolution When the judge, John T. Raulston, refused to allow scientists to testify on the scientific truth of evolution, Clarence Darrow used another way to make the case for evolution. He called William Jennings Bryan to the witness stand to expose the flaws in Bryan's arguments during cross-examination. According to many independent observers, Darrow succeeded in his efforts. This episode of the 11-day-long trial − more like a circus − became a highlight of the trial. Nevertheless, the jury found John Thomas Scopes guilty and the judge fined him $100. A successful film from 1960, *Inherit the Wind*, distributed by United Artists, is loosely based on this trial.

Most of the opposition to evolution can be linked to creationism. In this chapter, we explore the history of opposition from creationists to evolution with an emphasis on the opposition in the U.S. In doing so, we expose the real intentions behind and politics of this opposition, and its possible grave consequences at national level.

10.2 In the Beginning

In the beginning of the twentieth century, an intense change was underway at the international level in all fields of human life including social life, politics, economy, science, and technology. Newtonian physics was collapsing in the face of experiments being performed at the atomic level and as a result quantum laws were being discovered. At the same time, the theory of relativity was discovered. Mendel's work was rediscovered and as a result some important questions unanswered by the theory of evolution introduced by Darwin and Wallace were answered by using this work. The field of genetics was also taking shape. The change was not limited to educational and research institutes, the whole world was being rapidly transformed by a stream of new discoveries, inventions, and ideas hitting the public arena ranging from comic strips to radio, followed by movies and television - from suburbs to assembly lines, and from automobiles to airplanes.

The fact that people could suddenly run on wheels, fly in the sky, and travel faster than ever fostered an incredible sense of freedom. Advancements in the medicine industry such as the development of antibiotics (a very important part of modern medicine) based on the discovery of penicillin was enabling people to control disease better than ever.

As often happens during such times, these waves of intense change were out of phase and hence colliding with the outdated and inert establishments in all fields including social, political, and economical. The new world order was on the march and the old but established order was fighting tooth and nail. Labor unrest, riots around the world, and revolutions in China, Mexico, and Russia are some examples.

The U.S. economy was changing dramatically as the country was on the path of transformation from being a rural agriculture nation to becoming a leading industrialized nation. In this process of becoming a leading manufacturing country, the nation was going through and getting adjusted to urbanization. A lot of progress and accomplishments were being made in an environment filled with energy and hope, with a byproduct of unsettlement or unrest resulting from the change.

In the U.S., just like in other places of the world, intellectual experimentation and social change were flourishing in the backdrop of modernism; and the modernists and traditionalists were on a collision course. Creationists were a very important force among traditionalists in America.

10.3 What is Creationism Anyway?

In general, creationism is the religious belief according to which the Universe, including Earth and life on Earth, is the creation of a supernatural power. Within this common framework, various religions and belief systems have various accounts for this creation. For example, in the Christian and Jewish faiths, creationism is usually based on a literal interpretation of the biblical account of the beginnings of the Earth, life, and humans as described in the first two chapters of the *Book of Genesis*. The *Book of Genesis* is considered a sacred narrative in Judaism, Christianity, and Islam (their version). Hinduism and other religions have their own versions of the creation of the Universe and the life in it. However, there is one thing common to all these belief systems: not enough detail is provided as would be demanded from a scientific theory, and many concepts are left open to various interpretations. The positive aspect of this vagueness is that this leaves room for accommodating scientific discoveries and therefore reconciling them with the belief. For example, some Hindus find support for evolutionary ideas in their scriptures called the Vedas.

As shown in Figure 10.2, on June 18, 2004, a new landmark was unveiled at CERN, the European Center for Research in Particle Physics near Geneva, Switzerland — a two meter tall statue of the Indian deity Shiva Nataraja, the Supreme God in Hinduism. The statue, symbolizing Shiva's cosmic dance of creation and destruction, was given to CERN by the Indian Government to celebrate the research center's long association with India. A special plaque next to the Shiva statue at CERN explains the significance of the metaphor of Shiva's cosmic dance with several quotations from *The Tao of Physics,* an international bestselling book first published in 1975 and still in print in over 40 editions around the world. In this book, author Fritjof Capra presents the parallel between Shiva's dance and the dance of subatomic particles, and used this parallel as the central theme of the book. In choosing the image of Shiva Nataraja, the Indian government acknowledged the profound significance of the metaphor of Shiva's dance for the cosmic dance of subatomic particles, which is observed and analyzed by CERN's physicists; and so did the CERN administration, at least to some extent by accepting and embracing the gift in form of a landmark. In Fritjof Capra's words, "Modern physics has shown that the rhythm of creation

429

and destruction is not only manifest in the turn of the seasons and in the birth and death of all living creatures, but is also the very essence of inorganic matter," and that, "For the modern physicists, then, Shiva's dance is the dance of subatomic matter... Hundreds of years ago, Indian artists created visual images of dancing Shivas in a beautiful series of bronzes. In our time, physicists have used the most advanced technology to portray the patterns of the cosmic dance. The metaphor of the cosmic dance thus unifies ancient mythology, religious art and modern physics."

Figure 10.2 The statue, symbolizing Shiva's cosmic dance of creation and destruction, was given to CERN by the Indian Government to celebrate the research center's long association with India. Photo: Courtesy of Giovanni Chierico, CERN.

This is not to say that all Hindus agree with the Indian Government's gesture or with Fritjof Capra's interpretation; however, this is an example of reconciling religious beliefs with scientific discoveries and theories.

Most religious people, sooner or later, reconcile their beliefs with new scientific discoveries. However, there is always a set of creationists who find their beliefs threatened by certain scientific discoveries and theories about the Universe, Earth, and life. For example, a set of members of the International Society for Krishna Consciousness (ISKCON) has opposed the idea of evolution.

Islamic creationism is the belief that the Universe and life in it, including humans, were directly created by God as explained in the Qur'an. Like in other Abrahamic religions – or any religion for that matter – the creation doctrine of Islam is vague and allows for a wider range of interpretations. Several liberal movements within Islam generally accept the scientific positions on evolution and related issues such as the age of the Earth and the age of the Universe. Similarly, Judaism has a wide spectrum of views about creation, evolution, and their relationship.

In the case of evolution, reconciliation has been (and still is) particularly difficult in the United States, where the belief in religious fundamentalism has largely affected attitudes towards evolution more than believers anywhere else, except perhaps in Turkey. A big part of the reason for this has been that political partisanship in the U.S. is highly correlated with fundamentalist thinking, to such a degree as not found in Canada and in most countries of Europe, for example. So, to capture all the major elements of the struggle between creationism and evolution, we review how this story has unfolded in the U.S.

10.4 Evolution of Creationism: A Full Circle

In Western countries, the inclusion of evolution in the science courses of educational institutes has been mostly uncontroversial, except in the United States. This is such an irony, given that, in general, the United States is one of the most advanced countries in science and technology. In this country, religious fundamentalists have waged war against evolution for about a century now, which as you will see in this chapter, has come full circle with a very interesting history exposing the very character of this opposition in the process. This circle or cycle of the creationists' opposition efforts started in the beginning of the twentieth century by claiming that evolution was just a theory, not a fact and hence should be expelled from schools. Then creationists attempted to bundle creationism with evolution as an alternative scientific theory, and now are back to the attempt of forcing educators to mis-educate students about evolution by presenting it as just a theory and not a fact.

In the following sections, we explore different stages of this cycle of evolution versus creationism in the United States in order to review the history of opposition to evolution.

10.5 Expel Evolution: Defying the Scientific Truth

In the beginning, the strategy of the creationist movement was outright opposition of evolution: defy the theory of evolution and keep it out of the education system. The Scopes Trial is a good example to illustrate the story of this opposition.

Theory of evolution, after being first introduced by Darwin and Wallace in 1858, and then in 1859 with the publication of Darwin's *On the Origin of Species*, was widely accepted during the 1860s and 1870s. Furthermore, with developments in other fields such as astronomy and geology, it became impossible to ignore evolution. So, public schools began to teach evolution as part of science courses. This was reconciled with Christianity by most believers, but was considered by a number of early fundamentalists to be directly in conflict with the Bible.

In the U.S., creationists have battled for about a century against teaching evolution in public schools. Their strategy and argument against it has been changing in response to legal setbacks. In the late 1910s and early 1920s, more Americans became exposed to the concept of evolution than ever before as high school attendance during this period increased considerably. In the aftermath of World War I, the creationists leveraged the Fundamentalist-Modernist Controversy to develop a surge of opposition to the very idea or concept of evolution.

In the early 1920s in America, intellectual experimentation and social change was flourishing in the backdrop of modernism. After World War I, modernism was in full swing in America, when Americans were dancing to the tunes of the Jazz Age and debating abstract art and Freudian theories, while showing their contempt for alcoholic prohibition. For the younger modernist generation, the issue changed from *whether society would approve of their behavior* to *whether their behavior was in par with their intellect* (Linder, 2012). Traditionalists, the people married to older Victorian thought, worried as they saw the values important to them eroding right in front of their eyes. They viewed the traditional social patterns in chaos. As Professor Douglas O. Linder puts it, "in a response to the new social patterns set in motion by modernism, a wave of revivalism developed, which became especially strong in the American South."

Putting it into one sentence, the question at the center of all this was: Who would dominate American culture—the

And God Said: Let There Be Evolution

modernists or the traditionalists? So, the conflict between modern values and traditional values was the bigger context in which the courtroom drama of the *Monkey Trial* was unfolding in Dayton, Tennessee. There in Dayton, in the summer of 1925, journalists witnessed the inevitable showdown between the two forces that would shape the American future.

At face value, a science teacher was on trial for illegally teaching evolution, and his guilt or innocence would be decided by a jury. So, it should not be a big deal. However, it was the bigger context mentioned above that was making it a big deal. Even the possible challenge to the constitutionality of the law that was making the teaching of evolution illegal was a big deal only in the bigger context: the conflict between modern values and traditional values. What would it mean for the ongoing conflict between traditionalists and modernists, and who would shape the American future, traditionalists or modernists?

Figure 10.3 The King & Carter Jazzing Orchestra photographed in Houston, Texas, January 1921.

> **Decide it yourself!**
> Do you see the conflict between traditional social values and modern intellectual values even in today's America? For example, when looking professorial becomes a negative and "the guy you can have beer with" becomes a positive for presidential candidates.

William Bell Riley, head of the World's Christian Fundamentals Association, had lobbied state legislatures in several states to pass anti-evolution laws. He succeeded in Tennessee as the Butler Act was passed. Working with Riley was William Jennings Bryan, a populist and a three-time Democratic presidential candidate, who had been leading the fundamentalist crusade to expel the theory of evolution from the classrooms of America. In the words of columnist H. L. Mencken, who covered the Scopes Trial, Bryan had transformed himself into a "sort of Fundamentalist Pope." By 1925, Bryan, Riley, and their followers realized the fruit of their campaign: Fifteen states had introduced and were considering legislation to ban the teaching of evolution, and some states, including Tennessee, had already passed it into law. It was in February of 1925 when, by enacting a bill introduced by legislator John Washington Butler, Tennessee made it illegal "to teach any theory that denies the story of divine creation as taught by the Bible and to teach instead that man was descended from a lower order of animals."

> **Note.** William Jennings Bryan (1860 – 1925), a lawyer, was the Democratic Party nominee for President of the United States in 1896, 1900, and 1908, and he served as the 41st U.S. Secretary of State under President Woodrow Wilson. He was noted for a deep, commanding voice.

And God Said: Let There Be Evolution

The house passed the anti-evolution bill on a vote of seventy-one to five within seven days of Butler introducing it. There were those legislators who agreed with the bill and therefore voted for it. And there were also those who voted for it out of the fear of a backlash from their constituents in case they voted no. Not only were there no public hearings held before the vote, there was almost no discussion on the bill. When a house member made a request to hold the bill over for a debate, Butler responded, "I do not see the need for any further talk, as everyone knows what evolution means."

It is fair to acknowledge that not everyone in Tennessee agreed with Butler's view on the value of the debate. This was clear from the letters to the editors that swamped the offices of the state newspapers. Many of these letters compared the proposed Tennessee legislation to the position taken by the Catholic Church that led to the 1633 trial of a great physicist, Galileo Galilei, for publishing his allegedly heretical view that the Earth revolved around the Sun, rather than vice versa as the Bible seemed to suggest. As Professor Linder reports, one letter writer jokingly commented that there was "no better proof" to be found of the "truthfulness of Darwin's theory than to visit Capitol Hill and view some of its occupants."

When the American Civil Liberties Union (ACLU) offered to defend anyone who wanted to bring a test case against one of the anti-evolution laws, John T. Scopes, a high school biology teacher, accepted the offer. Scopes defied the Butler Act by teaching evolution in his Tennessee class. As reported in an article published in The *New York Times* issue of May 25, 1925, Scopes was indicted on a specific charge that on April 24, 1925, John T. Scopes "did unlawfully and willfully teach in public schools of Rhea County, Tenn., which said schools are supported in part and in whole by the public school funds of the State, certain theory and theories that deny the story of Divine creation of man as taught in the Bible and did teach thereof that man descended from a lower order of animals."

Going into the trial, the ACLU had no position on the issue of evolution: being neither for nor against it. The ACLU's original intended stand was to oppose the Butler Act on the grounds that it violated a teacher's individual rights and academic freedom, and was therefore unconstitutional. This strategy changed further as the trial progressed. In the beginning of the trial, the defense proposed that there was actually no conflict between evolution and the creation account in the Bible: a viewpoint later called theistic evolution. However, as the trial progressed, Clarence Darrow, the defense attorney, attacked the literal interpretation of the Bible as well as Bryan's limited knowledge of science and other religions. Only after a decision in favor of the prosecution, when the case went to appeal, did the defense return to the original claim that the prosecution was invalid because the law was essentially designed to benefit a particular religious group, which would be unconstitutional.

There was another aspect to the trial: the publicity. The trial was unfolding in the medial limelight with modernists and traditionalists pitted against each other over two visible issues: the teaching of evolution in schools and a fundamentalist interpretation of the Bible. Edward J. Larson, a historian, won the Pulitzer Prize for History for his book *Summer for the Gods: The Scopes Trial and America's Continuing Debate Over Science and Religion.* He notes, "Like so many archetypal American events, the trial itself began as a publicity stunt," and he goes on to cite some interesting facts from this event. The press coverage of the "Monkey Trial" was overwhelming. For example, the front pages of newspapers like the *New York*

And God Said: Let There Be Evolution

Times were dominated for days by news items covering this case. More than 200 newspaper reporters from all parts of the United States and two from London, England were situated right there in Dayton to cover the trial. It is often cited that twenty-two telegraphers sent out 165,000 words per day on the trial over thousands of miles of telegraph wires hung for this purpose. Furthermore, as Larson notes, the reporters set a new record by transmitting more words to Britain about the Scopes trial than for any previous American event. Chicago's WGN radio station had its own first on-the-scene reporting of a criminal trial: Their announcer Quin Ryan broadcasted via clear-channel broadcasts the first on-the-scene coverage of a criminal trial. Two reporters with movie cameras had their film flown out daily in a small plane from an airstrip specially prepared for this purpose. With trained chimpanzees performing on the courthouse lawn, Dayton was looking more like amidst a festivity rather than a criminal trial. Both modernists and traditionalists tried to use this highly-publicized event to express their side of the argument in any ways they could. The results of the Scopes trial were a mixed bag. Within the bigger context in which the trial was unfolding, both sides got something. It can be looked upon as a win-win without a complete victory for either side.

How did the modernists win? Even though modernists technically lost the trial, they made some significant gains in the bigger context of the conflict between traditional and modern values. The defense wanted to use this opportunity to bring out the truth about evolution and disseminate the theory of evolution, and they succeeded. To begin with, when the judge, John T. Raulston, failed this attempt by the defense by refusing to allow scientists to testify on the truth of evolution, Clarence Darrow, the defense attorney, used another tactic. He called William Jennings Bryan, the prosecuting attorney, to the

438

witness stand for cross-examination. According to many independent observers, during this cross-examination Darrow successfully exposed the flaws in Bryan's arguments and therefore in the arguments used by traditionalists or fundamentalists against evolution.

Some in the media such as H.L. Mencken criticized the prosecution and the jury which was "unanimously hot for Genesis." Mencken even mocked Dayton's inhabitants as "yokels" and "morons." He called Bryan, the prosecuting attorney, a "buffoon" and his speeches "theologic bilge." In contrast, he called the defense "eloquent" and "magnificent." Of course, there was an element of propaganda on both sides.

The cross-examination of Bryan by Darrow was portrayed nationally by the media; a play and a movie, *Inherit the Wind*, followed. This exposed, on the national stage, the weakness of the argument being used by the creationists to millions of Americans, and made religion-based opposition to the theory of evolution look ridiculous to many. This is how this trial became a critical turning point in the American creation-evolution controversy. Even though the traditionalists won the trial, the case overall provided a blow for fundamentalists in the form of eroding public support for their side of the argument. In the court of public opinion, creationism overall lost during this trial.

Five days after winning the case but getting bad publicity, Bryans died in his sleep on July 26, 1925.

How did the traditionalists win? Well, to begin with, traditionalists did win the legal case, technically. The law was heavily slanted against the modernists and for the traditionalists. Although the ACLU had taken on the trial as a cause, in the wake of Scopes' conviction they were unable to find any volunteers to take on the Butler law because they technically lost the case. Eventually by 1932, the ACLU had given up. It was not until 1965, that the anti-evolutionary legislation was challenged again. In the meantime, after William Jennings Bryans' death, his cause of trashing the theory of evolution and keeping it out of the public schools was taken up by other individuals and organizations including the Bryan Bible League, the Defenders of the Christian Faith, and the Klu Klux Klan.

Traditionalists did enjoy some tangible success out of the Scopes Trial. In the trail and afterwards, the Tennessee anti-evolution law, the Butler Act, prevailed or survived. Tennessee was not the first state to pass an anti-evolution law. Anti-evolution laws or measures were already in place in some states such as California, Florida, North Carolina, and Oklahoma, before the anti-evolution law in Tennessee was passed. After the Scopes Trial, some other states such as Arkansas, Louisiana, Mississippi, and Texas passed anti-evolution laws and measures. The high

Figure 10.4 The Rhea County Courthouse bell tower, Dayton, Tennessee; U.S. National Historic Landmark. John Thomas Scopes, a County High School teacher, was tried here in July 1925 for teaching evolution.

school biology texts most widely used in the second half of the 1920's (after the Scopes Trial) and the early 1930's bear witness to this anti-evolution crusade. There is only one of these books which lists evolution, although in the index and countered with biblical quotations.

So, the 1920s were marked with the prosecution and conviction of John T. Scopes for teaching evolution and thereby violating the Butler Act. Even though the case was appealed to the Tennessee Supreme Court, the decision was overturned on a technicality; the Butler Act and other anti-evolution laws and measures in other states survived and continued doing their job.

So, the evolution controversy among the public continued.

Note. The Scopes Trial, hatched as a publicity stunt, obviously brought publicity to the town of Dayton, Tennessee. The Rhea County Courthouse, where the trial was held, was placed on the National Register of Historic Places in 1972. The courthouse was then designated a National Historic Landmark by the National Park Service in 1976.

It is important to note here that between World War I and World War II, creationists on one hand supported eugenic policies, discussed in the previous chapter, and on the other hand opposed evolution. Even though creationists succeeded to expel evolution from biology textbooks, there was an emerging demand that science textbooks be written by scientists rather than educators or education specialists or under religious pressure.

However, after World War II, something else started unfolding on the international stage that would affect the ongoing cultural war in the U.S. Sometimes as a nation (or even as individuals), we can close our eyes and deny the existence of the truth. But the truth comes back to get us eventually. In this case, the truth came back in the late 1950s symbolized by a 184-pound metal ball called Sputnik launched during the Cold War between the Soviet Union and the United States, which started after the Second World War. This immediately affected the cultural battle in the U.S. between the traditionalists and modernists centered on evolution.

10.6 Wake up and See the Sputnik

On October 4th, in 1957, during the Cold War era, the Soviet Union successfully launched the first satellite, named Sputnik I, to orbit the Earth. It was the first in a series of satellites collectively known as the Sputnik program. This unanticipated event acted as a wake-up call for a relatively happy America feeling good about its future. This *Sputnik moment* precipitated a *Sputnik crisis* in the U.S. and heightened fears in the U.S. that the Soviet Union was achieving superiority over the U.S. in science and technology. The implications were that U.S. scientists were falling behind Soviet scientists and that the United States education system was falling behind that of the Soviet Union. This shook the American belief that the U.S. was superior in science and technology to all other countries. The response to Sputnik included the creation of the National Aeronautics and Space Administration (NASA) on October 1, 1958. During that same year, on September 2, 1958, another response had come in the form of the National Defense Education Act (NDEA), signed into law, which authorized funding to United States educational institutions at all levels. There was a critical need for state-of-the-art science textbooks to support the plan of cultivating top-notch homegrown scientists, and the National Science Foundation (NSF) was encouraged by the U.S. Congress to fund the development of the required educational materials.

One of the results of these efforts was that the Biological Sciences Curriculum Study (BSCS) was founded in 1958 by a grant from the NSF to the education committee of the American Institute of Biological Sciences (AIBS). The AIBS had realized the need for an extensive study and review of two things: the content of courses in biological sciences, as well as the entire process of teaching and learning for students of all ages. One specific task the BSCS was charged with was developing updated high school biology textbooks. The biology textbooks that the BSCS developed covered evolution as it is, the unifying principle in biology, which had been largely missing from textbooks after the Scopes Trial. The commercial publishers followed the suit.

Creationists, of course, were not pleased with these developments; so the reinvigoration of the education system in bio-sciences did not go unchallenged. For example, complaints were lodged with the State Textbook Commission in Texas where the backlash was the greatest. Church sermons and media were used to launch attacks on these developments in the education system. But the Sputnik crisis had set some change in motion that was unstoppable. It started or catalyzed the development of some trends in the U.S, which included increased interest in improving public education and legal separation of public education from religion. Continued urbanization in the South also added to the momentum of this change. Overall, these trends, in addition to the federal support triggered by the Sputnik crisis, was turning the tide of the public opinion or discussion in favor of teaching evolution. This national trend in favor of the change weakened the opposition in Texas and elsewhere, and led to the eventual repeal of the Butler Law in Tennessee in 1968.

As part of the educational change occurring in the country, the books developed by the BSCS were being widely used in U.S. high schools. This brought back once again the public controversy about teaching evolution in public schools into the lime light. Teacher Susan Epperson (Figure 10.5), a 10th grade biology teacher at the Little Rock Central High School, challenged the anti-evolution law. The adoption of a new biology textbook and curriculum that included evolution put her in a legal dilemma because it remained a criminal offense to teach the material in her state according to an anti-evolution law passed in 1928 modeled after the Butler Act of Tennessee. Epperson was not against teaching evolution. So with the unequivocal support of the Little Rock Ministerial Association along with backing from

Figure 10.5 Susan Epperson, the Arkansas teacher who successfully challenged her state's anti-evolution law in the 1968 Supreme Court case, *Epperson vs Arkansas.*

the Arkansas Chapter of the National Education Association and the American Civil Liberties Union, she filed a suit to test the constitutionality of the 1928 Arkansas state law prohibiting the teaching of evolution.

So, in case of *Epperson v. Arkansas*, the U.S. Supreme Court ruled in 1968 that laws barring the teaching of evolution in public schools were unconstitutional. After this blow from the U.S. Supreme Court, creationists changed their strategy or tactics to: banish the teaching of evolution by stopping funding.

10.7 Strangle Evolution: Stop the Life Supply

After the U.S. Supreme Court ruling of 1968 in case of *Epperson v. Arkansas* that laws barring the teaching of evolution in public schools are unconstitutional, creationists modified their strategy. To further their agenda of keeping evolution out of schools, they resorted to other means; one of those being to stop the public funding of teaching that involved evolution, such as publicly-funded biology textbooks. This strategy was on display in a suit filed in 1973 by an evangelical opponent of evolution named William Willoughby. This case, known as *Willoughby v. Stever,* was brought against the director of National Science Foundation, Horton Guyford Stever, and the Board of Regents of the University of Colorado. These officials were accused of using taxpayer money to fund textbooks developed by the BSCS which included evolution content. The creationists' goal in this case was to have the material on evolution in the textbooks legally recognized as an unconstitutional establishment of religious secularism. Their attempts failed however when the case was dismissed as meritless. As a setback to the creationists in this continued American cultural and political battle over teaching evolution in schools, this groundbreaking decision set another legal precedent against the efforts of keeping evolution out of public schools.

However, the battle continued. After this blow, creationist strategy took a very interesting turn: piggyback creationism with evolution.

10.8 Bundle Creationism with Evolution

After having failed to keep evolution out of the classrooms in American schools, creationists tried to get creationism into the classrooms. They tried to make the sale of creationism as science under scientific sounding names such as creation science or scientific creation. The strategy was to present creationism as a scientifically credible alternative to the theory of evolution. As a result of this nationwide effort, in the 1970s and early 1980s, legislations calling for equal time for "creation science" as an alternative to evolution were introduced in at least 27 states. Many states including Kentucky, Louisiana, and Tennessee passed such laws.

Let's consider the example of the 1981 Arkansas Act 590, named *Balanced Treatment for Creation-Science and Evolution-Science Act*. The Act mandated that "creation science" be given equal time in public schools along with evolution. Where did the legislation for the Act originate? The legislation was based on education, research, and media promotion of creation science and Biblical creationism. It was put forth in the legislature by a Christian fundamentalist on the basis of a request from the Greater Little Rock Evangelical Fellowship. Here are some very interesting facts about the Act: The Act was opposed by many religious organizations along with non-religious groups. Moreover, when the Act was challenged in court, the plaintiffs were led by Methodist minister William McLean. In this case, known as McLean vs Arkansas, Judge William Overton issued a ruling on January 5, 1982, that "creation-science" as defined in the Arkansas Act 590 "is simply not science." Because it was a U.S. District Court ruling, schools outside the Eastern District of Arkansas were not bound to it.

Judge Overton's decision provided a good, quick education on this issue for anybody with open mind. Quoting from the decision of the judge:

More precisely, the essential characteristics of science are:
(1) It is guided by natural law;
(2) It has to be explanatory by reference to nature law;
(3) It is testable against the empirical world;
(4) Its conclusions are tentative, i.e. are not necessarily the final word; and
(5) It is falsifiable.

Note that this definition of science is consistent with what you learned about science, scientific theory, and the scientific method in Chapter 2. Judge Overton found that "creation science" failed to meet these essential characteristics. Some of the reasons offered by Judge Overton to support this statement are listed in the following:

1. Sudden creation "from nothing" is not science because it depends upon a supernatural intervention which is not guided by natural law; it is not explanatory by reference to natural law; it is not testable; and it is not falsifiable.

2. The "insufficiency of mutation and natural selection in bringing about development of all living kinds from a single organism," is an incomplete negative generalization directed at the theory of evolution.

3. The "changes only within fixed limits of originally created kinds of plants and animals" fails to conform to the essential characteristics of science for several reasons:

    • There is no scientific definition of "kinds."

    • The assertion appears to be an effort to establish outer limits of changes within species, while there is no scientific explanation for these limits which is guided by natural law and the limitations; whatever they are, cannot be explained by natural law.

4. The "separate ancestry of man and apes" is a bald assertion. It explains nothing and refers to no scientific fact or theory.

5. The "explanation of the earth's geology by catastrophism, including the occurrence of a worldwide flood." This assertion completely fails as science because catastrophism and any kind of Genesis Flood depend upon supernatural intervention, and cannot be explained by natural law.

6. "Relatively recent inception" has no scientific meaning, is not the product of natural law; not explainable by natural law; nor is it tentative.

7. The methodology employed by creationists is another factor which is indicative that their work is not science. A scientific theory must be tentative and always subject to revision or abandonment in light of facts that are inconsistent with, or falsify, the theory. A theory that is by its own terms dogmatic, absolutist, and never subject to revision is not a scientific theory.

8. The creationists' methods do not take data, weigh it against the opposing scientific data, and thereafter reach the conclusions...Instead, they take the literal wording of the Book of Genesis and attempt to find scientific support for it.

Recognizing that the creationists do not follow the scientific method, Judge Overton noticed:

"The method is best explained in the language of Morris in his book (Px 31) *Studies in The Bible and Science* at page 114:

... it is ... quite impossible to determine anything about Creation through a study of present processes, because present processes are not creative in character. If man wished to know anything about Creation (the time of Creation, the duration of Creation, the order of Creation, the methods of Creation, or anything else) his sole source of true information is that of divine revelation. God was there when it happened. We were not there ... Therefore, we are completely limited to what God has seen fit to tell us, and this information is in His written Word. This is our textbook on the science of Creation!

The very fact that the plaintiffs in this case were being led by a Methodist minister, William McLean, was a very bitter pill for the creationists to swallow. It was a stark reminder to them that they had lost in the court of public opinion and failed to keep evolution out of the American classroom. Because this ruling was issued by a district court judge, it had no effect on other places where teachers were being forced to teach creationism along with evolution. One such place was the state of Louisiana, where in 1981 Republican governor David Treen signed into law a bill passed by the Louisiana State Legislature called the *Louisiana Balanced Treatment Act,* also known as the *Creationism Act.* The Act required teachers to teach creationism if they taught evolution.

In the face of this law, many biology teachers in Louisiana would rather have avoided teaching evolution than teach creationism with it. However, Don Aguillard was one of fewer than ten biology teachers who agreed to challenge the law, and he became the

Figure 10.6 Dr. Donald Aguillard, then a biology teacher who as a lead plaintiff in the U.S. Supreme Court case Edward vs Aguilard successfully challenged the state of Louisiana against mandating teaching of creationism along with evolution.

lead plaintiff in the lawsuit known as Edwards v. Aguillard. During this case, seventy-two Nobel prize-winning scientists, seventeen state academies of science, and seven other scientific organizations filed amicus briefs which described "creation science" as being composed of religious tenets. On June 19, 1987, six years after the passing of the law, the Supreme Court issued its decision affirming the following:

1. The Act is facially invalid as violative of the Establishment Clause of the First Amendment, because it lacks a clear secular purpose.

And God Said: Let There Be Evolution

2. Because the primary purpose of the Creationism Act is to endorse a particular religious doctrine, the Act furthers religion in violation of the Establishment Clause.

3. It is unconstitutional to mandate or advocate creationism in public schools, for creationism is a religious idea.

4. The Act does not further its stated secular purpose of "protecting academic freedom." It does not enhance the freedom of teachers to teach what they choose, and fails to further the goal of "teaching all of the evidence." Forbidding the teaching of evolution when creation science is not also taught undermines the provision of a comprehensive scientific education.

5. Louisiana's "balanced treatment" law is unconstitutional because it "impermissibly endorses religion by advancing the religious belief that a supernatural being created humankind."

6. The contention "a basic concept of fairness" for requiring the teaching of creation science is "without merit."

Although the ruling offered a crushing blow to the American creationist movement, it only affected state schools. Independent schools, home schools, Sunday schools, and religious schools such as Christian schools, were left free to still teach creationism as a scientific theory.

Yet, the ruling was a national setback to the efforts of creationists. However, within two years following this ruling, creationists adapted quickly to repack creationism as intelligent design.

10.9 Repack Creationism: Here Comes Intelligent Design

The U.S. Supreme Court ruled in the 1987 case of *Edwards v. Aguillard* that to require the teaching of "creation science" alongside evolution was a violation of the Establishment Clause of the First Amendment, which prohibits state endorsement of a religion. The Supreme Court in the same decision also held that "teaching a variety of scientific theories about the origins of humankind to school children might be validly done with the clear secular intent of enhancing the effectiveness of science instruction." This latter assertion by the court is not new to scientists; this is already part of scientific culture. However, this assertion was used by a group of American creationists to circumvent the court rulings, attempting the backdoor entry of creationism into the schools under the reformulation of intelligent design. This reformulation of creationism was introduced in what was presented as a supplementary textbook titled *Of Pandas and People,* produced and published by the Texas-based Foundation for Thought and Ethics, which styled itself as a Christian think tank. Intelligent design is advertised by its proponents as not based on any sacred texts and therefore not requiring any appeal to a supernatural being. So what did they replace the supernatural intervention of creationism with? According to them, now the designer may be a space-alien, a time-traveling cell biologist, or it might be God. Yet, given that teaching creationism in public schools is unconstitutional, they reject any characterization of intelligent design as creationism. However, it is no coincidence that almost all the leading proponents of intelligent design are associated with the Discovery Institute, a politically conservative think tank that makes no bones about believing and asserting that the designer be the God of Christianity. Furthermore, upon a close examination, the claim that intelligent design is not a presentation of creationism does not hold water.

Creationists pushed to sell intelligent design as a scientific theory through the religious members of school boards. They got some success when a local school district in Pennsylvania, the Dover Area School District, in a decision taken on October 18, 2004, required that a disclaimer be read aloud in science classes when evolution was taught alleging that "evolution is a theory...not a fact," that gaps in the theory exist for which there is no evidence, and that intelligent design as presented in *Of Pandas and People* is a credible scientific alternative to evolution. However, the success was short lived. Two of the nine school board members resigned as a protest against this decision. Almost all the biology teachers refused to implement the decision in the classrooms as according to them intelligent design was a religious doctrine and a not a scientific theory.

Furthermore, eleven parents of the students filed suit in federal district court arguing that the policy of the school district was unconstitutional. The case, known as *Kitzmiller v. Dover Area School District,* was tried in a bench trial from September 26, 2005, to November 4, 2005, before Judge John E. Jones III, a conservative Republican appointed in 2002 by Republican president George W. Bush.

The expert witness testimonies presented at the trial were devastating to the claim that intelligent design was scientific. As mentioned earlier, intelligent design was established to be just another version of creationism. For example, a biochemist, Michael Behe, one of the expert witnesses in the trial, testified that no articles had been published in the scientific research literature that "provided detailed rigorous accounts of how intelligent design of any biological system occurred." And now read this carefully: Behe was testifying in defense of the school board's policy.

On December 20, 2005, Judge Jones issued his ruling that the Dover mandate was unconstitutional and barred intelligent design from being taught in the public school science classrooms of the Dover school district. He wrote, "In making this determination, we have addressed the seminal question of whether [intelligent design] is science. We have concluded that it is not, and moreover that [intelligent design] cannot uncouple itself from its creationist, and thus religious, antecedents."

Think About It!
Teaching creationism in the name of intelligent design or under any other name along with evolution would be the closest equivalent of teaching what with astronomy?

Answer
Teaching astrology with astronomy.

Having failed to prove the scientific credibility of their ideas, creationists came full circle to their initial position and attempted again to directly challenge evolution to expel it from the American classroom. However, recall that they had already been there and had been defeated both in the court of public opinion and in the court of law. What could be worse than a Christian minister (William McLean) leading the case against them? So now their original position reduced to creating doubts about the scientific theory of evolution by misrepresenting it.

10.10 Misrepresent Evolution

As mentioned in the previous section, after having failed to bundle creationism with evolution as a credible scientific alternative, creationists are back to singing their old song:

evolution is just a theory. This current policy reflects, for example, from the strategy of the Texas-based Institute of Creation Research (ICR), which promotes biblical creationism. After the U.S. Supreme Court struck down the Louisiana Balanced Treatment Act in 1987, the ICR took on the strategy that "school boards and teachers should be strongly encouraged at least to stress the scientific evidences and arguments against evolution in their classes…even if they don't wish to recognize these as evidences and arguments for creation."

When creationists say *evolution is just a theory and not a fact,* they are doing it as a tactic. Through this phrase, in a sneaky way, they are trying to propagate the idea that evolution is just a conjecture or a speculation. However, by doing this they are either playing plain ignorant of what theory means in science, or they are being dishonest to misuse a term to mislead the public. We addressed this issue of a scientific theory and scientific method in Chapter 2. However, this policy of the creationists to create confusion and doubt about evolution did bear some fruit. For example, giving in to the pressure of local creationists, the Cobb County Board of Education in Georgia made a decision to require a warning label on biology textbooks: "evolution is a theory, not a fact." The sticker read:

This textbook contains material on evolution. Evolution is a theory, not a fact, regarding the origin of living things. This material should be approached with an open mind, studied carefully, and critically considered.

Approved by
Cobb County Board of Education
Thursday, March 28, 2002

The sticker, of course, has an enormous historical value in that it signifies the amazing documented fact that how even in the twenty-first century, a school board in the United States of America can put on display such ignorance.

On November 8, 2004, five parents in the county filed suit in federal district court arguing that the Board's policy was unconstitutional. In this case, known as *Selman vs. Cobb County School District,* the trial judge agreed with the parents. Ultimately, due to a technicality the case was settled outside of the court in favor of the plaintiffs. As a result, Cobb County school officials would not order the placement of "any stickers, labels, stamps, inscriptions, or other warnings or disclaimers bearing language substantially similar to that used on the sticker that is the subject of this action" and would not undermine science education in the future. The school district also agreed to pay $166,659 towards attorneys' fees in the case.

Meanwhile, after the 1987 Supreme Court decision of Edwards v Aguillard, the Discovery Institute, the hub of the intelligent design movement, had started a campaign of *Teach the Controversy* to the public, education officials, and public policymakers. Their strategy has been to advertise evolution as a controversial theory. Intelligent design proponents, through the Discovery Institute, employed a number of specific political strategies and tactics to further their goals and agenda. Although their immediate goal was and remains to undermine the credibility of evolution in the public mind, their ultimate goal remains the same: replace evolutionary theory with creationism or mandate the teaching of intelligent design (creationism) along with evolution. One of their strategies has been to populate the municipal, county, and state school boards with intelligent design proponents.

In addition to *Teach the Controversy,* the other slogans creationists have used in their campaign to undermine evolution by misrepresentation include the catchphrases "critical analysis" and "academic freedom." Such phrases are part of their efforts to coat their dogmatic non-scientific ideas with a flurry of scientific or academic sounding phrases, labels, and slogans. This strategy of the creationists materialized into a series of so-called academic freedom bills introduced in the U.S. state legislatures between 2004 until the writing of this book. All these bills are written from the same creationist template.

Here is an example of one such legislation that actually turned into law. The catchphrase "academic freedom" was part of the *Louisiana Academic Freedom Act of 2008*, which was renamed (or changed) to The *Louisiana Science Education Act* and signed into Law by Republican Governor Bobby Jindal on June 25, 2008. Note the phrases "academic freedom" and "science education" coming right from the strategy book of intelligent design supporters. At face value, the law looks innocuous when its text suggests that it's intended to foster critical thinking, calling on the Louisiana State Board of Education to "allow and assist teachers, principals, and other school administrators to create and foster an environment within public elementary and secondary schools that promotes critical thinking skills, logical analysis, and open and objective discussion of scientific theories." Who would object to that? However, perhaps the Louisiana Legislature were unaware (or ignorant) of the fact that critical thinking, logical analysis and objective thinking are already part of the framework and curriculum of science. These elements are integral parts of science by definition and do not need to be legislated. As always, the devil however here is in the details. Actually the law creates a backdoor for the entry of non-scientific theories such as intelligent design by allowing local school boards to approve supplemental classroom materials specifically to undermine the teaching of scientific theories that the creationists don't like such as evolution. This creates possibilities, in the words of molecular biologist and science editor John Timmer, "to allow poorly-informed board members to stick their communities with Dover-sized legal fees." According to the bill, fostering a critical thinking environment includes providing "support and guidance for teachers regarding effective ways to help students understand, analyze, critique, and objectively review scientific theories being studied." Which scientific theories? The bill is remarkably selective in its suggestion of topics that need critical thinking, as it cites scientific subjects "including, but not limited to, evolution, the origins of life, global warming, and human cloning." If you see this in the whole context of the organizations that lobbied for this bill, the target is obviously the scientific theory of evolution and the topics related to it, and the goal is to create confusion about evolution in the minds of the students and to create a backdoor entry for non-scientific material to be presented as an alternative to evolution.

Maybe they improved the chances of saving their jobs by pandering to a block of voters, but the Louisiana Legislature and Governor Bobby Jindal did not do any favors to science or to the people of Louisiana, especially to its children, by passing this bill. The American Association for the Advancement of Science told Jindal that the law, "would unleash an attack against scientific integrity." The National Association of Biology Teachers in their appeal to the Louisiana Legislature to defeat the bill, pleaded that, "the state of Louisiana not allow its science curriculum to be weakened by encouraging the utilization of supplemental materials produced for the sole purpose of confusing students about the nature of science." The American Institute of Biological Sciences had warned that by passing this bill, "Louisiana will undoubtedly be thrust into the national spotlight as a state that pursues politics over science and education."

But all these warnings went unheard in Louisiana. However, such anti-evolution bills written from the same template in the disguise of critical thinking and academic freedom introduced in a number of other states such as Alabama, Florida, Missouri, Oklahoma, South Carolina, and Texas died by the end of 2011.

Another state that has followed suit to Louisiana is Tennessee when it went back to the future when its state legislature passed their version of the so-called academic freedom bill in the first half of 2011. The law encourages teachers to, "present the scientific strengths and scientific weaknesses of existing scientific theories covered in the course being taught." But the only examples given in the bill of "controversial" theories are "biological evolution, the chemical origins of life, global warming, and human cloning." As mentioned earlier, these so-called academic freedom bills in all different states are written from the same creationist template in order to implement the current strategy of creationism against evolution: *Teach the Controversy.*

In Tennessee, this legislation became law when Governor Bill Haslam decided neither to sign the bill nor to veto it. As we have mentioned earlier, these bills are being pushed to implement the policy of creationists to create confusion about scientific theories such as evolution and to create backdoor entry for non-scientific theories as alternative to scientific theories. They do not add anything new to science education because critical thinking is already a part of science. Governor Haslam sounded in agreement with some of these points in the statement he issued:

458

I have reviewed the final language of HB 368/SB 893 and assessed the legislation's impact. I have also evaluated the concerns that have been raised by the bill. I do not believe that this legislation changes the scientific standards that are taught in our schools or the curriculum that is used by our teachers. However, I also don't believe that it accomplishes anything that isn't already acceptable in our schools. The bill received strong bipartisan support, passing the House and Senate by a three-to-one margin, but good legislation should bring clarity and not confusion. My concern is that this bill has not met this objective. For that reason, I will not sign the bill but will allow it to become law without my signature.

At best this stand represents political opportunism and lack of leadership on part of Governor Haslam. This new law does not ban the teaching of evolution as the Butler Act did, under which John Scopes was convicted, but it does possibly open the backdoor for the entry of non-scientific theories such as intelligent design to be presented side by side with scientific theories such as evolution and thereby confuse students and undermine their scientific growth. Most of such bills introduced in many states are already dead for lack of enough public support. However, creationists continue their efforts, pushing these anti-evolution bills forward wherever and whenever they can.

Here is a word of caution about the "academic freedom" that creationists are trying to exploit: it does not mean the freedom of teaching whatever you want. For example, in physics, we know that classical theories known as Newtonian physics fail in the atomic scale, where only the quantum theory can explain physical entities and their behavior. Now, academic freedom does not mean that a teacher can present Newtonian physics as the physics of the atomic world as well because he/she believes so. Similarly academic freedom does not mean to create confusion about the scientifically-credible theory of evolution or to present a non-scientific theory such as intelligent design as a credible alternative to evolution. However, that is exactly how creationists are trying to use the concept of academic freedom. For them critical thinking means to undermine the scientifically-proven credibility of evolution with a belief system and by academic freedom they mean the freedom to present creationism as science; and the Louisiana State Legislatures and the Governor went along with them on this anti-science mission. Call it a political opportunism or plain ignorance, in the end it is all the same because the results are the same.

This history of the legal battle of creationists with evolution exposes their real intentions by distinguishing their goal from the tactics and strategy applied to achieve that goal.

10.11 Chronology of the Legal Battle between Evolution and Creationism

Table 10.1 summarizes the brief history of the creationism versus evolution battle in terms of court cases in the U.S.

Judgment year	Case	Judgment
July 21, 1925	*The State of Tennessee vs. Scopes.* High school biology teacher John Scopes was accused of violating the state's Butler which prohibited teaching evolution.	John Scopes was found guilty for violating the Butler Act by teaching evolution, but the verdict was overturned on a technicality and he was never brought back to trial.

December 12, 1968	***Epperson v. Arkansas.*** Biology teacher Susan Epperson filed the suit to test the federal constitutionality of the 1928 Arkansas state law prohibiting the teaching of evolution.	The U.S. Supreme Court ruled that the laws barring the teaching of evolution in public schools are unconstitutional.
1973	***Willoughby v. Stever.*** Evangelist William Willoughby sued National Science Foundation director H. Guyford Stever and the Board of Regents of the University of Colorado for using taxpayer money to fund textbooks developed by the Biological Sciences Curriculum Study (BSCS) because they included evolution instruction.	Affirmed: Public funding of evolution education in textbooks support science, not religious teachings. States may not require teaching be tailored to satisfy religious beliefs.
January 5, 1982	***McLean vs Arkansas.*** A lawsuit filed in the U. S. District Court for the Eastern District of Arkansas by various parents, religious groups and organizations, biologists, and others against the Arkansas state law known as the *Balanced Treatment for Creation-Science and Evolution-Science Act (Act 590),* which mandated the teaching of "creation science" in Arkansas public schools to balance the teaching of evolution.	The Arkansas *Balanced Treatment Act* of 1981 requiring schools balance the teaching of evolution with the teaching of creation science violated the Establishment Clause of the First Amendment to the U. S. Constitution

And God Said: Let There Be Evolution

June 19, 1987	**_Edward vs Aguillard._** A legal case heard in the U.S. Supreme Court regarding Louisiana's _Creationism Act_, which forbade the teaching of the theory of evolution in public schools unless accompanied by instruction in "creation science," which basically was biblical creationism.	The law requiring that creation science be taught in public schools along with evolution is unconstitutional because the law was specifically intended to advance a particular religion.
December 20, 2005	**_Kitzmiller v. Dover Area School District._** The first direct challenge brought to the U.S. federal courts against a public school district that required the presentation of intelligent design as an alternative to evolution in explaining the origin of life.	Teaching intelligent design in public school biology classes violates the Establishment Clause of the First Amendment to the U.S. Constitution because intelligent design is not science and "cannot uncouple itself from its creationist, and thus religious, antecedents."

This table presents a very brief history of the historical legal battles between creationists and evolution and is not complete by any means. Nevertheless, it does expose the fact that the real goal of creationists has been to banish evolution from schools or to replace it with creationism. To achieve this goal, they applied different tactics and strategies and hid their real intentions under different covers when they had to.

The Acts of the legislators of Louisiana and Tennessee are for now the last paragraphs in about a century long history of creationist opposition to evolution in the U.S. Glancing over the history briefly discussed in this chapter, one sees three underlying principles in the creationists' overall strategy.

10.12 The Triangular Strategy of Creationism

Since the times of the Scopes Trial, during their long journey against evolution, creationists have developed a three-pronged central rhetoric theme as part of their strategy to attack evolution. These three underlying principles or claims which are part of creationists' overall strategy, also called the three pillars of creationism, are the following:

Evolution is just a theory. By this mantra, creationists imply that the theory of evolution is untested, unproven, and has no substantial evidence in its support; which is a plain lie, as every student in biology knows. Creationists try to imply that evolution is unsupported or even in conflict with facts in science. These incorrect claims that at best present evolution as shaky science are constantly fed to the general public by the creationists by using media such as websites, videos made for presentations at gatherings, and popular books. This pillar, however, collapses in the face of all the accounts and evidence presented in this book. As you see in this book, an incredible body of evidence supports the theory of evolution, and evolutionary science is a dynamic field which continues to develop and expand from its already strong base of evidence.

Evolution equates to atheism. Creationists claim that teaching evolution is harmful; it threatens religion, morality, and society. The basic goal here is to imply that the acceptance of evolutionary science is incompatible with religious faith by definition. Creationists use this fear to drive much of the opposition to evolution. However, this pillar is only as strong as were the arguments against Galileo Galilei in opposition to his scientific observation that the Earth revolves around the

And God Said: Let There Be Evolution

Sun, and thereby Earth is not the center of the Universe. Now, we take this fact for granted and we still have our faiths.

Appeal to fairness and academic freedom. The creationists' far-cry here is: It is only fair to teach both sides; be fair, exercise academic freedom, and teach creationism along with evolution. This claim can sound appealing to the public because it invokes fairness, a value cherished in America and many other cultures. However, this third pillar is bound to fall because it's based on the false foundation that creationism is science, which it is not. Fairness does not warrant teaching religious faith in the name of science. Doing so will be unfair to our children because we will be confusing them and undermining their scientific foundation.

Creationists continue to preach the fallacy that there are only two alternatives, evolution or creationism, and that any evidence against evolution is evidence for creationism, and therefore discrediting evolution credits creationism. This thinking is obviously wrong because you can only compare a scientific theory with a scientific theory and creationism is not a scientific theory. Even between two scientific theories, the fall of one theory does not necessarily mean the rise of the other theory. Each theory has to prove itself by facing scientific scrutiny on its own. So, even if the theory of evolution falls one day, the candidate theory to replace it would not be any form of creationism because *intelligent design* and any other form of creationism do not even qualify to be scientific theories as they are not falsifiable. For this reason, they don't even qualify to be scientific hypotheses.

In the U.S., creationists are intentionally or unintentionally playing the dangerous game of damaging the scientific future of this country.

Accepting the reality of evolution does not rob your life of purpose or meaning and does not necessarily turn you into an atheist. Neither does it promote immorality; nor does it endorse the mentality of Stalin or Hitler. To the contrary, you saw in Chapter 9 how through eugenics, creationists and Hitler became connected in some sense. In the following section, you will see how the opponents of evolution connect with the Stalin mentality.

10.13 Look Who Is On the Other End of the Rope

Nobody can irritate a religious fundamentalist more than an atheist or even worse a communist, and vice versa. However, ironically, it is very interesting to note that on many issues the extreme positions taken by ideologically diametrically opposed people act like two ends of a rope which can easily meet. In the twentieth century, it was not only religious fundamentalists but also a faction of communists that felt threatened by the theory of evolution. For example, Joseph Stalin (1878-1953), a communist dictator of the Soviet Union, did not like the idea that genes were determining certain characteristics (traits) of organisms. According to communist (or Marxist) dogma, most of people's traits are due to the external environment and they can be changed by changing the environment. Following this line, Stalin's Director of Agriculture, Trofim Denisovitch Lysenko, who was also a biologist, went a long way to run a hate campaign against genes, chromosomes, and the inheritance laws discovered by Mendel, discussed in Chapter 3.

Lysenko's campaign eliminated throughout the Soviet Union all research and study that involved Mendelian genetics. The campaign also resulted in the expulsion, imprisonment, and death of hundreds of scientists. For instance, one of the

And God Said: Let There Be Evolution

best geneticists in the Soviet Union was Nikolai Vavilov (1887-1943), who had travelled in Europe from 1913 to 1914 and studied plant immunity in collaboration with British biologist William Bateson, the founder of the discipline of genetics and who coined the very term *genetics*. Vavilov was arrested in 1940 and interrogated for a long time. He was found *guilty* of belonging to a "rightist conspiracy, spying for England, and sabotage of agriculture" in a trial that lasted for five minutes. He died of starvation in a prison camp in 1943.

In 1962, nine years after Stalin passed away, three of the most prominent Soviet physicists, Yakov Borisovich Zel'dovich, Vitaly Ginzburg, and Pyotr Kapitsa, presented a case against Lysenko that his work was false science. Furthermore, in 1964, famous nuclear and particle physicist and human right activist Andrei Sakharov spoke out against Lysenko in the General Assembly of the Academy of Sciences:

> He is responsible for the shameful backwardness of Soviet biology and of genetics in particular, for the dissemination of pseudo-scientific views, for adventurism, for the degradation of learning, and for the defamation, firing, arrest, even death, of many genuine scientists.

Encouraged by this presentation, the Soviet press took on the responsibility of examining the case, and the press was soon filled with articles pointing to the damaging role Lysenkoites had played. A very strong voice was raised in the media in support of restoring scientific methods to all fields of biology and agricultural science. Subsequently, in 1965, Lysenko was dismissed from his position as Director of the Institute of Genetics at the Academy of Sciences, followed by the dissolution of the Institute itself. He was confined to an experimental farm in Moscow's Lenin Hills.

So, what linked Stalinists and creationists, the two ends of the rope? It was the desire to replace scientifically validated theories and ideas with their own ideological and religious dogmas.

In a nutshell, whether we further our religious beliefs such as creationism or serve ideological dogma such as that of Stalin and Lysenkoites in the name of science, the results will be damaging in many ways.

10.14 Fatal Consequences of Creationism in Opposition to Evolution

So where are we in the battle between modernists and traditionalists that began with the Monkey Trial in Dayton, Tennessee? Scientific and legal scrutiny has confirmed the fact that creationism and its various versions such as intelligent design are not scientific theories. If you do not believe the scientific community and judges, then believe the creationists: The fact that creationism is not science clearly reflects in the Wedge Document, a secret policy manifesto leaked out of the Discovery Institute, which clearly aims at pushing their religious agenda in the name of science. For example, it states, while describing a religious goal, to "reverse the stifling dominance of the materialist worldview, and to replace it with a science consonant with Christian and theistic convictions."

The fact that *creationism is not science* does not mean that there is anything bad about it; it just means that it is not a scientific theory, so it must not be taught as such. Nobody is arguing against teaching it in a course on religion, along with other religious beliefs, naturally, or just by itself if the course is designed that way. Teaching it as a scientific theory will result in confusing students, undermining their scientific foundations, and thereby damaging the scientific future of this nation.

To keep pushing these beliefs as an alternative to or complementary of the theory of evolution is a dangerous game. The creationists' propaganda along with bills they have been pushing through legislatures or have succeeded to pass, such as the Louisiana Science Education Act and the Tennessee Academic Freedom Act, have catastrophic potential consequences. Such laws are double-edged swords: on one hand they provide license to a section of religiously-biased teachers to further their religious agenda in the name of science, and on the other hand such laws can be used to bully the teachers who are not creationists by creationist education administrators. According to an informal survey by the National Science Teachers Association conducted in 2005, 31 percent of science teachers in the U.S. reported pressure to include non-scientific alternatives to evolution such as "creation science" or "intelligent design" in their teaching, whereas 30 percent reported pressure to downplay evolution and related topics in their science curriculum.

The steady propaganda and misrepresentations of evolution by creationists has already taken its toll. According to the National Survey of High School Biology Teachers conducted in 2011, only 28 percent of teachers consistently and "unabashedly" introduce evidence that evolution has happened, and build lesson plans with evolution as a unifying theme linking different topics in biology. Furthermore, 13 percent of teachers explicitly endorse creationism or intelligent design, and spend at least one hour of class time presenting it in a positive light. Five percent of the surveyed teachers reported that they support creationism in passing or when answering questions from the students. According to the same survey, about 60 percent of high school teachers avoid taking sides. Most of these teachers have not adequately studied or taken courses in evolutionary biology themselves; so they lack the confidence to answer questions from skeptical or hostile students and parents.

Only creationists can take this scenario as a blessing. For the rest of the nation, it should sound off an alarm.

As we have shown so far in this book, evolution is inarguably a unifying principle in biology and therefore has strong implication for all health sciences, including medicine. There are always aspects of scientific theories, even well-established theories such as Newtonian physics and quantum physics, that scientists are still working on or arguing about. This does not make the theory a controversial theory. The textbooks and curriculum used in the schools present basic principles which are uncontroversial and scientifically well settled; the same is true with evolution. It is as valid a scientific theory as Newtonian physics, also called classical physics, and as quantum physics. "A theory not a fact" sticker on a biology text is only as true as the same sticker on a physics or chemistry text. How about a sticker on an airplane:

This plane is designed after Newtonian physics which is just a theory and not a fact. At the atomic scale, this theory has been replaced by quantum physics.

And how about teaching astrology in an astronomy class? I can keep going, but the point of this discussion is that by creating confusion about evolution in public schools, we will be weakening the scientific foundation of our next generations, thereby undermining the scientific future of this country. Students who are deprived of the proper understanding of evolution will not achieve a basic level of scientific literacy in biological sciences, including health sciences, the same way as students who are deprived of proper understanding of classical physics and quantum physics will not succeed in the fields of physical sciences and engineering.

10.15 Summary

The 1925 trial of a science teacher, John Scopes, in Dayton, Tennessee, for teaching evolution declared a battle between modernists and traditionalists in the United States. During this battle, creationists, who were at the forefront of the traditionalists' side, exhibited a three-pronged self-contradictory action strategy: Oppose evolution to protect creationism; support the eugenics movement, which they believes was based on evolution, to support their race-oriented dream about the future of America; and later when the eugenics movement ended up in disaster and failed miserably, use it against evolution by associating it with evolution.

Even though creationists continuously modified their strategy in response to legal defeats, their goals have always been to keep the theory of evolution away from schools, replace evolution with creationism, or present creationism or one of its derivatives such as intelligent design or creation science as an alternative to evolution.

Even though creationists technically won the famous John Scopes Trial of 1925, they lost in the court of public opinion, and have kept losing ground since then. Unfortunately for the advancement of evolution, creationists have steadily been fortunate enough to find enough ignorance (or science illiteracy), political opportunism, or both among legislators to get their agenda materialized into anti-evolution laws. The ultimate fate of these laws has almost always been defeat in courts of law. In response to these defeats, creationists have been continuously changing their strategy and tactics to fulfill their ultimate goal of replacing evolution with creationism. This century-long battle can be broadly divided into three stages marked by the John Scopes Trial followed by two landmark decisions by the U.S. Supreme Court:

The Post John Scopes Trial Period: 1925-1968. During this period, the party line of creationists was this: Evolution is just a theory and not a fact; so it should be expelled from schools. The John Scopes Trial brought defeat to the creationists in the court of the public opinion as the whole nation witnessed that the creationists' anti-evolution argument could not stand against the scientific logic of the modernists. Nevertheless, during this period, creationists enjoyed a great success in keeping evolution out of the schools with the help of the law.

The Post Susan Epperson Period: 1968-1987. In the landmark case of *Epperson v. Arkansas*, the U.S. Supreme Court ruled in 1968 that laws barring the teaching of evolution in public schools were unconstitutional. In response to this defeat, creationists changed their strategy to what could be deemed piggybacking creationism with evolution, that is, using the law to force teachers to teach creationism as an alternative theory to evolution. Creationists used fancy terms such as creation science, equal-time, and balanced treatment to sell their agenda or implement their policy, which was materialized into law in many states including Arkansas, Kentucky, Louisiana, and Tennessee. Also during this period, creationists made unsuccessful attempts to legally stop public funding for textbooks or teachings that included evolution.

The Post Don Aguillard Period: 1987-Current. The 1987 landmark ruling of the U.S. Supreme Court in *Edward vs Aguillard* put a stop to creationists' efforts to mandate teachers to teach creationism with evolution. In response to this legal setback, creationists adopted a two-pronged strategy: misrepresent evolution as just a theory and not a fact (*Teach the Controversy*) and repackage creationism with non-religious sounding names such as intelligent design to create a legal backdoor for a repackaged creationism to enter the classroom in the name of academic freedom. They have been pushing acts through legislatures of various states that go something like the Academic Freedom Act or the Science Education Act. They have even succeeded to get such acts passed and turned into law in the states of Louisiana and Tennessee. Efforts in other states continue.

During the long and continuous course of success with the state legislatures and ultimate defeats in courts of law, creationists have also been losing in the court of public opinion and among their own ranks and files as well. For example, the creationists' efforts to keep the scientific theory of evolution out of the schools have been challenged by not only the scientific community but also by religious leaders and organizations who understand the importance of religion and science having their own places. It was not just a strange event or coincidence that Susan Epperson, who successfully challenged the law that prohibited the teaching evolution, was supported by the Little Rock Ministerial Association, or that William McLean, who successfully challenged the teaching of creationism in the biology classes, was a Christian minister.

According to surveys, the continuous efforts of the creationists supported by the science illiterate (ignorant) or politically opportunist state legislatures have already started adversely affecting the science education in schools. This is equivalent to undermining the future of our coming generations in this increasingly competitive world of science and technology. So it is neither creationists versus scientists, nor creationism versus evolution; it is some fundamentalist creationists versus the future of our children.

In a Nutshell

Here are the three biggest takeaways from this chapter.

Creationism is not science. It is not creationism versus evolution because creationism is not a scientific theory. A scientific theory must be verifiable and falsifiable, which creationism is not because it's a religious belief, which does not lend itself to scientific scrutiny. The claim that creationism or its derivatives such as intelligent design are credible scientific alternatives to evolution has also not survived legal scrutiny.

Theory of evolution is a well-tested, sound theory supported by facts. The theory of evolution is as sound as other well-established scientific theories such as Newtonian physics and quantum theory.

Any misrepresentation of evolution in public schools is an attack on the scientific foundations of our future. Because evolution is a unifying principle of biology, presenting a religious belief as an alternative to evolution, or miseducating our children by misrepresenting evolution in any way in the public schools is equivalent to depriving them of a sound scientific foundation.

10.16 Behold

In this chapter, we explored how creationism belongs to a belief system, whereas evolution is a natural process that occurs according to verifiable natural laws, described by the theory of evolution. Belief systems are not science. A major factor that distinguishes a scientific law from a belief is that a scientific law can be verified on its own without referring to divinity, and once understood, can often be put into practical use in many cases. In other words, applications can be developed based on scientific laws. Most of the devices and facilities today that are part and parcel of our daily lives, such as cars, airplanes, televisions, cellular phones, and computers, are applications of laws of physics.

So, what are the applications of the theory evolution? We explore the answer to this question in the next chapter.

ELEVEN

Applications of Evolution

And God Said: Let There Be Evolution

Galileo was no idiot. Only an idiot could believe that science requires martyrdom - that may be necessary in religion, but in time a scientific result will establish itself.

David Hilbert

11.1 Once Upon a Time

If historians have to determine a specific date for the public victory of the theory of evolution, one candidate date would be October 23, 1998 when Dr. Richard J. Schmidt was convicted of attempted murder.

According to the story unfolded by this case, Schmidt was a philanderer who has been carrying on an extramarital affair for 10 years with Janice Trahan, a nurse. Year after year, Schmidt promised to leave his wife, but never did. His primary interest in Janice was purely sexual. After having realized this, Janice had tried many times to break off their affair, but in vain. He would always force her to continue the relationship by using multiple tactics such as threaten to kill her or himself, blackmail her that he would post erotic photos of her on the Internet, and threaten to expose the fact that he helped her cheat on exams. Janice finally gathered the courage to break off her ten-

Figure 11.1 Statue of Lady Justice, the symbol of justice, at the Shelby County Courthouse, Memphis, Tennessee.

year-long relationship with Schmidt in July 1994. Her life, however, took a terrible turn in the evening of August 4, 1994, when she woke up to find Schmidt standing in her bedroom holding a hypodermic needle in his hand. However, his visit was expected and she had even

And God Said: Let There Be Evolution

left the front door unlocked for him, as since before their breakup, Schmidt had been giving her B-12 vitamin shots that he had prescribed for her chronic fatigue. He had phoned her earlier to let her know that he would be coming over to give her another shot. In spite of her resistance, Schmidt delivered the shot quickly before she could react. As it turned out, this was not just another B-12 vitamin shot. She felt intense pain upon receiving the shot; it was agonizing. On December 20, 1994, Janice found out that Schmidt had injected her with death, when she tested positive for HIV and Hepatitis C.

So, on October 23, 1998, Dr. Richard J. Schmidt was found guilty and subsequently convicted of attempted second-degree murder by injecting his former girlfriend with blood or blood-products obtained from an HIV type 1 (HIV-1) infected patient under his care. This conviction happened about three quarters of a century after the conviction of biology teacher John Thomas Scopes on July 21, 1925, for teaching evolution. Where is the connection between the two convictions, you ask, and what does this have to do, if anything, with evolution? Well, here is the connection. The evidence used against Dr. Schmidt was based on phylogenetic analysis, which itself is based on the theory of evolution. The defense unsuccessfully appealed to the Louisiana Supreme Court and then to the U.S. Supreme Court. In the spring of 2002, the U.S. Supreme Court rejected the appeal and the guilty verdict stood. This was the first victory of the application of evolutionary science within the U.S. court system.

The only purpose of telling this story here is to point out that this case marked the first time that phylogenetics, an evolutionary analysis, was admitted as evidence in a U.S. criminal court. Phylogenetics is an example of the many invaluable tools and technologies that theory of evolution or the evolutionary science has provided us in its application including genetic markers, molecular clocks, DNA analysis of pathogen evolution, metagenomics, and evolutionary medicine. You will become more familiar with all of these terms and concepts as you make your way through this chapter.

11.2 Applications of Evolution

As you know from previous chapters, the marvelous complexity and diversity of living things is largely attributed to mutations and natural selection, two essential elements of the theory of evolution. Mutations produce changes in the genetic code and natural selection selects adaptations, that is, characteristics that improve the fit of the organisms with the environment. The natural products of evolution can be seen at all complexity levels of life: from the amazing diversity of life at the level of organisms; all the way down to cells, the basic units of life; and to individual protein molecules made in the cell. One feature of science that also distinguishes it from belief systems is that once we learn something from nature through a theory, a physical law, or a principle that scientists discover, we can often apply it to real-world situations indiscriminately. For example, the laws of classical physics, also called Newtonian physics, are applied to mechanics in the macroworld, including the basis of aviation and machines that run on land and water. The whole field of consumer electronics including televisions, cellular phones, computers, and the Internet is an application of our scientific understanding of matter at atomic level.

As emphasized earlier in this book, a theory in science continuously goes through the test of experimentation. It is kept as it is, improved, or thrown away and replaced with some other theory depending on if it fits the observed facts and experimental results. The ultimate validity of a theory, however, comes through its applications. If we can build something based on a theory, that is, apply the theory to the real world, it would mean it must have some truth to it, otherwise it wouldn't work. Theory of evolution has gone through vigorous scientific scrutiny and tests and resulted in improvements and additions since the times of Darwin and Wallace when it was first introduced. So, one can ask what are the application of the theory, if any?

In this chapter, we discuss the applications of the theory of evolution, starting with phylogenetics, an evolutionary tool used in the case of Dr. Richard J. Schmidt, as mentioned earlier.

11.3 Phylogenetics: The Study of Evolutionary Tree

Phylogenetics is the study of evolutionary relatedness among various groups of organisms such as species. In modern evolutionary science, these relationships are discovered by analyzing the genetic relationships because evolution at its root is defined as genetic change. Therefore, genetic relationships are of primary importance in deciphering evolutionary relationships. Phylogenetics uses the phylogenetic tree, which is a graphical representation of the evolutionary relationships among three or more organisms or genes. In case of genes, the analysis is also called molecular phylogenetic analysis. So, molecular phylogenetic analysis involves constructing and studying the evolutionary tree of genes showing how the genes evolve. So, in the case of Dr. Schmidt, scientists constructed evolutionary trees for genes from three sources: the patient whose HIV genes were allegedly injected into the victim, the victim, and another infected individual (to be used as reference point). The evolutionary tree analysis of the HIV genes from the patient and victim as compared to those from the third party clearly showed that the victim's genes were derived from the patient's genes.

We have just presented an application of evolution in *forensics*, a scientific field that addresses questions and issues that are of interest to the legal system. Another example of a tool developed from evolutionary science is genetic markers, a fancy name given to the genes or just DNA sequences with a known physical location on a chromosome.

11.4 Genetic Markers

A genetic marker is a DNA sequence (chain of nucleotides) which may or may not be a gene but has a known location on a chromosome. The fact that the location of the sequence is known is what makes it a marker that can be used to identify cells, individual organisms, or species. First, let us ask a question: Who introduced us to genetic markers? The answer is evolutionary science. In the early 1980s, geneticists discovered that some regions of human DNA evolve very rapidly. Soon after, scientists found use of these fast evolving DNA sequences or genes as genetic markers: the unique identifiers of individuals like fingerprints but only with much greater detail and hence accuracy.

> **Note.** In general, a genetic marker is an easily identifiable piece of genetic material, usually DNA, that can be used in the laboratory to tell apart cells, individuals, populations, or species.

How do genetic markers help identify individuals? First, genetic markers can be described as a variation or alteration that may arise due to mutation in a gene or even in a DNA sequence that is not a gene. Because these variations can happen many ways, the markers can vary among individuals within the population of a species. This combined with the fact that the locations of markers on the chromosomes are known facilitates the identification of individuals.

481

And God Said: Let There Be Evolution

The markers can also be used to identify genes. As mentioned earlier, the marker itself may be a part of a gene or may just be a DNA sequence with no known function. DNA segments close to each other on a chromosome tend to be inherited together. Markers can be used to track the inheritance of a nearby gene that has not yet been identified but whose approximate location is known. Furthermore, genetic markers can help link an inherited disease with the responsible genes. If the analysis of a large number of individuals with a genetic disease demonstrates that the individuals who inherit a genetic disease always (or nearly always) inherit a specific marker, then that marker is linked to the disease, that is, to the gene that causes the disease. The marker in this case is either part of the disease gene or very close to it on the chromosome. Now if we see that marker in a chromosome of another individual, the probability is high that the individual has inherited the disease gene and therefore the disease. Genetic markers have been used to identify genes for several genetic diseases and are also used to diagnose many diseases. They are used in paternity testing as well.

So, you can see that genetic markers have their applications in medical science. For example, as Figure 11.2 illustrates, a genetic marker, Prostate Cancer Gene 3 (PCA3), has been proposed to be used to detect the prostate cancer. Prostate cancer (PCa), the most common cancer in men, does not cause any symptoms in most cases. This is because only a minority of cancer cells progress rapidly, but they are potentially lethal. These cells need to be detected before they spread out from the prostate. Previously, the only way to diagnose PCa was a *prostate biopsy* which has many drawbacks such as it is invasive.

Detection of Cancer Cells in Normal Cell Background

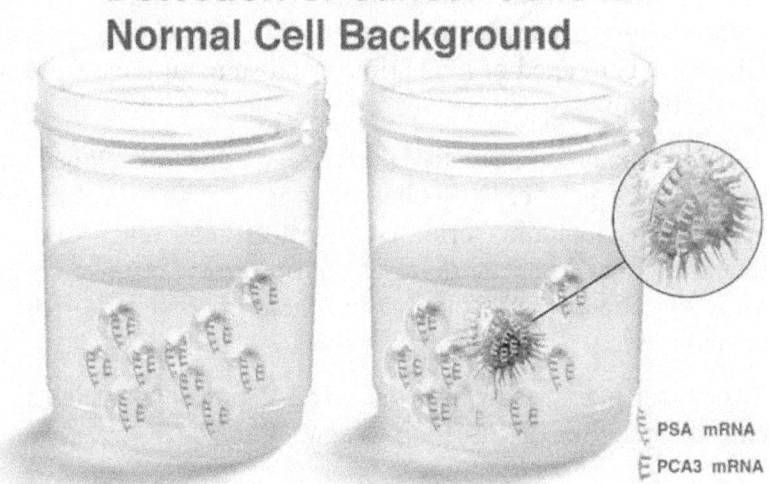

PSA mRNA

PCA3 mRNA

Figure 11.2 On the left: A background low-level expression of the cancer gene PCA3 present from the benign prostate cells in urine. On the right: A single cancer cell shown to greatly over-express the gene, and hence with the help of genetic marker allowing the detection in a urine specimen of an abnormal quantity of PCA3 relative to the normal background. Image: courtesy of Dr. Leonard S. Marks, Urological Sciences Research Foundation.

The PCA3 test based on genetic marker technology is now already becoming routinely available in European countries including Belgium, France, Germany, The Netherlands, and the United Kingdom.

In forensics, specific genetic markers can be used to link suspects with crime scene evidence such as a human hair that the suspect might have left behind, saliva left on cigarette butts, urine, or blood. It is not only proving the crime, but a suspect's innocence can also be proven by demonstrating a non-match of the suspect's markers with those of the crime scene evidence.

According to the report of a public policy organization named the Innocence Project, since 1989, non-matching of genetic markers has exonerated more than 220 individuals, most of who were convicted of rape crimes, and some of them were on the death row.

To summarize, genetic markers, a tool developed by evolutionary science, is being used in a wide spectrum of applications including criminal investigations, diagnosis and identification of genetic diseases, and paternity testing. Throw away the concept of evolution and there is no genetic marker technology.

Genetic markers are part of a DNA molecule, which is a part of a genome, discussed in Chapter 4.

11.5 Applications of the Research on Genome Evolution

Genomes of various species are being sequenced (deciphered) and compared with one another to study the evolution of genomes. The knowledge obtained from these studies is giving rise to various applications. For example, from this comparison, scientists can determine which genes have been conserved (are the same or unchanged) between two species. This is very useful information, which can be used in many ways. For instance, certain genes in yeast have their counterparts in humans which cause disease in humans. Scientists study the yeast genes similar to these human disease genes with the goal of finding a cure for these diseases. It is much safer and wiser to experiment on yeast than on humans.

During genome research, in the context of evolution, scientists unveiled that the genetic code is universal. This means that many processes in cells of different species are identical. In Chapter 4, we discussed the central dogma of molecular biology, the process used by a cell to make proteins based on the information in genes. The fact that this process works in all organisms has enormous applications. For example, a condition called Type 1 diabetes develops in humans because the body cannot produce a protein called insulin. A treatment of this disease involves injecting insulin into the patient's body, and the insulin is produced by bacteria that have the genes to produce insulin. This would not be possible if there was no evolutionary relationship between bacteria and humans and no commonality between their genome.

One of the most important concepts underlying the applications of evolutionary science is that of molecular clocks.

11.6 Molecular Clocks

One of the practical problems for studying evolution is that its effects become apparent over many generations. For certain organisms such as bacteria and fruit flies, however, you can collect data for many generations in the lab in a reasonable amount of time because their lifetime is short and they regenerate quite frequently. Subsequently, you can use this data to study evolution. The question arises then, how do you study the evolutionary relationships or history of organisms whose lifetime is not that short, such as humans? Techniques such as radioactive dating, fossil studies, and molecular clocks are used in such cases.

Molecular clock is a technique in evolutionary science which estimates the time taken for an evolutionary change by using the fact that some regions of the genome change at a constant or known rate. It can be used to deduce the time in evolutionary history when two or more species or groups of organisms diverged from their common ancestor group. The molecular data used for such analysis are usually the nucleotide sequences of DNA or amino acid sequences of proteins. Molecular clocks are based on this hypothesis: For a given DNA sequence, mutations in all evolutionary lineages accumulate at a constant rate. Of course, the hypothesis adds approximation, but it facilitates the extrapolation of the evolutionary tree in past times for which data may not be available. In other words, molecular clocks facilitate analyzing molecular evolution in particular sets of lineages to establish the approximate dates of phylogenetic events such as divergence of living groups, including the events for lineages not documented by fossils. Molecular clocks are sometimes also referred to as *evolutionary clocks* or *gene clocks*.

To summarize, molecular clocks assume that a given stretch (sequence) of DNA accumulates mutations at a regular rate. Once we determine this rate for the DNA sequence of closely-related species from the existing data, for example, from genes and fossils, we can work our way back in time to mathematically extrapolate when these species diverged from a common ancestor.

Understanding evolution has also helped us in our war against infectious diseases caused by harmful microbes.

11.7 Winning the Microbial War

An infectious disease is an illness resulting from the presence of pathogenic microbes such as bacteria, virus, fungi, and parasites. Just by being what we are, humans, makes us potential hosts for and hence vulnerable to a huge variety of these pathogenic microbes. During an infection, a pathogen multiplies in the cells or tissues of its host, and disease results unless our defenses such as the immune system halt this multiplication process. It looks as if infectious diseases, just like other evils in this world such as crime and war, will always be facts of life that we must continuously face and deal with.

A disease epidemic is nothing less than an invasion of human society. Its effects can be as devastating as those of a conventional war or sometimes even more. For example, recall the outbreak of severe acute respiratory syndrome (SARS) also known as Yellow Pneumonia, carried by the SARS virus, which began on November 16, 2002 in the Guangdong Province of China. Worldwide, mostly in Asia, 8273 people were reported infected, out of which 774 died. In many countries, it targeted vital social infrastructures such as hospitals and medical personnel, caused transportation systems to shut down in many countries, and brought education to standstill as schools and universities were closed. As a result, the economy was hit hard; about a $30 billion (US) production loss was reported, mostly in Asia. "In this region we are more likely to be invaded by microbes than by a foreign army," said Chua Jui Meng, then Health Minister of Malaysia.

> Note. A pathogen is any agent, such as a bacterium, that infects a host organism (or organ) and causes disease by disrupting the host function. However, not all bacteria are pathogens; most of them are helpful and play very important and useful roles in maintaining life on our planet.

As another example, in 1918, the Spanish Flu epidemic was estimated to have killed between 20 and 40 million people. The death toll from a possible influenza pandemic today is expected to be in the millions across the world in a very short time. The human and economic consequences would be severe. In 2003, a highly pathogenic avian flu virus broke out in Southeast and Northeast Asia. According to the World Health Organization

And God Said: Let There Be Evolution

(WHO), 150 million birds have either died from the disease or been culled so far. Although this specific virus at present has a limited capacity to infect human beings, the virus does not travel from human to human—only from animals to human. Scientists are worried that it is only a matter of time before the virus acquires the ability to transmit easily from human to human; that is, the situation that can trigger a pandemic.

> Note. Influenza, commonly referred to as the flu, is an *infectious disease* caused by RNA viruses of the family Orthomyxoviridae (the influenza viruses), that affects birds and mammals. *Infectious* means capable of spreading from one person to another by infection. Such viruses are worse than the normal flu because they are much more virulent, where virulence is the relative ability of a pathogen to cause diseases.

I can overwhelm you by citing examples of regular diseases in action such as acute respiratory infections that kill 4.7 million per year, tuberculosis that kills 3.1 million per year, and AIDS that kills 2.6 million per year. But you get the picture: we are at microbial war with pathogenic agents. And here is the most important news from the battlefield: evolutionary science is our strongest ally in this war.

Here is how evolutionary science has been helping us in this war. By using evolutionary science tools such as phylogenetic analysis of DNA, we can determine the genealogy, that is, gene linage ancestry or evolutionary history of pathogens. This information from genealogy of a pathogen then gives us clues about its means of reproduction and transmission in addition to information about its preferred habitats. For example, in the framework of evolution, close evolutionary relatives are more likely to share the traits related to inherited genes than distant relatives. This evolutionary information is then used to attack pathogens at the correct targets, for example, making recommendations on how to minimize their transmission capabilities and how to enhance immunity of the potential hosts to them. For example, from these evolutionary studies, biologists have discovered a process called horizontal gene transfer. In addition to inheriting genes from a parent cell, pathogens, by using horizontal transfer, can exchange genes among the cells of the same or different species in the same generation. This knowledge has resulted in seeking new kinds of antibiotics that would block the replication and transmission abilities of such genetic entities.

Note. Figuring out the evolutionary mechanism of pathogens involves identifying the causes of mutations and role of natural selection, and their relationship, for example, why some specific heritable changes (or mutations) persisted.

From the phylogenetic analysis of influenza virus, strains such as H5N1 have been identified, and scientists have learned that wild birds are a primary source of this virus, and that domestic pigs often act as intermediary hosts in transferring the virus from birds to humans. Based on this knowledge, some preventive actions were recommended including that people in certain regions of higher risk keep their poultry and pigs in separate enclosed facilities in order to avoid contact with wild birds. Furthermore, coupling phylogenetic (evolutionary) history of a virus with geographic sampling enables scientists to identify candidates for use in vaccine development, and also to predict the spread of diseases. For example, in 1997, scientists convinced authorities in Hong Kong to get rid of all domestic fowls, an evolutionary lineage or subset of domestic birds, which were determined to be the local source of the H5N1 virus. This way, scientists thwarted by a hair's breadth a potentially catastrophic outbreak of a disease called *bird flu* caused by this virus.

As a bonus, during the evolutionary studies of pathogens, scientists have learned about a very important phenomenon, the evolution of drug-resistant pathogens. We have learned that by killing pathogens with drugs, we give selective advantage in evolution to the pathogens that are resistant to the drugs. Because the lifespan of a generation is very short for most pathogens, drug treatments help the drug-resistant pathogens, which are initially in extremely small numbers, to multiply rather quickly. For example, the first antibiotic drug widely used against pathogens was penicillin, which saved many lives during World War II. This gave rise to an optimism of eliminating the threat of bacterial diseases all around the world. This optimism was however challenged by the discovery of the evolution of penicillin-resistant bacteria called *Streptococcus pneumoniae*, which first appeared in 1967 in Australia, and about 10 years later in South Africa. Today about 50% of all strains of this pathogen are drug-resistant. It is commonly transmitted among children for example at day care centers causing chronic middle-ear infections and pneumonia. At least we know from genetic and evolutionary studies that these strains arise by spontaneous mutations or by horizontal gene transfer discussed earlier in this section; this knowledge can be used to deal with them.

In nutshell, while the risk of future disease epidemics are looming on us, the actions being taken based on the evolutionary source and mechanism of the responsible microbes is helping to minimize the risks.

Think About It!

True or False: Some anti-HIV drugs have become ineffective because their use has created drug resistance in the HIV virus.

Answer

False.

Anti-HIV drug here is the environmental factor for the virus, and the environment selects some traits against others, but it does not create and hand out the traits. The drug-resistant virus was already there in the HIV virus population but in small numbers, and it was multiplied over generations because the non-resistant virus (which was initially in greater number) was killed by the drug.

Because evolution has played a vital role in the process of us becoming what we are today as humans, therefore it obviously has an important role to play in our health and medicine.

11.8 Evolutionary Medicine

First, let me make it clear that Darwin and Wallace, who pioneered the theory of evolution, did not discuss the implications of their theory for medicine. Evolutionary medicine is the result of applying modern evolutionary theory (that we call the standard theory of evolution in this book) to understanding our susceptibilities to diseases and offering cures, and promoting health. Because we have billions of years of history, the evolutionary history inside of us in form of genes, understanding this history helps us to understand some diseases, and hence helps in finding their cures. Evolutionary medicine adds another dimension to medical science.

And God Said: Let There Be Evolution

As advanced our current medicine is compared to ancient times, it is still very primitive in many respects. When a part of our body is sick, we often attack the whole body with drugs. In other words, there is very little spatial and temporal control over the drug delivery to our body. For instance, when we take a pill, not only do the parts of the body that need it get the medicine, but also the parts that don't need it; and that may have adverse effects or cause collateral damage. Also, because we do not have control over the exact time when the drug will be delivered to a specific part of the body, its effectiveness is not optimized.

Note. You learned in Chapter 5 how quantum physics is more accurate physics and the only physics that works in the microworld, and that classical physics is an approximation of quantum physics, the approximation that works well at the macro-scale.

This problem is partly related to the fact that although quantum physics is already about a century old, our medicine model is still based on classical physics, also called Newtonian physics. Disease agents such as pathogens work at molecular levels, which can be better understood and handled by using the laws of quantum physics. However, our medicine model, still based on the classical approach as opposed to quantum approach, looks at diseases largely as anatomical (structural) changes in parts of the body such as organs, cells, and even genes, brought directly by external agents such as pathogens. As a result, medicine assumes the determinism of classical physics, but ironically it goes totally un-deterministic in attacking the diseases in the body as mentioned earlier: a dilemma. This is consistent with the slow speed of biology in keeping up with the developments in physics as compared to some other fields such as chemistry and engineering.

But good news is around the corner. Medical science has two new interrelated avenues open for it: evolutionary medicine and quantum (or nano) medicine. Both are related because both have a common origin: the microworld. The origin of evolution, as you already know from the previous chapters, is the mutations in genes, which due to their small size are quantum mechanical entities. Scientists are already working on using quantum mechanical objects such as buckyballs, a carbon structure of nanosize (Figure 11.3), to make drug deliveries to specific places in the body using nanotechnology. It

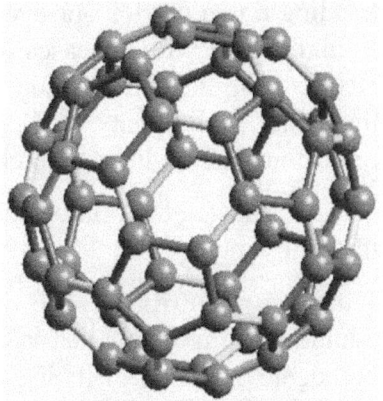

Figure 11.3 A Buckyball molecule that due to its structure and size is capable of holding drugs for their delivery to a precise location in the body.

sounds like common sense that drug molecules should hit the correct spot in the body at the right time to be effective: anti-depressants should go to the brain and anticancer drug molecules should go straight to the cancer tumor sites. The problem arises with the incompatibility between the human body and the size drug molecules. Current efforts in the drug delivery field with the help of nanotechnology are focused on implementing this common sense: development of targeted delivery methods that facilitate activating the drug only in the targeted area of the body over a pre-determined period of time. There is a name for this issue: *bioavailability*.

How does evolution help with this? Within the framework of evolution and theory of evolution, we know that our personalized evolutionary history is written on our genes. We also know that our growth starts from a single cell fused from two half cells, one from each of our parents. That cell and all the following cells that run our life are run by the genetic code handed down to us in our first cell from our parents. This handing down carries our evolutionary history and to consider it in dealing with diseases is a matter of commons sense. Because the genetic makeup of no two individuals is exactly the same, evolutionary medicine will give rise to personalized or customized medicine, which will be more effective.

The aspects of evolution relevant to medical science include the following:

Pathogen evolution. The important aspects of pathogen evolution from the medical perspective include the evolution of pathogens in terms of their degree of resistance to antibiotics, virulence (relative ability to cause diseases), and the ability to subvert the host's immune system. The success of any pathogen depends upon its ability to play hide and seek with the host's immune system. For example, some pathogens have evolved to gain the ability to infect a host and then hide to evade detection and destruction by the immune system of the host. The evasion methods gained through evolution include hiding in the host cells, hiding within a protective capsule, misdirecting the response from the immune system by secreting the appropriate compounds, binding to their antibodies, and changing surface markers or masking them with the host's native molecules. Understanding of the evolution of these capabilities helps to devise a strategy to combat them.

Aspects of human evolution. The aspects of human evolution useful in medical science include the processes involved in human evolution along with constraints and trade-offs made during the selection of various traits. For example, humans have a higher rate of birth problems than that of other primates because the evolution of female pelvis size has not stayed consistent with the evolution of infant brain size which has been naturally selected to be larger.

Furthermore, early human populations made some adaptations, that is, natural selection selected some traits that helped them fit better with the existing environment at that time. How do those adaptations now affect us with the changed environment in terms of factors such as diet, degree of physical exercise, and life expectancy? Understanding such issues has significant implications in medical science.

Early human adaptations. This means to explore how the past adaptations of early humans supported in the environment of that time differ from the modern lifestyle and how these differences affect the health-related aspects of today's human such as diet, hygiene (preservation of health), life expectancy, and physical exercise. For example, the modern diet generally contains more fat, salt, and refined carbohydrates (sugar) than the diet of the early humans, making the modern human more vulnerable to related diseases. Furthermore, modern lifestyle discourages humans from getting engaged in physical exercise as compared to the lifestyle of early humans which included physical exercise as the mechanism for collecting food and getting around. This also has bad health implications for the modern human.

Understanding these issues helps to understand some diseases and devise their treatments.

Response evolution. This means the evolution of the response system of the human body that enables it to protect, heal, and recuperate from infections and injuries. Examples of processes involved in accomplishing this include immunity and sickness behavior. For example, diarrhea, fever and vomiting actually facilitate purging microbial infections. The immune system is a fascinating defense mechanism against diseases. Understanding how it works at molecular level is of immense significance to medical science; and how it evolved is a critical part of understanding it.

The point here is that modern medical science only focuses on where we are, and evolutionary medicine adds another dimension: how we got here. To develop the most effective medical system, both the current status and the evolutionary history of organisms must be taken into account.

For some people, the problem in comprehending evolution lies in the fact that it is a slow process that occurs over generations; so who knows? The beauty of scientific theories as opposed to belief-based conjectures or doctrines is that once we understand the theory we can apply it to different contexts. For example, once we understand the big bang theory, the creation of the Universe billions of years ago, we can re-create that big bang on a much smaller scale in experiments such as those at the European Organization for Nuclear Research (CERN) and Stanford Linear Accelerator Center (SLAC), and test some aspects of the theory. Similarly, after we understand the theory of natural evolution, which generally occurs slowly, we can facilitate accelerated evolution in the lab, called directed evolution.

11.9 Directed Evolution: Evolution in the Lab

Recall from Chapter 4 that proteins are the workhorses of life in an organism. Directed evolution is a method that applies elements of evolutionary theory to evolve proteins with enhanced desirable functions. It is emerging as an important technique in the field of protein engineering, the field of developing useful or valuable proteins.

As you learned in Chapter 4, cells produce proteins according to the instructions chemically coded in the genes. Because the evolved genes have the code to generate proteins, the proteins are the direct product of evolution. Proteins play a vital role in the development and maintenance of living organisms; nearly each dynamic function of a living organism from bacteria to humans depends on proteins. Therefore, proteins, which account for 50% of the dry mass of most cells, are instrumental in everything an organism does. They speed up chemical reactions (enzymes); provide structural support such as making hair, nails, and horn; facilitate transportation such as transporting oxygen through blood from lungs to other parts of the body; support movement such as contraction of muscles; and defend the body against diseases and combat pathogens such as harmful bacteria and viruses. Proteins are highly structured molecules with multiple layers of structure: primary, secondary, and tertiary. This gives rise to thousands of types of proteins, each type responsible for a unique function defined by its unique shape or structure.

Because the making of proteins is coded in the genes (or DNA) of a given organism, which the organism inherited from its parents, only those proteins with specific functionalities that are coded in the DNA will be created. Your genome, the set of genes which you got from your parents, is a result of evolution through generations. The lack of a protein or the presence of a wrong protein can give rise to illnesses and disabilities. If your body needs a protein for which your genome does not have a gene, you are out of luck. That may mean that you are stuck with your illness, condition, disorder, or disability. However, thanks to our understanding of evolution, help is on the way.

And God Said: Let There Be Evolution

Scientists are using evolution in the lab, that is, directed evolution, to generate proteins with enhanced desired properties which might be missing in an organism as a result of natural evolution. The following are the main steps in the process of directed evolution:

1. **Diversification through mutation.** The genes that have the code for the proteins of interest are diversified through mutations in a controlled environment by using one of several algorithms. The large collection of gene variants produced this way is called a library of genes or DNA library.

2. **Expressions and production.** Genes are allowed or induced to express themselves to produce proteins coded in the gene.

3. **Selection.** The functional performance of the produced proteins is measured. Based on the performance results, sets of top performers (mutants or gene variants) are selected by using some algorithm.

4. **Amplification.** The selected mutants (genes) are multiplied many times by replication. The large quantity of genes obtained through replication enables scientists to understand the selected mutations by techniques such as DNA sequencing.

These steps define a cycle or a round of directed evolution. The selected genes are put back into the cycle starting with further diversification through mutations. Repeating these cycles a great many times creates impressive results in generating the best genes for producing the most effective proteins you need.

Think About It!

Why do you need genes to produce proteins? Can't you just buy protein supplements from the shelf?

Answer

The so-called proteins that you consume through food, including supplements, are actually the raw material for proteins. Even if you eat a real protein, it would be broken into its building blocks by your digestion system, that is, into the raw material. That raw material is necessary but not sufficient to produce real proteins, which are produced from this raw material by the cells according to the instructions in the genes. If the right gene to produce a protein is not there, all the raw material in the world won't help.

Uses of directed evolution are obvious: we can evolve the desired properties and capabilities that are not available to the organism from the inherited DNA that evolved through natural evolution. Furthermore, directed evolution can be used to deal with diseases. For example, a vaccine against human papillomavirus has been developed using directed evolution. Another example of the success of directed evolution is the improvement of hepatitis C vaccines against the hepatitis C virus which is responsible for infectious disease adversely affecting the liver. The improvement is not marginal: The new vaccine is 250,000 times more effective than the traditional one. This is accomplished by shuffling around the DNA segments at the molecular level based on our evolutionary understanding. As another example, researchers are working on improving the capability of some proteins to suppress tumors.

Note. Human papillomavirus is a virus that causes papilloma responsible for human genital warts: a small, benign epithelial tumor consisting of an overgrowth of cells on a core of smooth connective tissue, which in a minority of cases can lead to cancer.

When a theory in a science field enjoys enormous success, the next step of its success involves other fields harnessing the benefits of the theory by designing its indirect applications.

11.10 Indirect Applications of Evolution

Researches in various fields outside of biology are using elements of the theory of evolution to develop algorithms in order to design optimal solutions for complex problems. Examples of such applications range all the way from designing effective robot control systems, electronic circuits, and bridges to determining optimal combinations of medicines; to accurately forecasting weather; to balancing stock portfolios, and so on. These goals are accomplished through computer programs that use evolutionary algorithms. A typical evolutionary algorithm includes the following steps:

1. **Compile a population.** Put together a group (population) of candidate solutions.

2. **Measure the fitness.** Evaluate the fitness, that is, effectiveness of each solution in the population.

3. **Select the better solutions.** If a solution in the population meets all the pre-determined target criteria, select it and end the process; or else select a relatively fit set of candidate solutions and rank them as parents, and go to step 4.

4. **Reproduce through heredity.** Subject the selected parent solutions to mutations (changes) and combine them (equivalent of sexual recombination) by combining their functions or traits to generate a set of new candidate solutions; and repeat the process by sending this new set of candidate solutions to step 1.

This kind of programming, also called genetic programming, designs solutions which are hard to design manually due to the complexity and enormity of computations involved. Here is an example: The orbits of communication satellites affect the loss of signals received by the ground-based receivers. Genetic algorithms based on the theory of evolution have been successful in configuring sophisticated orbits for constellations of satellites that out-perform the usual symmetric arrangement configured manually by designers.

Note that these evolutionary algorithms are not just theoretical constructs. They are based on the theory of evolution and the evolutionary study of real life. Relatively better performance of these algorithms or the solutions is further evidence for the validity of the theory of evolution and the very concept of evolution.

We can count all the applications of a theory, but the maximum benefits can only be reaped if society and politics are in harmony with the truth of the theory and with the related scientific community.

11.11 Consequences of Denial and Misrepresentation

Unfortunately, there still exists a considerable amount of denial and opposition based on misunderstanding and misrepresentation of the theory of evolution. We as a society cannot make the best use of a scientific discovery if we are not on par with it. For instance, accepting or rejecting the reality of evolution has a direct bearing on how governments design their public policy and how citizens make some choices in running their daily lives. For example, overfishing selects for smaller fish and may result in higher prices at the supermarket; and excessive use of antibiotics naturally helps in developing (or evolving) drug-resistant bacteria, as discussed earlier. These are two examples of how regulators and legislators could do their job and citizens could make their choices more effectively if they understood evolution.

At personal level, many diseases of our times such as autoimmune disorders, diabetes, and obesity are partly the result of the mismatch between our genes and an environment that changes more quickly than the human genome can evolve. Understanding of this fact can help governments to adopt the right policies regarding these issues. This understanding can also help to convince citizens to make dietary choices that conform better to the demands of their genetic heritage in order to combat disease.

To summarize, evolution is a fact of life that can be understood in terms of the theory of evolution, which has a profound explanatory and predictive power. As suggested earlier, its implications and uses go beyond the discipline of biology. In other words, evolution is a theory based on facts and tested through scientific scrutiny such as experiments; and its analytical framework can be used to explore a diverse array of subjects from medicine to literature, from engineering design to construction, from crops to mate choice, and from biotechnology to religion. Furthermore, it is a way of looking at the world and at the future ahead of us. Therefore, it is of paramount importance that we make the theory of evolution an essential part of general education curriculum at secondary and higher education levels without any reservations such as imposed by the recent state laws in Louisiana and Tennessee in the name of academic freedom, mentioned in Chapter 10.

11.12 Summary

As you have seen throughout this book, theory of evolution is supported by an overwhelming body of evidence. Ultimate validity and public acceptance of a theory comes from its real-world applications. Based on the very concept of evolution, the theory of evolution, and the evolutionary study of life, applications are already being developed in various fields including forensics, medical science, and engineering. Tools and techniques such as genetic markers, molecular clocks, and evolutionary algorithms have been developed based on evolutionary concepts. These tools have enabled scientists to determine the origin of viruses that could potentially cause epidemics, and thereby helps to stop the epidemic from occurring. These tools help us to diagnose and treat diseases more effectively. Furthermore, the framework of evolution helps us to understand pathogens better and therefore combat them better, which will help us tremendously in winning the war against pathogens.

Understanding natural evolution has helped scientists to use evolutionary process in the lab, called directed evolution, for various purposes such as developing cures for diseases and developing optimal solutions for problems in many fields.

In a Nutshell

✓ The tools and techniques developed within an evolutionary framework help to identify the origin of viruses such as the one that causes the bird flu and thereby helps to stop potential pandemics from occurring.

✓ Genetic markers, developed in the framework of evolution, are already used in diagnosing cancer.

✓ Medical science is in the process of developing personalized medicine by using the evolutionary history coded into the genes of an individual.

✓ We are already using some medicines such as some vaccines which have been developed by using evolution in the lab, directed evolution. These medicines are proving tremendously more effective than traditional ones.

✓ Genetic algorithms based on the theory of evolution are being devised and used to develop (evolve) optimal solutions in many fields. For example, such algorithms have been successful in configuring sophisticated orbits for constellations of satellites that out-performed the usual symmetric arrangement configured manually by designers.

✓ We as a society can only make the best use the scientific theory of evolution if we understand it and appreciate it. Therefore, it is of utmost importance that we make the study of evolution an essential part of our general education curriculum.

11.13 Behold

Great is the power of applications developed based on a theory. In our right minds, we cannot take a plane ride and still refute Newton's laws of motion on which the planes are designed. Similarly we cannot take a vaccine developed by using evolution to defend ourselves against some virus and then attack the very concept of evolution. Of course, in general, you can attack (test) any scientific theory within the scientific framework discussed in Chapter 2. Here by attack I mean the kinds of attacks that have been coming from creationists to keep the teaching of evolution out of the public schools, as discussed in Chapter 10.

One of the reasons for those attacks on evolution or for the denial of the reality of evolution is that some religious folks feel that the very concept of evolution is an attack on their faith. However, evolution is not the first theory where faith came between the reality of a theory and its acceptance by a set of people. During the history of science and the progress of human civilization on this planet, we have been there before more than once. We discuss this issue in the next chapter.

And God Said: Let There Be Evolution

TWELVE

In the Beginning: Back to the Future

And God Said: Let There Be Evolution

There is a grandeur in this view of life, with its several powers, having been originally breathed into a few forms or into one; and that, whilst this planet has gone cycling on according to the law of gravity, from so simple a beginning endless forms most beautiful and most wonderful have been, and are being evolved.

Charles Darwin

> *Divinity sleeps in stones,*
> *breathes in plants,*
> *dreams in animals,*
> *and awakens in human beings.*

An Asian Indian Proverb

When there is a conflict between science and religion, religion should change its opinion

Dalai Lama

12.1 Once Upon a Time: The Galileo Affair

As celebrated physicist and author Stephen Hawking put it, "Galileo, perhaps more than any other single person, was responsible for the birth of modern science." Albert Einstein called Galileo the father of modern science. However, his scientific work did not fit well with those who believed in the literal interpretation of the Bible, just like the creationists of today.

a)   b)

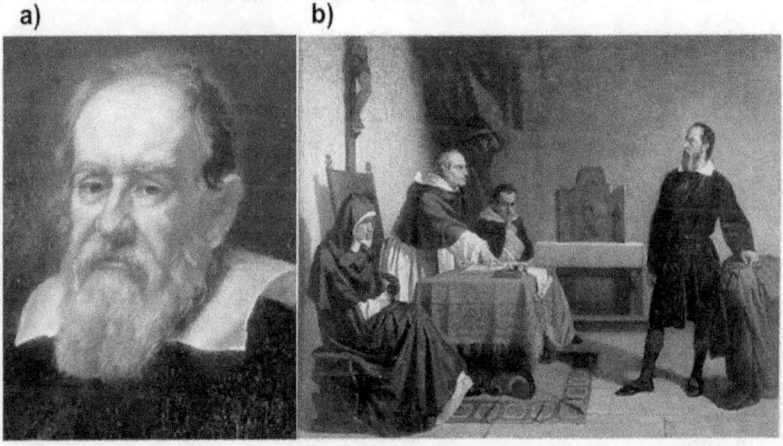

Figure 12.1 a) Portrait of Galileo Galilei by Giusto Sustermans. **b)** Galileo facing the Roman Inquisition; 1857 painting of Cristiano Banti.

And God Said: Let There Be Evolution

Most proximately due to his book *Dialogue Concerning the Two Chief World Systems* and generally due to his overall scientific work, Galileo was ordered in 1633 to Rome to stand trial on suspicion of heresy, "for holding as true the false doctrine taught by some that the Sun is the center of the world..." He was tried before the Roman Inquisition, a judicial system established by the papacy in 1542 to regulate church doctrine. As a result of the trial, Galileo was found guilty, and the sentence decision of the Inquisition was in three essential parts:

1. **Heresy.** Galileo was found "vehemently suspect of heresy," specifically for holding the following opinions:

   - The Sun, and not the Earth, rests at the center of the Universe, and the Earth is not the center of the Universe and it moves.

   - An opinion that has been declared and defined contrary to the Holy Scripture, Galileo dared to treat it, defend it, and show it as probable.

2. **Life Imprisonment.** Galileo was sentenced to imprisonment. However, due to his age (70 at the time of this trial), the sentence was later turned into the house arrest for the rest of his life.

3. **Ban.** Galileo's offending book the *Dialogue Concerning the Two Chief World Systems* was banned. Not only that, although not included in the trial judgment, publication of any of Galileo's current or future works was prohibited.

Note.

The Inquisition was a Roman Catholic tribunal for discovery and punishment of heresy established in 1233 by Pope Gregory IX.

Heresy in the Roman Catholic Church means the willful and persistent rejection of any article of faith. In general, it is any belief or theory that strongly varies from the established beliefs or customs.

Galileo, under pressure, agreed to plead guilty in return for a lighter sentence. Ultimately, he was sentenced by Pope Urban VIII to be put under house arrest for an indefinite period. It was during his house arrest that Galileo put his earlier work into a book titled *Two New Sciences*, which worked as the foundation of the modern science. These sciences are now called *kinematics* and *strength of materials*. Also during his house arrest, Galileo became blind and died on January 8, 1642 at the age of 77.

12.2 In the Beginning

In this chapter, we show how opposition to scientific discoveries is not limited to evolution. Whenever a scientific breakthrough challenges the traditional way of thinking, it faces opposition. The very idea of the ancient Greek philosophers that the world around us can be understood in terms of consistent natural principles was opposed by the pre-Christian religious believers of that time. We explore some aspects of and reasons for such opposition to science by focusing on the opposition to evolution. History is a witness that this opposition begins with strong rejection of the scientific discovery and ends up at reconciliation with it.

Take the example of gravity, which plays a crucial role in keeping our solar system running. Today even children know, and it is obvious to adults, that the Earth revolves around the Sun and the seasons are due to this motion of the Earth, and that Earth is not at the center of the Universe. Creationists are neither lobbying to get legislations passed against teaching this scientific knowledge nor are they pressuring to put warning stickers against gravity on physics books that contain theory of gravitation, which causes this motion. But this was not so when this scientific fact was discovered by scientists like Galileo. I In the seventeenth century, the creationists in power responded basically the same way as the creationists have responded to the theory of evolution about four centuries later (will we ever learn?).

There have been two major theories about our place in the Universe. According to the geocentric model, also called *geocentricism* or the *Ptolemaic* theory of the Universe, the Earth is the center of the Universe and all other astronomical or celestial objects, such as the Sun, revolve around it. In the absence of sound scientific knowledge at the time, this theory had prevailed from antiquity partly because it was also consistent with the literal interpretation of the Bible and other religious books. However, there also existed in parallel a non-prevailing theory, called *heliocentric* theory or *heliocentricism*, which was brought into limelight by Copernicus (Figure 12.2) in 1543 in his publication *De revolutionibus orbium coelestium (On the Revolutions of the Heavenly Spheres)*. According to *heliocentricism*, Earth and other planets revolve around the Sun, and therefore the implication is that the Earth cannot be the center of the Universe. The hypothesis that the Earth revolves around the Sun was first presented in the third century BC by Aristarchus of Samos, a Greek astronomer and mathematician, born on the island of Samos, in Greece. The idea that he Universe is not earth-centered and therefore not human-centered proved to be a milestone in understanding the cosmos. However, the idea was forgotten and not picked up again until the Renaissance period. So, it was not until the sixteenth century that a rigorous mathematical model predicting the heliocentric system was presented by the Renaissance astronomer, mathematician, and priest Nicolaus Copernicus (1473-1543), eventually leading to the Copernican Revolution.

Figure 12.2 *Astronomer Copernicus: Conversation with God,* by Matejko. In background: Frombork Cathedral.

Note. After its publication in 1543, it took about 200 years for the Copernicus model or geocentric theory to replace the Ptolemaic theory.

It often happens that groundbreaking scientific ideas are not accepted right away because they don't fit well to existing mainstream thinking within the society and sometimes within the scientific community as well. Copernius' model, based on the heliocentric theory from third century, was not very popular even when Copernicus died in 1543. There were only about ten Copernicans between 1543 and 1600 and most of them worked outside the university environments at places such as princely, royal, or imperial courts. The most famous of these supporters of the Copernican model were an Italian physicist, mathematician, astronomer, and philosopher, Galileo Galilei (1564-1642); and a German astronomer and mathematician, Johannes Kepler (1571-1630). Each of these scientists had his own set of reasons different from that of any other scientist for supporting the Copernican model. Such is the character of science culture and scientific method. This kind of approach is healthy because it covers a wide spectrum of ideas and concepts and when filtered through the scientific method, the best science comes through it. After all, each hypothesis, model, or theory has to go through continuous scientific scrutiny through observations and experimentation.

Galileo played a great role in testing all the models about the Universe at that time. In addition to his contributions to pure science, Galileo also made contributions to applied science such as his work on strength of materials, invention and improvement of military compasses, and improvements to the telescope, which worked as a tool to test the models of the Universe. With his improved telescope, Galileo began looking at the night sky in 1609; by March of 1610 he was able to publish his findings in a small book, *The Starry Messenger* (*Sidereus Nuncius*). More observations followed. These discoveries provided strong support for the validity of Copernicus' model and were not consistent with the Ptolemaic model.

These findings did not fit well with the establishment of the time.

12.3 We Have Been There Before: Science on Trial

In 1632 Galileo published his book, *Dialogue Concerning the Two Chief World Systems* (Figure 12.3), in Florence, Italy after receiving formal authorization from the Inquisition. In this book, Galileo used the dialogue form, a communication genre common in classical philosophical works, to demonstrate the truth of the Copernican system over the Ptolemaic system. He masterfully proved for the first time, that the Earth revolves around the Sun. Most proximately due to this book and generally due to his overall scientific work, Galileo was ordered to Rome to stand trial on suspicion of heresy in 1633, "for holding as true the false doctrine taught by some that the Sun is the center of the world..."

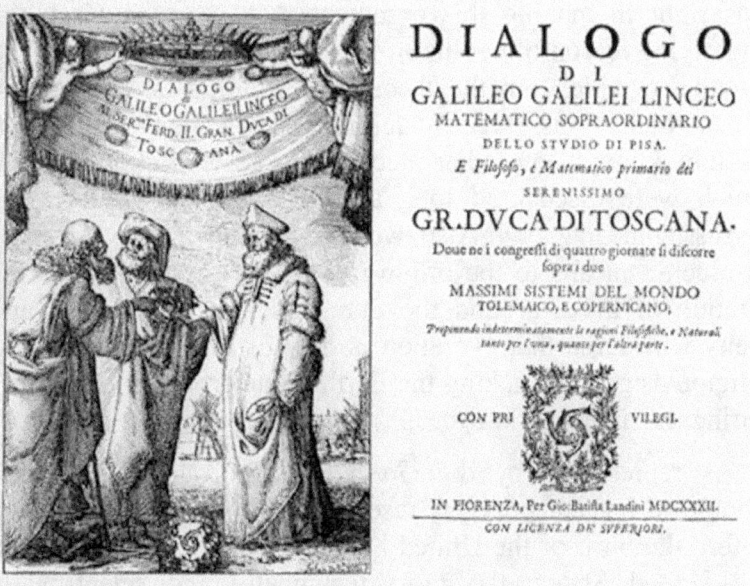

Figure 12.3 Title page of *Dialogo* (means the dialogue), the book for which the religious establishment of his time rewarded Galileo, then 70 years old, with house arrest for the rest of his life.

It was during the term of Pope Urban VIII, when as a result of this trial, Galileo was put under house arrest for the rest of his life. It was during this house arrest that Galileo became blind and died on January 8, 1642 at the age of 77.

515

And God Said: Let There Be Evolution

> Note. According to popular legend, under pressure during the trial, after recanting his theory that the Earth moved around the Sun, Galileo allegedly muttered the rebellious phrase "And yet it moves." However, there is no definite evidence that this actually happened.

A good scientist is always careful about the scope and limits of a discovery. For example, being careful about the limits of his discovery, Darwin, in the first edition of *On the Origin of Species*, wrote: "I am convinced that natural selection has been the main but not the exclusive means of modification." Darwin was right in making this comment in two ways: First, the underlying reason of evolution was found in genes and genetic mutations, which put the theory of evolution on even much firmer foundations. Even though natural selection is still valid, we now know a few other mechanisms for evolution as well, which were discovered long after Darwin and Wallace, as discussed in Chapter 4. Also, we know by now that evolution is as much a means to the origin of species as gravity is to the rotation of Earth around the Sun. Also, in my opinion and many will agree that evolution is only as much of a threat to religion as gravity making the Earth rotate around the Sun was during the times of Galileo.

As represented by the *Galileo Affair*, Europe had their trouble with science largely resolved in general long ago, even before the birth of the United States of America. It is sad that the United States, the most technically and scientifically advanced nation in the world, had to play this evolution drama that spanned over most of the twentieth century and made its way even into the twenty-first century.

As argued in Chapter 10, even if the theory of evolution falls one day, the candidate theory to replace it would not be *intelligent design* or anything like this because the so-called theories like intelligent design do not even qualify to be scientific theories because they are not falsifiable. For this reason alone, they don't even qualify to be scientific hypotheses. That being said, ideas like creationism or intelligent design may have a place, but not in science. Many of us would agree that personal conviction often has tremendous value in our lives but it should not be confused with a scientific theory. If a personal conviction is not scientific, it does not mean that it's bad; it just means that it's not scientific. It becomes bad only when we try to claim it as scientific.

12.4 Holding on to Your Conviction and the Reality: Finding Your Way through the Maze of Politics, Religion, and Science

I'm using the word *religion* here in a broader sense that goes beyond the limits of organized religions; in the sense of bond, faith, or conviction to the extent that even a non-religious or atheist person may have some faith or conviction in some belief system. As shown in Figure 12.4, politics, religion, and science influence one another in a triangular relationship, which in turn affects the whole society. For example, my religious views may influence on my political views and may affect my relationship with science. Similarly, my scientific background may influence my religious and political beliefs, even though scientific method

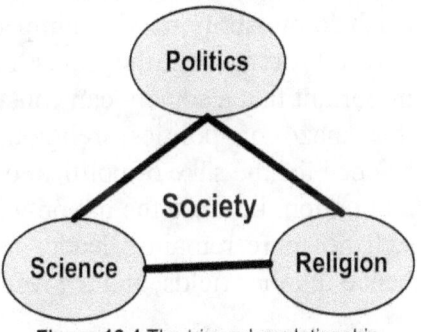

Figure 12.4 The triangular relationship between politics, religion, and science.

and theories have nothing to do with politics or religion as such. But scientists are social beings, and doing scientific work does influence society and is influenced by it. Here we are only talking about the relationship of science with the other two aspects of the society: religion and politics. All these three aspects of life may feed upon each other. The influence of these relationships could be positive or negative. Therefore, the relation management among these three areas at personal level and at national level is very crucial. Mismanagement of these relationships has destroyed nations; history is witness.

For example, when motivated (or pressured) by a religious movement to save their political standing or make political gains, a politician like Bobby Jindal of Louisiana or Bill Haslam of Tennessee (Chapter 10) signs a bill even under the fancy names of science education or academic freedom, opening up the possibility of circumventing a scientific theory, it is a bad mix and a bad management of the relationship between politics, science, and religion. The same can be said about the policy regarding stem cell research adopted by the Bush administration, which impeded progress in science and as a result in finding the cures for diseases in the U.S. It is important that a society can constructively find its way through this maze of politics, religion, and science. To victimize science for the sake of political or religious gains is suicidal for any nation, because the nation will be left behind; it eventually will no more remain a leader in science and innovation, and hence in other fields, and the rest of the world won't wait.

We must realize that science by definition is a systematic study of nature limited to a subset of the world: only that which is observable directly with naked eyes or indirectly with tools such as optical microscopes, electron microscopes, and more sophisticated tools such as particle colliders at CERN.

Questions such as *"why do I exist?"* or *"what is my purpose in life?"* or *"does God exist or not?"* are out of the scope of science. Answers to such questions are mostly subjective. This does not mean that subjective answers have no value; it just means that they are not scientific. In order to function, any human society has to establish and share some standards for making aesthetic, ethical, moral, and philosophical judgments even if they are subjective. Even though these standards vary from society to society and may vary from individual to individual, they help people to decide what is important and good for them as a society and as individuals. All this helps give meaning to what we do.

History is the witness that despite the clear scope of science, scientists and scientific theories do cause controversy when they discover a natural explanation for something that was either supposed to be unexplainable or if the explanation does not fit the traditional thinking or belief system. One of the reasons for this may be that the moral standards of the society at that time are interwoven with traditional interpretation which does not fit well with the scientific theory.

This is what happened in Europe when Nicolaus Copernicus proposed, based on his studies of the heliocentric theory, that Earth revolves around the Sun, and Galileo's observations confirmed it. This generated a storm for some people in the religious establishment because the prevailing religious belief was that Earth was at the center of the Universe and therefore could not revolve around the Sun. But today we take heliocentric theory as an obvious fact, and not many creationists are fighting to "put the Earth back at the center of the Universe."

Here is the point I'm trying to make: Scientific investigations sometimes, when they challenge our traditional

519

view of the natural world, can be misinterpreted as questioning morality, faith, or even challenging the existence of God. That is exactly the angle from which creationists including intelligent designers have been responding to evolution. And history (for example, *the Galileo Affair*) is the witness that these are the intelligent designers or creationists that would ultimately do more damage to religion than the theory of evolution. First, understand the fact that scientists, as individuals and as a community, are not any less moral, less lawful, less compassionate, or even less religious than anyone else; even though religion, especially organized religion, is not the sole source of morality or ethics. Before evolution was discovered and the theory of evolution was established supported by evidence, most of the scientists, just like almost everyone else, were creationists. They may not have accepted everything written about creation in every religious book, but most of them pretty much believed that living species were originally created by the Creator pretty much the way they appear today.

Nevertheless, scientists change their beliefs based on evidence. Also do not assume that all scientists who believe in evolution are atheists. Theory of evolution explains how species evolved; it is about the origin of species and not about the origin of life. Scientists follow a certain standard called scientific method, explained in Chapter 2, in their studies: explanations and hypothesis must be testable in the natural world in ways that anyone can repeat the tests.

Looking at all the conflicts between science and religion that occurred during the history of the human species on Earth, one cannot help but ask this question: Why do some of us almost always collide with science when a breakthrough sheds light on a previously unknown aspect of matter or life and challenges an established understanding or belief in doing so? By listening to the arguments presented by creationists against evolution and by studying the history of such conflicts, I have concluded that by and large it is not due to our faith or religion, and it is not due to our concept of God either; rather it is due to the wrong-headed approach that we sometimes take to our faith. Here is that approach: Assume we rest our faith in the dark of unknown rather than in the light of knowledge or belief, and nourish it (the faith) or part of it on the gaps in science or scientific knowledge, on the mysteries that the science has not solved yet but could solve in the future. When the science does solve these mysteries, does shed light on the dark where we have rested our faith or part of it, we run into trouble and act the way we do. This is true whether it is the Catholic Church and creationists of Galileo's times or it is the creationists of today opposing evolution. So from this perspective, it is a self-infected problem for which we cannot blame scientists, religion, or God, but only ourselves. This wrong-headed approach and the problem created from it, however, unfortunately affects the progress of the whole society. And when politicians begin feeding this approach for their short-term political gains, the national loss multiplies.

Here is a simple example to illustrate this point of resting our faith on the gaps in science. Fifty years ago, biologists did not know why flowers bloomed. The blooming phenomenon, like many other phenomena not understood, on one hand triggered research work and on the other provided another seat for creationists to lean on against evolution or science. They pointed to this gap in science as further proof for the invalidity of science and for the existence of God. A half century later, scientists have come to understand the blooming phenomenon: A cluster of leaves is instructed by the plant's genes to bloom; blooming is a gene expression. Pollen is scattered in springtime as a way to reproduce according to the instructions in the genes. Now, the creationists will drop this gap from their attack tools, but will continue resting their criticism of evolution on some other gaps, which may be explained tomorrow. The key to religious peace of mind with evolution is not to trash the theory of evolution or scientists but to discard the assumption that science and religion rule each other out.

If any scientific discovery could destroy religion, it could have been done with the discovery that Earth is not the center of the Universe, that it is Earth that rotates around the Sun due to gravity and not the other way around. If gravity could not destroy religion, neither could evolution; so evolution should not be taken as a threat to religion. Because evolution is as much a fact as gravity, all of us have to consider at our own personal level or institutional level what it means to us. Some of us, like creationists, would attack the concept of evolution; however by doing that we would be lifting a rock that would fall on our feet, as history is witness. We could not destroy gravity in the case of Galileo, and now we cannot defy evolution.

The approach of putting our faith in the way of science is a self-defeating approach; it is hard to see how it can serve any religion or the concept of God. Instead it is undermining the future of our children by weakening their science education. The other approach is to accept the scientific reality and reconcile.

12.5 Accepting the Scientific Reality

As discussed in Chapter 6, evolution is the unifying principle of biology, which has found its roots in genetics, and an emerging body of evidence is indicating that quantum physics principles may also be operating at the origin of evolution. Let

522

me be clear: the point here is not that Darwin was right. The point is also not that everything he said in *On the Origins of Species* is right; he was not. For example, scientists have observed that at times evolution can move in bursts that can be measured in years rather than in bits which can only be measured over ages. This observation can be thought of as a challenge to Darwin's emphasis on slow and steady progression. But that is not the point either. The point is that the core of the theory of evolution originally introduced by Darwin and Wallace has been validated by discoveries and scientific scrutiny. Furthermore, the theory has grown or developed further with more discoveries such as those of genes; and the theory has been put on even firmer foundations as a result of these discoveries and scrutiny. Scientists have discovered more mechanisms of evolution in addition to natural selection. Evolution is certainly not synonymous with Darwinism, as discussed in Chapter 9, for example, in Section 9.10.

That said, overall, we can say now that the following statement is scientifically true:

Genes with different versions (alleles) and mutations continue to provide the intrinsic variability of traits among living organisms, which in turn provides the raw material for a process in which, in Darwin's own words, "...endless forms most beautiful and most wonderful, have been, and are being evolved."

As we have seen all through this book, evolution is a scientific reality, a fact of life. Reality and believing in reality can be two different things, however. Do we accept the reality of evolution? Why won't we? As said earlier, our perception of and approach toward our religion and faith may discourage us from accepting this reality. Other reasons could be lack of proper awareness and information.

Even though, in the U.S., Christianity is the religious background of an overwhelming majority of creationists who had trouble with evolution, it is important to note that almost all religions had to deal with it in their own ways. To be fair here, even in Christianity, not all the creationists oppose evolution by believing in the literal interpretation of scripture. Just like it happened in the case of any other shock from science such as gravity, reconciliation has been happening in case of evolution as well and in all religions. One byproduct of reconciliation is creating blended versions of creationism and evolution. For example, Islam has its own school of *Evolutionary creationism or Theistic evolutionism,* which holds that the scientific principles of the origin of the Universe are supported by the Qur'an, Islam's holy book. There are many Muslims who believe in evolutionary creationism, especially among Liberal movements within Islam.

Even though Jewish views on creationism and evolution vary widely, a number of Jewish denominations accept evolutionary creationism. Major Jewish denominations, including many Orthodox Jewish groups, have their own versions of evolutionary creationism. Although Conservative Judaism does not seem to have an official view on the subject, many conservative Rabbis believe in theistic evolution. To its credit, Conservative Judaism does generally embrace science while recognizing it as a "challenge to traditional Jewish theology."

There is no substitute for experience. It is partly due to its experience including the *Galileo Affair* that Europe has come to terms with scientific discoveries such as evolution or stem cells better than U.S. has. There is no doubt that opposition to evolution has been exceptionally louder in the U.S. The belief in religious fundamentalism in the U.S. is much more likely affecting attitudes towards evolution than it is for believers elsewhere, except perhaps in Turkey.

Why do we care if we accept evolution or not? First, it is nice to accept the physical reality of the world around us because it is the first step toward doing the right thing and making progress. Second, if as a society we deny to accept a scientific reality as crucial as evolution, we are bound to fall behind the rest of the world. So, how are we doing in accepting this reality of evolution?

The August 11, 2006 issue of the journal *Science* reported the results of an interesting study. The researchers consolidated public survey data on evolution collected over 20 years, that is, between 1985 and 2005, in the United States, Japan, and 32 European countries. In each of these surveys, adult participants were asked whether they thought the statement, "Human beings, as we know them, developed from earlier species of animals," was true, false, or if they were unsure. The study found that over these 20 years, the percentage of U.S. adults who accepted evolution declined from 45 to 40 percent. Comparing with peoples' views in other countries, the United States ranked near the bottom in the public acceptance of evolution; only Turkey, where religious fundamentalism is prevalent, ranked lower with 75 percent rejecting evolution. In other European countries such as Iceland, Denmark, Sweden, and France, 80 percent or more of adults accepted evolution, whereas in Japan, 78 percent of adults accepted evolution.

To identify the factors shaping attitudes toward evolution, researchers used 10 independent variables in their study, which included religious belief, political ideology, and the understanding of evolution-related concepts such as genetics. The comparison of the data on these 10 variables collected from adults in the U.S. and in European countries revealed an interesting fact: Americans with fundamentalist religious beliefs were more likely to reject evolution than Europeans with similar beliefs. Fundamental religious belief means the belief in substantial divine control and frequent prayer. The researchers attributed this discrepancy to the differences in the interpretation of the Bible by the Christian fundamentalists in America and in other countries. According to the co-author of this study, Jon Miller of Michigan State University, "American Protestantism is more fundamentalist than anybody except perhaps the Islamic fundamentalism, which is why Turkey and we are so close." According to the researchers, fundamentalist protestant Christians in the U.S. tend to interpret the Bible literally and as a result view Genesis as a true and accurate historical account of creation, whereas mainstream Protestants in both Europe and the United States take Genesis as rather metaphorical.

By studying the political ideology variable, the researchers noticed political ideology and confusion about evolution are considerably more correlated in the U.S. than in Europe. Researchers agreed that unlike in Europe, politics is contributing heavily to the widespread confusion about evolution in the U.S., where mainstream politicians have been making opposition to evolution a prominent part of their political campaigns to attract conservative votes. After reading Chapter 9 and 10, this result should come as no surprise to you. This is something that does not happen in Europe or in some Asian countries such as Japan on this scale. Paul Meyers, a biologist at the University of Minnesota who was not involved in the study, quoted in Live Science, said: What politicians should be doing is saying, "We ought to defer these questions to qualified authorities and we should have committees of scientists and engineers who we will approach for the right answers."

Researchers also found a considerable correlation between the understanding of some biological concepts and acceptance or rejection of evolution. From this study, a contributor to America's low confidence in evolution is the poor grasp of biological concepts, especially genetics, by American adults. This should also not come as a surprise to you because by reaching this point of this book, you have already realized the intrinsic relationship between genetics and evolution. Understanding genetics enables us to appreciate the unity behind diversity of life and our relationship to other forms of life.

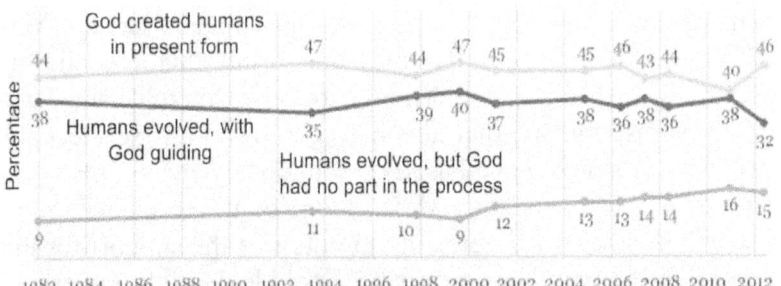

Figure 12.5 Views of Americans about creationism and evolution: A trend from 1982 to 2012. Courtesy: GALLUP Polls.

These results are also generally upheld by more recent surveys. Gallup's analysis based on its polls in 2009 in the U.S. also found the dependency of the confusion about evolution on the educational level and religiosity. According to this analysis, religiosity even outweighed educational level in shaping views on evolution. Overall, those with the most education are far more likely to support evolution than those with the least. Look at this: Only 21 percent of respondents who had up to a high school level of education believed in evolution, whereas 74 percent of those with postgraduate degrees did. What effect of religiosity was discovered by this poll? Among weekly churchgoers, only 24 percent believed in evolution, whereas 41 percent did not, and 35 percent had no opinion. Compare these numbers with those who seldom or never attend church: 55 percent accepted the reality of evolution, whereas only 11 percent did not, and 34 percent had no opinion.

Figure 12.5 shows the trend in views about evolution among Americans from polls conducted by GALLUP between 1982 to 2012. As this figure shows, a slight majority of Americans believe in evolution with or without God's guidance, whereas a significant number continues to disbelieve in the evolution of humans from other life forms. The most recent survey (May 2012) has also upheld the previous result that disbelief in evolution is directly proportional to belief in religion. Two-thirds of respondents that attend weekly religious services did not believe in science-based evolution whereas only one-quarter of the attendants who seldom or never go to church share that view. It is interesting to note that the support for this non-scientific view varies over political ideology. For example, 58 percent of republicans, compared with 41 percent of democrats and 39 percent of independents believe that God created humans in their present form within the last 10,000 years.

In a nutshell, from these survey studies it is clear that there is a strong correlation between religious fundamentalism and opposition to evolution. Lower education level and lack of understanding for some biological topics such as genetics also contribute to the non-acceptance of evolution. Furthermore, political partisanship entangled with religious belief has been playing a dangerous role in this battle. As said earlier, the correlation between political partisanship and religious fundamentalism is not as significant in Europe as in the U.S. That said, many Christians around the world have reconciled with their faith in the light of evolution; they accept evolution as the most likely explanation for the origins of species by not accepting a literal interpretation of Genesis.

Why am I giving you all this information? Why is this so important? Let me put it this way: If researchers found that four in ten people in the U.S. still did not believe in gravity, that Earth revolves around the Sun and that Earth is not the center of the Universe; or if four in ten people did not accept that atoms exist, it would be a big shakeup to the education system. We would be embarrassed. Now that more than four in ten people do not accept evolution, should it shake up the education system; should we be embarrassed? The answer is definitely yes, for the same reasons as we would be in the case of gravity and atoms.

In summary, according to researchers, the major factors contributing to America's low score on accepting the reality of evolution include poor understanding of biology, especially genetics; the politicization of science; and the literal interpretation of the Bible by a small but vocal group of American Christians. Back to Jon Miller, the co-author of one of the studies we discussed in this section: "Whether it's the Bible or the Koran, there are some people who think it's everything you need to know. Other people say these are very interesting metaphorical stories in that they give us guidance, but they're not science books." Before you call the authors or researchers of these studies biased, hold on until we learn from the next section that these results are not much different from what Christian scholars reached centuries ago.

529

12.6 Spare Evolution from Being the Only Challenger

We can spare evolution from being the only challenger to creationism based on the literal interpretation of Scripture. Neither is it the first challenger. Here is a very interesting fact: Long before Darwin and Wallace discovered evolution, there was a school of thought right within Judaism and Christianity that considered the account of creation in Scripture as an allegory rather than a historical account. As a general example of allegory, shown in Figure 12.6, the *Allegory of Music* is a popular theme in painting. In this example,

Figure 12.6 Allegory of Music by Filippino Lippi.

Filippino Lippi uses symbols that were popular during the High Renaissance. Many of these symbols refer to Greek mythology.

Note. An allegory is a figurative mode of communication or representation, as opposed to literal mode, used to convey meaning. An allegory uses the communication technique of symbolic representation by using symbolic figures or actions.

Let's get back to my comment that there was a school of thought right within Judaism and Christianity that considered the account of creation in Scripture as an allegory rather than a historical account. Here are two notable examples: Jewish neoplatonic philosopher Philo of Alexandria from the first century, and Saint Augustine (354-430), the bishop of Hippo

Regius who was also a former neoplatonist. According to the writings of Philo, it would be a mistake to think that creation happened in six days, or in any set amount of time. Driven by his own reasoning, Saint Augustine, one of the most influential theologians of the Catholic Church, argued against the literal interpretation of the account of creation in Scripture. For instance, he argued that everything in the Universe was created by God at the same moment in time, as opposed to six days as the literal interpretation of Genesis states. It can be argued that one of the implications of the six-day creation could be that God is not omnipotent. It appears that both Augustine and Philo felt uncomfortable with this implication of the literal interpretation of Genesis. It was Saint Augustine, and not an evolutionary scientist, who first suggested that the accounts in the Bible should not be interpreted literally if they contradict what we know from science, our experience, and our God-given reasoning. I cannot say it more eloquently than Saint Augustine said in his book *The Literal Interpretation of Genesis* (early fifth century, AD):

> It not infrequently happens that something about the earth, about the sky, about other elements of this world, about the motion and rotation or even the magnitude and distances of the stars, about definite eclipses of the sun and moon, about the passage of years and seasons, about the nature of animals, of fruits, of stones, and of other such things, may be known with the greatest certainty by reasoning or by experience, even by one who is not a Christian. It is too disgraceful and ruinous, though, and greatly to be avoided, that he [the non-Christian] should hear a Christian speaking so idiotically on these matters, and as if in accord with Christian writings, that he might say that he could scarcely keep from laughing when he saw how totally in error they are. In view of this and in keeping it in mind constantly while dealing with the book of Genesis, I have, insofar as I was able, explained in detail and

set forth for consideration the meanings of obscure passages, taking care not to affirm rashly some one meaning to the prejudice of another and perhaps better explanation. (From *The Literal Interpretation of Genesis*, 1:19–20, Chapter 19, AD 408)

Here is more from Saint Augustine in the same book:

With the scriptures it is a matter of treating about the faith. For that reason, as I have noted repeatedly, if anyone, not understanding the mode of divine eloquence, should find something about these matters [about the physical universe] in our books, or hear of the same from those books, of such a kind that it seems to be at variance with the perceptions of his own rational faculties, let him believe that these other things are in no way necessary to the admonitions or accounts or predictions of the scriptures. In short, it must be said that our authors knew the truth about the nature of the skies, but it was not the intention of the Spirit of God, who spoke through them, to teach men anything that would not be of use to them for their salvation.

Here is another interesting fact. Upon the publication of the theory of evolution through Darwin's book *On the Origin of Species*, its earlier enthusiastic supporters included Frederick Temple, an archbishop of Canterbury; Charles Kingsley, an author of several volumes of sermons; and the French Jesuit priest and geologist Pierre Teilhard de Chardin who saw evolution as confirmation of his Christian beliefs. It is fair to state the fact here that Pierre Teilhard de Chardin received condemnations from Church authorities for his theories.

The main points of Saint Augustine or other scholars discussed above are examples of reconciliation with the scientific reality as opposed to ignoring it or opposing it.

12.7 Reconciling with Evolution

We discussed in Chapter 10 how some people in the Hindu religion find support for evolutionary ideas in their scriptures called Vedas, and how India supports scientific research by reconciling it with its religious symbols. Similarly, leaders of the Anglican and Roman Catholic churches have already made statements in acceptance of the theory of evolution. The approach of the Catholic Church toward the theory of evolution has gradually changed in favor of evolution since the publication of Charles Darwin's *Origin of Species* in 1859. Officially or unofficially it has gone through all the three stages: opposition, neutrality (no position), and acceptance. For about 100 years, although there was opposition at time, there was no authoritatively official pronouncement or stand on the subject. By 1950, Pope Pius XII had moved to support the academic freedom for studying evolution to the extent that such a study was not anymore deemed to directly violate Catholic dogma. It is fair to say that overall, since the late twentieth century, the Church's attitude has been one of rather great tolerance for evolution.

Fast forward to the present time. Today, the Catholic Church's unofficial position is very much consistent with that of the proponents of *theistic evolution*: faith and scientific discoveries about human evolution are not in conflict. At the same time, however, it regards humans as a *special creation*, and requires the existence of God to explain both the spiritual component of human origins and monogenism – the belief that human species on Earth originated with a single ancestor or ancestral couple.

On October 22, 1996, during his address to the Pontifical Academy of Sciences, founded in 1936 by Pope Pius XI, Pope

533

And God Said: Let There Be Evolution

John Paul II, updating the Church's position in order to accept evolution of humans, said:

> In his encyclical *Humani Generis* (1950), my predecessor Pius XII has already affirmed that there is no conflict between evolution and the doctrine of the faith regarding man and his vocation, provided that we do not lose sight of certain fixed points....Today, more than a half-century after the appearance of that encyclical, some new findings lead us toward the recognition of evolution as more than a hypothesis. In fact it is remarkable that this theory has had progressively greater influence on the spirit of researchers, following a series of discoveries in different scholarly disciplines. The convergence in the results of these independent studies – which was neither planned nor sought – constitutes in itself a significant argument in favor of the theory.

It is fair to mention that in the same address, Pope John Paul II rejected any theory of evolution that provides a materialistic explanation for the human soul:

> Theories of evolution which, because of the philosophies which inspire them, regard the spirit either as emerging from the forces of living matter, or as a simple epiphenomenon of that matter, are incompatible with the truth about man.

One can extrapolate that once the scientific research on what we call consciousness, soul, or spirit makes significant progress, the Church will need to reconcile with those scientific discoveries as well. It is important to note here as well that as in other countries, Catholic schools in the United States teach evolution as part of their standard science curriculum.

Saint Augustine and the Pope said it well. It is not within the scope of biology, for example, to tell if Christ arose from the dead or not or if Lord Krishna will come back or not. By the same token, it is not within the scope of religion to pass judgments on a scientific theory such as evolution from a

534

religious viewpoint. Scientists are not asking for stickers to be placed on religious books. If a religious person wants to disprove a scientific theory they will need to do so following the scientific method just like a non-religious person. If anyone (religious or non-religious) wants to present a theory as an alternative to theory of evolution, that theory first will be required to be qualified as a scientific theory, that is, testable and falsifiable. Then it will need to go through the same scientific investigation as the theory of evolution (or any other scientific theory for that matter) has gone through. Isn't that fair enough?

We have just briefly discussed the reconciliation by the Catholic Church with the theory of evolution. How about other denominations of Christianity? Looking at the International scene, while most contemporary Christian leaders including scholars from mainstream churches, such as Anglicans and Lutherans, endorse Genesis as spiritual meaning of creation, at the same time they reject the notion reading it can shed light on the physics or biology of creation. This is very much consistent with the view of Saint Augustine discussed earlier in this chapter.

So, is the reconciliation complete?

12.8 Wake Up and Smell the Sun: Are We There Yet?

Using religious terminology here, the human species is blessed with the capability of examining the physical world around us and acquiring knowledge about it. In other words, we have the capability of discovering the physical reality. However, it seems like we are also cursed with our very strong appetite for ignorance and dogmas. We have the capability of constructing our own environments of the mind and create our own image of the world and life, which may be in stark contrast with the physical reality.

Therefore, there will always be some people who under creationism or some other banner will oppose theory of evolution and any other scientific theory that does not fit with their version of their religion or belief. At this point, the issue is not limited to evolution, but it extends itself to the whole of science. This issue involves two problems: poor education and understanding of science among some sections of society, and vested interests of some people regardless of their understanding of science. The first problem is solvable to some extent, and the second problem can only be fought against.

As evidence of what I am saying, even today there are some creationists who not only oppose evolution but also still believe that Earth is the center of the Universe and it does not move, much less revolve around the Sun. For example, it is not difficult to find websites or web pages that propagate such views. One such example is: www.fixedearth.com.

Here is an example how their opposition is not limited to evolution but expands to the whole of science:

"Today's cosmology fulfills an anti-Bible religious plan disguised as "science". The whole scheme from Copernicanism to Big Bangism is a factless lie. Those lies have planted the Truth-killing virus of evolutionism in every aspect of man's "knowledge" about the Universe, the Earth, and Himself."

---From the website: www.fixedearth.com, which was accessible at the time of writing this book.

Here is the irony: The fact that Earth is moving and revolves around the Sun is tied to the theory of gravitation. It is a fact that no creationists can practically defy gravity, for example, jumping from a mountain top believing to safely land. Airplane flights are designed on the assumption that Earth is moving and on the validity of theory of gravitation, and some creationists do take these flights. They may also, if they haven't already, use medical treatment based on evolution. As is clear from the above quotation, some creationists tie evolution to cosmology and the whole of science at an emotional level and seemingly respond by attacking the whole of science. But I have a better connection for them at factual level. As is clear from Chapter 5, evolution is connected to physics including quantum physics, which explains the microscopic world and is the basis for the whole of physics and hence the whole of science. So, attacking evolution is attacking the whole of science, which these creationists are literally doing on the mentioned website anyway. To be true to their ideology, they should not be selling their anti-evolution ideas or anti-science literature on the Internet because without science there would be no computer, rest aside Internet. They should not be driving cars because without Newtonian physics that they consider a part of conspiracy against their religion, the car could not be designed. They should not go see a doctor because today's medicine is based on medical science. I can go on, but you get the point: they, just like the rest of us, are using material and facilities in their daily lives that are based on the very scientific facts that they are opposed to. So, you can see that at this point, the unscientific opposition to evolution such as the one by creationists just becomes a mere case of ignorance, illiteracy, and extreme dogmatism.

Opposition to science by Christianity and other religions followed by reconciliation did not begin with evolution; it goes back to the very origin of science. As mentioned in Chapter 1, the very idea that nature follows consistent principles which can be understood started with the Greek philosopher Thales of Miletus (624-546 BC). This scientific approach of the Greek philosophers dominated Western thought for about 2000 years. The Greek's Christian successors (at least a subset of them), however, dismissed the idea that the Universe can be understood by the laws of nature, that the Universe is in fact run by indifferent laws of nature. For instance, under the instructions (or blessings) of Pope John XXI, Bishop Stephen Tempier of Paris published and condemned in a list of 219 heresies on 1277. One of those heresies was the idea that nature follows laws, because this idea conflicted with God's omnipotence. Ironically, a few months later, the roof of the Pope's palace collapsed on him, and he was killed by the gravitational law of nature.

Let me make it clear that creationists are not limited to the U.S. or Christianity. They come from a variety of religions and create their basis throughout the world including Australia, Turkey, and the United Kingdom. At some level and to some extent, the battle against evolution has been the battle between superstitions and rationality, between ignorance and knowledge, and between the drive for truth and the drive for vested interests; and that part of the battle will perhaps never end. When such a battle loses its interest in evolution, it will move to the next topic.

Some ignorance, lack of education, denial of physical reality, and the willingness to be driven by our vested interests rather than by scientific truth will perhaps stick with us forever. Furthermore, opposition to evolution as part of opposition to science will always remain, as it is for many other scientific theories. After accepting this truth, we can safely say that the battle between the opponents and the proponents of the theory of evolution is practically over. The results are already in and the dust is settling down. Evolution (and its theory, which will continue improving) will be part of common sense for coming generations if it is already not, just like gravity is to us today.

12.9 The Final Call

As mentioned earlier in this chapter, reconciliation of religious beliefs with evolution comes in terms of concepts such as *evolutionary creationism* and *theistic evolution*. Basically these concepts assert the belief that religious teachings about God are compatible with the modern scientific understanding about biological evolution. Some scholars with religious orientations such as John Polkinghorne, a former Cambridge University professor and an ordained priest, who was awarded the $1 million Templeton Prize for "exceptional contributions to affirming life's spiritual dimension," argued that evolution is one of the principles through which God created living beings.

In nutshell, theistic evolutionists believe:

✓ God exists.

✓ God created the material universe and all life within the Universe.

✓ Evolution is simply a natural phenomena built into process of creation that was put into motion by God.

So, according to the theistic evolutionary view, evolution is simply a tool that God employed to develop human life. Scientists have no problem if somebody holds such a belief as long as the belief remains belief and is not presented as scientific theory or an alternative to a scientific theory such as evolution. This is because theistic evolution is not a scientific theory, but a particular view about how the science of evolution relates to religious beliefs and interpretation. In this sense, it should be welcome as one of the reconciliation tools for those of us who are religious.

Reconciliation has been happening and is bound to continue to happen. This is because no matter what kind of world we create in our minds in the comfort of our beliefs, we cannot escape the physical reality of the natural world. For example, think of the applications of evolution discussed in Chapter 11. It is not hard to realize that times are arriving quickly when some people will be speaking against evolution while a medicine based on evolution will be doing its job in their body.

I can almost hear it, a voice from the near future. May be not so literally, but the Catholic Church has already spoken it in some form. And as far as the Protestant Church goes, there is not any other exit left either. I will not be surprised to see a pastor enthusiastic about his faith, life, and the truth, saying during his sermon something like this:

And God said: Let there be evolution!

And there is evolution!

Amen.

The Beginning

Glossary

adaptation Inherited characteristic (trait) of an organism that enhances the likelihood of its survival and reproduction in specific environments. Also refers to the remarkable fit between organisms and their environment.

adaptive evolution The evolution that improves the match between a population and its environment.

adaptive radiation The process in which a lineage in the evolutionary tree rapidly gives rise to many new species through adaptations, which enable the new species to fill different niches offered by the environment.

allele One of two or more versions of a gene. All alleles of a gene produce phenotypic effects which are distinguishable from one another such as blue eyes versus green eyes versus brown eyes.

archaea One of the two prokaryotic groups (domains) of organisms; the other being Bacteria.

atom The smallest particle of an element.

ATP (adenosine triphosphate) An energy molecule that is made by the cells from the food that organisms eat, and in return it powers cellular work.

autosome Any chromosome which is not a sex chromosome.

bacteria One of the two prokaryotic groups (domains) of organisms; the other being Archaea.

biogeography Study of the geographic distribution of organisms and their groups such as species.

biology The scientific study of life.

biosphere The sum total of all the parts of the Earth that are inhabited with life; includes water, air, crust etc. Includes all the ecosystems of our planet.

blending hypothesis A common belief held among biologists before Mendel's work according to which the genetic material of offspring was a uniform blend of that of the two parents, giving rise to the phenotypes of the offspring intermediate to that of the parents.

bya Billion years ago.

catastrophism The hypothesis that states changes in the past occurred suddenly and were caused by geological mechanisms that were different from those operating today.

cell Smallest living entity; the structural and functional building block of all living organisms.

cell differentiation The process by which a cell becomes a specialized cell type such as blood cell, brain cell, etc.

central dogma of molecular biology A process used by cells to produce proteins based on the information coded in the genes.

chromosome A structure that carries a single large DNA molecule in the cell. Different species of animals and plants carry different numbers of chromosomes.

community A set of all populations of all species living in a specified area.

convergent evolution Evolution of similar traits in different evolutionary lineages due to adapting to similar environments in similar ways. Example: bats, birds, and flying insects, although evolved along different lineages, have acquired similar traits that help them to fly.

deoxyribonucleic acid (DNA) A molecule that contains the hereditary information about an organism which acts as a blueprint for its development and functioning.

directed evolution A method that applies the elements of evolutionary theory to evolve proteins with enhanced desirable functions.

DNA sequencing A process to determine the order of nucleotides in a DNA molecule. Many techniques of DNA sequencing are available.

ecosystem All the organisms in an area and the biological environment with which they interact. May contain one or more communities interacting with the environment.

element A pure substance that cannot be decomposed into other pure substances by chemical methods. All atoms of an element are identical.

embryo An organism in the early stages of its development within the womb; for example in humans up to the end of second month of pregnancy.

embryology The study of the development of an embryo.

entropy A degree of randomness and disorder in a system.

eukarya One of the three domains of life; the others being Archaea and Bacteria. It includes all eukaryotes.

eukaryotes Organisms made of eukaryotic cells, the cells that contain membrane-bound internal structures called organelles including a nucleus.

evolution The concept that living different species descended from common ancestral species through modifications.

fossil A preserved remain, impression, or trace of an organism that lived in the past, such as a skeleton or a footprint, etc. Most fossils are found in the rocks called sedimentary rocks.

gamete A sexual half-cell, called a reproductive haploid cell such as a sperm or an egg in animals, and pollen and egg in a plant. Two haploid cells such as a sperm and an egg fuse into a diploid zygote.

genealogy The gene linage ancestry or evolutionary history of an organism or a group of organisms.

genetic factors Entities now known as genes or versions of genes (alleles) were predicted by Mendel to explain his experimental results and were referred to as factors of genetic factors.

genetic marker A DNA sequence which may or may not be a gene but has a known location on a chromosome.

genome The complete genetic material in a cell of an organism; the set of DNA molecules with each molecule packaged inside a structure called chromosome. For example, each cell of your body, except mature red blood cells, contains 23 pairs of chromosomes; each chromosome of a pair is inherited from each parent.

gene A unit of heredity; a region in the DNA molecule.

gene pool The sum total of all the alleles of all genes in all organisms of a population.

genetic drift Change in allele frequencies of a population due to chance events.

genetic marker A DNA sequence, which may or may not be a gene but has a known location on a chromosome.

genetics The scientific study of heredity, and similarities and differences among organisms based on heredity.

genotype Set of specific alleles carried by an individual organism corresponding to a phenotype; also called genetic makeup.

horizontal gene transfer Gene transfer between the genome of one organism to the genome of another organism when the two organisms do not have parent-offspring relationship.

heredity Transmission of a genome from parents to offspring. This is how the heritable traits are passed from one generation to the next.

hypothesis Testable and falsifiable explanation of a phenomenon.

inheritance Transmission of a genome from parents to offspring. This is how the heritable traits are passed to the offspring.

macroevolution Evolutionary change that appears at organism or species level such as appearance of a new traits in organisms and the appearance of a new species. Macroevolution has its roots in microevolution.

metabolism The sum total of all biochemical reactions that cells use to manage their energy resources.

microevolution Evolutionary change in the gene pool of a population that occurs over generations.

molecular clock A technique in evolutionary science which estimates the time taken by an evolutionary change by using the fact that some regions of the genome change at a constant or known rate.

molecule A structure composed of two or more atoms (of the same or different elements) bonded together by some kind of electromagnetic force.

morphology External form of an organism defined by visual parameters such as dimensions, shape, size, and mass.

mutation A change in the sequence of nucleotides in a DNA molecule of an organism.

Mya or Ma Million years ago.

natural law A generalized conclusion or a set of generalized conclusions based on observations of a physical entity, set of physical entities, or behavior. Laws are at the highest of scientific maturity levels.

natural selection A process in which organisms with certain inherited traits that fit the environment better have better chances to survive and reproduce; this process always improves the fit between the population and its environment.

organ A body structure of an organism composed of multiple tissues that interact to accomplish a set of tasks. Examples: heart, liver, kidney.

organism An individual entity that consist of one or more cells. Examples: cow, palm tree, and humans.

paleontology The branch of science that studies fossils.

phenotype A physical or physiological trait of an individual organism determined by the corresponding genotype (or genetic makeup).

photosynthesis A cellular process used by the plants, alga, and certain prokaryotes to use light, carbon dioxide, and water to produce carbohydrates and oxygen.

phylogenetics The study of phylogeny.

phylogeny The evolutionary history of groups of organisms such as species or groups of species.

population A group of organisms of a given species living in an area.

prokaryote Any single-celled organism made of a prokaryotic cell, which is a cell that has no nucleus or any other membrane-enclosed internal structures called organelles. Prokaryotes have two groups or domains: Archaeans and Bacteria.

protocell The primitive version of a living cell that contained an internal biochemical environment different from the external environment and protected by a simple fatty acid membrane.

ribonucleic acid (RNA) A type of nucleic acid molecule that helps produce proteins based on the information encoded in genes.

scientific method A technique used in science for acquiring new knowledge and amending the existing knowledge through verification.

sister chromatids. Two replicated copies of a duplicated chromosome attached to each other during the process of meiosis.

species A group of organisms that have the potential of interbreeding and producing fertile and viable offspring. It may consist of several populations living at different places.

speciation A process in which one species forks into two or more species.

sympatric speciation. The process in which a new species forms within the same population without a geographic split.

sedimentary rock A rock that is formed from the mud and sand settled down to the bottom of lakes, seas, and swamps. Most fossils are found in sedimentary rocks,

theory An explanation that is broader in scope and mature in experimental testing than a hypothesis, and may include several well-tested hypotheses and laws.

trait A characteristic variant of an organism such as blue eyes or green eyes. Traits may belong to form, function, or behavior. A heritable trait is called a phenotype.

uniformitarianism The principle that change occurred gradually due to geological mechanisms that are constant over time.

vestigial traits Leftover traits which have no or marginal use in the current species but were useful in the ancestral species.

virus A small particle that acts as an infectious agent that can replicate itself but only inside the living cell of an organism. It's not a complete cell; just a package of genetic material.

And God Said: Let There Be Evolution

Art Credits and Acknowledgements

Unless otherwise acknowledged, all pictures and illustrations in this book are the property of Infonential, Inc. We have made our best effort to trace and acknowledge the ownership of the following items. In the event of any question or issue arising from the use of any of these items, we will be pleased to make the necessary corrections in the future printings.

Public Domain: Figures 1.1, 1.4, 1.5, 1.6, 1.8, 2.1, 4.1, 4.9, 6.5a, 6.5b, 6.6, 6.12, 6.13, 6.14, 7.1, 7.2, 7.10, 10.1, 10.3, 10.4, 10.6, 11.3, 12.1, 12.2, 12.3, and 12.6. Images of Jean Baptiste Lamarck and Alfred Wallace in Chapter 1; Gregor Mendel in Chapter 2; Godfrey Hardy and Wilhelm Weinberg in Chapter 4; Max Delbrück and Erwin Schrödinger in Chapter 5.

GNU Free Documentation License: Figures 2.1, 6.3, 6.7, 6.9, 8.6, 8.8, 9.1. Images of Jean Charles Darwin in Chapter 1.

Figures: 2.3 courtesy of CERN; 4.6 courtesy of Daniel Horspool; 4.7 courtesy of National Human Genome Research Institute; 5.1 courtesy of U.S. Department of Energy; 5.3 courtesy of John Cairns et al, Nature, 1988, September; 6.11 and 6.17 courtesy of University of California Museum of Paleontology, The National Center for Science Education, The National Science Foundation, The Howard Hughes Medical Institute; 8.1 courtesy of Jiuguang Wang; 8.4 courtesy of Ancheta Wis; 8.5 courtesy of Derek Ramsey; 8.7 courtesy of Jon Sullivan; 10.5 courtesy of www.law.umkc.edu; 11.1 courtesy of Einar Einarsson Kvaran.

And God Said: Let There Be Evolution

Bibliography, Further Reading, and References

Charles Darwin (1859), *On The Origin of Species,* Dover Publications, New York, June 2006. Originally published: London: J. Murray, 1859.

Erwin Schrödinger (1944), *What is Life?*, Cambridge University Press, Cambridge, U.K., 1992.

Johnjoe McFadden, *Quantum Evolution*, W.W. Norton & Company, New York, 2002.

J.B. Reece et al., Campbell Biology, Benjamin Cummings, San Francisco, 2010.

C. Starr et al., Biology, Brooks Cole/Cengage, Independence, KY, U.S., 2007.

Special Issue on Evolution, Scientific American, **300** (1) (January 2009).

Website of the University of California Museum of Paleontology. http://www.ucmp.berkeley.edu/

Chapter 1

Armstrong, S., Fog, wind and heat: life in the Namib desert; New Scientist; **127**: 46 – 50 (14 July 1990). *The head-standing beetle.*

Alfred Russel Wallace website: http://wallacefund.inf

Chapter 4

Genome Publications:

Human
International Human Genome Sequencing Consortium, Initial sequencing and analysis of the human genome, *Nature* **409**: 860-921 (15 February 2001)

Rat
Rat Genome Sequencing Project Consortium, Genome Sequence of the Brown Norway Rat Yields Insights into Mammalian Evolution, *Nature* **428**: 493-521 (1 April 2004)

Mouse
Mouse Genome Sequencing Consortium, Initial sequencing and comparative analysis of the mouse genome, *Nature* **420**: 520 -562. (5 December 2002)

Fruit Fly
M. D. Adams, et al., The genome sequence of *Drosophila melanogaster*, *Science* **287**: 2185-95 (24 March 2000).

Arabidopsis - First Plant Sequenced
The Arabidopsis Genome Initiative, Analysis of the genome sequence of the flowering plant *Arabidopsis thaliana*, *Nature* **408**: 796-815 (14 December 2000).

Roundworm - First Mutlicellular Eukaryote Sequenced
The *C. elegans* Sequencing Consortium, Genome sequence of the nematode *C. elegans*: A platform for investigating biology, *Science* **282**: 2012-8 (11 December 1998).

Yeast
A. Goffeau, et al., Life with 6000 genes, *Science* **274**: 546, 563-7 (25 October 1996).

Bacteria - *E. coli*
F. R. Blattner, et al., The complete genome sequence of *Escherichia coli* K-12, *Science* **277**: 1453-1474 (5 September 1997).

Bacteria - H. influenzae - First Free-living Organism to be Sequenced
R. D. Fleischmann, et al., Whole-genome random sequencing and assembly of *Haemophilus influenzae* Rd, *Science* **269**: 496-512 (28 July 1995).

Genome similarities of human with:

Chimpanzees
NIH/National Human Genome Research Institute. www. nih.gov; www.genome.wellcome.ac.uk/doc_WTD020730.html. Comparing the chimp and human genomes; 31 August 2005. Based on research published in the 1 September 2005 issue of *Nature*.

Cats
Joan U. Pontius et al., Initial sequence and comparative analysis of the cat genome, www.genome.cshlp.org/content/17/11/1675.full, *Genome Res.* 2007, **17**: 1675-1689.

Cows
Christine G. Elsik et al.; The Genome Sequence of Taurine Cattle: A Window to Ruminant Biology and Evolution; *Science* **324**: 522-528 (24 April 2009).

http://www.sciencemag.org/content/324/5926/522.full

Mouse
Deanna M. Church et al, Lineage-Specific Biology Revealed by a Finished Genome Assembly of the Mouse, www.plosbiology.org/article/info:doi/10.1371/journal.pbio.1000112, PLOS Biology, May 2009 Issue.

Fruit Fly
Background on Comparative Genomic Analysis, December 2002, www.genome.gov/10005835.

And God Said: Let There Be Evolution

Chicken
Science Daily, Dec 10, 2004;
http://www.sciencedaily.com/releases/2004/12/041208230523.htm

Source: National Institute of Health and National Human Genome
Research Institute.

Chapter 5

Paul Sanghera, *Quantum Physics for Scientists and Technologists*,
Wiley-Interscience, April 2011.

Fisher, R. A. *The Genetical Theory of Natural Selection,* Clarendon
Press, 1930, ISBN 0-19-850440-3.

[Luria, SE, Delbrück, M, Mutations of Bacteria from Virus
Sensitivity to Virus Resistance, *Genetics* **28**:491-511 (1943).
http://www.genetics.org/content/28/6/491.

Newcombe, H.B, Origin of Bacterial Variants, *Nature* **164**:150-151
(1949).

J. Cairns, J. Overbaugh, S. Millar, The Origin of Mutants, *Nature*,
335(6186):142-5 (8 Sep 1988).

Chapter 6

Rice, W.R. and Salt, G.W., Speciation via disruptive selection on
habitat preference: experimental evidence, *The American Naturalist*
131: 911 917 (1988).

Rice, W.R. and Hostert, E.E, Laboratory experiments on speciation:
What have we learned in forty years?, *Evolution* **47** (6): 1637–1653
(1993); http://jstor.org/stable/2410209.

Dodd, D.M.B., Reproductive isolation as a consequence of adaptive
divergence in *Drosophila pseudoobscura*, *Evolution* **43**:1308–1311
(1989).

Rice, W.R., Salt, G.W., The evolution of reproductive isolation as a
correlated character under sympatric conditions: experimental
evidence, *Evolution*, **44** (5): 1140-1152 (August 1990).

Chapter 9

Selden, S. (1999, Fall), Popularizing Eugenics, www.eugenicsarchive.org, Image Archive on the American Eugenics Movement, an online archive of materials from national databases relating the popular eugenics movement in America. Sponsored by a grant from the Ethical, Legal, and Social Influences (ELSI) division of the Human Genome Project.

http://vector.cshl.org/eugenics/, Cold Spring Harbor Laboratory's DNA Learning Center.

Ronald L. Numbers, 222-223 (1992). *The Creationists.* New York: Knopf. *William J. Tinkle*, Section 9.7, paragraph 1.

Chapter 10

Original evolution-related trial transcripts and documents based on those.

Douglas O. Linder, State v. John Scopes ("The Monkey Trial"), website www.law2.umkc.edu/faculty/projects/ftrials/ftrials.htm

Larson, Edward J., *Summer for the Gods: The Scopes Trial and America's Continuing Debate Over Science and Religion*, Basic Books, ISBN 0-465-07509-6, 1997.

The Salem Republican, June 11, 1925.

Larson, Edward J., *Evolution*, Modern Library, ISBN 0-679-64288-9, 2004.

"The BSCS Story", http://www.bscs.org/ecommunity/bscsstory.html

Arnold Grobman, "The Biological Sciences Curriculum Study," AIBS bulletin, April, 1959.

McLean v. Arkansas Board of Education, Decision by U.S. District Court Judge William R. Overton; January 5, 1982.

Grabiner, J.V. & Miller, P.D., Effects of the Scopes Trial, *Science*, New Series, **185** (4154):832-837 (September 6, 1974).

And God Said: Let There Be Evolution

Moore, Randy, The American Biology Teacher, **60** (8):568-577 (October, 1998).

"The Wedge Strategy," Center for the Renewal of Science and Culture published, 1998, published on the website: www.antievolution.org/features/wedge.html

Chapter 12

St. Augustine, *The Literal Interpretation of Genesis*, 1:19–20, Chapter 19, AD 408.

Ker Than, U.S. Lags World in Grasp of Genetics and Acceptance of Evolution, posted on 10 August 2006, www. livescience.com.

The Galileo Affair: A Documentary History, edited and translated by Maurice A. Finocchiar, University of California Press, Berkeley and Los Angeles, California, 1989.

Judaism and Evolution, Jewish Virtual Library, a Division of The American-Israeli Cooperative Enterprise; www.jewishvirtuallibrary.org

Guided Evolution: Proof from Punctuated Equilibrium, Ataul Wahid Lahaye and Zia H Shah; www.alislam.org

Appendix A

TIMELINE OF THE THEORY OF EVOULTION

610-540 B.C. A Greek philosopher proposes that all life forms evolved from fish through a process of modification and migration to land.

1735. The foundations of taxonomy are laid down by the publication of the first volume of *Systema Naturae* by Carl Linnaeus.

1831-1835. Darwin collects data during the voyage of the Beagle that makes the foundation for his theory of evolution.

1844. Based on the collected data, Charles Darwin writes a long essay on the descent with modification and its underlying mechanism called natural selection. He did not publish the essay.

1848. Alfred Wallace sets off by ship from Liverpool to Pará (Belém), the first of his many journeys to collect data that would make the basis for the discovery of his law of Natural Selection.

1858. Law of Natural Selection is presented to the Linnean Society of London through a paper by Wallace and excerpts from Darwin's unpublished paper of 1844.

1859. First edition of *On the Origin of Species* by Charles Darwin is published.

And God Said: Let There Be Evolution

1866. Gregor Mendel, an Austrian monk and scientist, publishes his laws of inheritance based on a decade-long experimental study that he conducted on pea plants. His work, despite its huge implications for the theory of evolution by Darwin and Wallace, goes largely unnoticed for the next 35 years.

1900-1901. Mendel's work is rediscovered. Hugo de Vries makes the connection between Mendelian genetic factors, genes, and theory of evolution by Darwin and Wallace.

1925. The Scopes Trial in Tennessee, U.S., tries high school biology teacher John Scopes based on a state law called Butler Act that made it illegal to teach evolution.

1944. Based on experimental results, Oswald Avery, Maclyn McCarty, and Colin MacLeod announced that DNA is the genetic material.

1953. Francis Crick and James Watson discover the double-helix structure of DNA largely based on the results from experiments performed by Rosalind Franklin and Maurice Wilkins. This catalyzes the field of molecular biology including genetics.

1968. In the court case *Epperson v. Arkansas*, the U.S. Supreme Court declared Arkansas' law prohibiting the teaching of evolution unconstitutional because the motivation behind the law was based on a literal reading of Genesis, not science.

1987. In the court case *Edwards v. Aguillard*, the U.S. Supreme Court invalidated Louisiana's *Creationism Act* that required teaching creationism along with evolution in public schools. According to the Court, the law violated the Establishment Clause of the Constitution by attempting to advance a particular religion.

1953-Now. Genetic studies support evolution, and improve the theory of evolution by answering some questions not answered by the theory by Darwin and Wallace.

And God Said: Let There Be Evolution